AF847709

Rice Blast: Interaction with Rice and Control

Rice Blast: Interaction with Rice and Control

Proceedings of the 3rd International Rice Blast Conference

Edited by

Shinji Kawasaki

National Institute of Agrobiological Sciences,
Ibaraki, Japan

KLUWER ACADEMIC PUBLISHERS
DORDRECHT / BOSTON / LONDON

A C.I.P. Catalogue record for this book is available from the Library of Congress.

ISBN 1-4020-1228-4 (HB)
ISBN 0-306-48582-6 (e-book)

Published by Kluwer Academic Publishers,
P.O. Box 17, 3300 AA Dordrecht, The Netherlands.

Sold and distributed in North, Central and South America
by Kluwer Academic Publishers,
101 Philip Drive, Norwell, MA 02061, U.S.A.

In all other countries, sold and distributed
by Kluwer Academic Publishers,
P.O. Box 322, 3300 AH Dordrecht, The Netherlands.

Printed on acid-free paper

All Rights Reserved
© 2004 Kluwer Academic Publishers
No part of this work may be reproduced, stored in a retrieval system, or transmitted
in any form or by any means, electronic, mechanical, photocopying, microfilming, recording
or otherwise, without written permission from the Publisher, with the exception
of any material supplied specifically for the purpose of being entered
and executed on a computer system, for exclusive use by the purchaser of the work.

Printed in the Netherlands.

TABLE OF CONTENTS

Preface ix

Overview on the chemical control of rice blast disease
I. YAMAGUCHI 1

CHAPTER I PATHOGEN: MOLECULAR BIOLOGY

RNA silencing in the blast fungus *Magnaporthe grisea*
H. NAKAYSHIKI, N. KADOTANI, Y. TOSA, S. MAYAMA 15

SAGE (Serial Analysis of Gene Expression) in *Magnaporthe grisea*
-Profiling of cAMP-inducible genes involved in appressorium formation-
T. IRIE, H. MATSUMURA, R. TERAUCHI, H. SAITOH 23

Approach to understand metabolic networks involved in appressorium function of *Colletotrichum lagenarium*
G. TSUJI, S. FUJII, N. FUJIHARA, S. TSUGE, Y. KUBO 29

Proteomics of *Magnaporthe grisea*: liquid chromatography mass spectrometry for the identification of extracellular proteins
S.-C. WU, J. JOHNSON, A.G. DARVILL, P. ALBERSHEIM, R.ORLANDO 39

Identification and characterization of secreted proteins from *Magnaporthe grisea*
G. LU', C. FILIPPI, D. LI, D. EBBOLE 47

Repeat-induced point mutation (RIP) in *Magnaporthe grisea*: evidence for the presence of sexual cycle in nature
K. IKEDA, H. NAKAYASHIKI, Y. TOSA, S. MAYAMA 57

Relationship between two avirulence genes, *Avr-hattan 3* and *Avr-piks*
N. YASUDA, K. HIRAYAE, N. HAYASHI, Y. FUJITA, M. TSUJIMOTO, T. NAKAJIMA 65

CHAPTER II HOST: PLANT RESISTANCE

Genome evolution and function of resistance genes
K. HIRANO, S. KAWASAKI 71

Broad-spectrum resistance genes *Pi2*(t) and *Pi9*(t) are clustered on chromosome 6
G. Liu, S. Qu, B. Zhou, L. Zeng, G.-L. Wang 79

Fine genetic mapping and physical delimitation of the rice blast resistance gene *Pi5*(t) to a 70-kb DNA segment of the rice genome
J.-S. Jeon, D. Chen, G.-H. Yi, G. L. Wang, S. Kawasaki, P. C. Ronald 87

Transposon-insertion lines of rice for analysis of gene function
A. Miyao , H. Hirochika 107

Quantitative trait loci (QTL) reactions to rice blast isolates from Japan and the Philippines
Y. Fukuta, M. J. T. Yanoria, D. Mercado- Escueta, L. A. Ebron, Y. Fujita, E. Araki , G. S. Khush 113

Selection of criterional varieties for evaluation of blast partial resistance of rice in the Tohoku region of Japan
T. Kataoka, H. Kato, T. Takita, N. Yokogami , M. Yamaguchi 123

Genetic dissection and mapping of genes conferring field resistance to rice blast in Japanese upland rice
S. Fukuoka, T. Shimizu, M. Yano, K. Okuno, T. Nagamine 131

Studies on partial resistance to rice blast in the tropics
H. Kato, H. Tsunematsu, L.A.Ebron, M.J.T.Yanoria, D.M.Mercado , G.S.Khush 137

Microarray analysis of gene expression in rice treated with probenazole, a resistance inducer, in special reference to blast disease
M. Nishiguchi', M. Shimono, Y. Eguchi, H. Okuizumi, J. Yazaki, K. Nakamura, F. Fujii, K. Shimbo, Z. Shimatani, Y. Nagata, A. Hashimoto, T. Ohta, Y. Sato, S. Honda, M. Iwano, K. Yamamoto, K. Sakata, T. Sasaki, N. Kishimoto, S. Kikuchi 145

Cytology of infection and host resistance in rice blast disease
H. Koga , O. Nakayachi 155

Probenazole (Oryzemate®) - a plant defense activator
M. Iwata, K. Umemura , N. Midoh 163

CHAPTER III HOST: RESISTANCE BREEDING

Genetic engineering for blast disease resistance in rice, using a plant defensin gene from *Brassica* species
M. KAWATA, T. NAKAJIMA, K. MORI, T. OIKAWA, S. KURODA 173

Transgenic rice expressing wasabi defensin gene exhibit its enhanced resistance to blast fungus (*Magnaporthe grisea*)
H. KANZAKI, S. NIRASAWA, H. SAITOH, M. ITO, M. NISHIHARA, S. YAMAMURA, K. SUZUKI, R. TERAUCHI , I. NAKAMURA 179

Improve breeding for resistance to blast disease
D. THARREAU , C. KAYE , I. FUDAL , H. BÖHNERT , M.H. LEBRUN , Y. WANG, X. ZHU , Y. SHEN , Z. LING , J. XU , L. ZHU , A. PRICE 185

Durable control of rice blast disease with Multilines
S. KOIZUMI, T. ASHIZAWA , K. S. ZENBAYASHI 191

Rice blast control with Sasanishiki multilines in Miyagi prefecture
M. SASAHARA, S. KOIZUMI 201

Development and utilization of isogenic lines *Koshihikari Toyama BL*
Y. KOJIMA, T. EBITANI, Y. YAMAMOTO, T. NAGAMINE 209

Studies on the rice blast pathogen, resistance genes, and implication for breeding for durable blast resistance in Colombia
F.J. CORREA-VICTORIA, D. THARREAU, C. MARTINEZ, M. VALES, F. ESCOBAR, G. PRADO, G. ARICAPA 215

Development of differential varieties for blast resistance in IRRI-Japan collaborative research project
Y. FUKUTA, E. ARAKI, M. J. T. YANORIA, T. IMBE, H. TSUNEMATSU, H. KATO, L. A. EBRON, D. MERCADO-ESCUETA, G. S. KHUSH 229

Identification of blast resistance genes in elite indica-type varieties of rice (*Oryza sativa* l.)
L. A. EBRON, Y. FUKUTA, T. IMBE, H. KATO, M. J. T. YANORIA, H. TSUNEMATSU, G. S. KHUSH, M. YOKOO 235

Resistance of some Chinese hybrid rice, conventional early *indica* and late *japonica* rice to *Magnaporthe grisea*
Y. SHEN, H. ADREIT, XD. ZHU, J. MILAZZO , HQ. CHEN , D. THARREAU 241

CHAPTER IV PATHOGEN: POPULATION ANALYSIS

Triticale and barley: new hosts of *Magnaporthe grisea* in São Paulo, Brazil - relationship with blast of rice and wheat
A.S. URASHIMA, T. DIAS MARTINS, C.R.N.C. BUENO, D.B. FAVARO, M.A. ARRUDA, Y.R. MEHTA 251

Genetics of fertility and mating type in *Magnaporthe grisea*
ANIL S. KOTASTHANE, J. KUMAR, U.S. SINGH & R.S. ZEIGLER 261

Mating type alleles, female fertility and genetic diversity of *Magnaporthe grisea* populations pathogenic to rice from five Asian countries
Y. SHEN , J. MILAZZO , XP. YUAN, H. ADREIT , YL. WANG, JL. NOTTEGHEM , D. THARREAU 271

CHAPTER V PLANT PROTECTION

Efficacy of carpropamid against mutants of *Magnaporthe grisea* at codon 75 on scytalone dehydratase
T. SHIGYO, Y. KUCHII, Y. ARAKI, H. SAWADA 281

Computer simulation approaches for rice blast disease forecasting in Japan
F. NEMOTO, K. ISHIGURO 289

Simulation models of rice blast epidemics:
-From a theoretical approach to a major component of IPM-
K. ISHIGURO 297

PREFACE

For the researchers of rice blast disease and plant blast resistances, IRBC is an important chance to exchange information and discussion on rice blast, which has been a rather minor topic on international plant pathological meetings. Especially for those in Asia, where the blast is one the most important agricultural concerns, IRBC was a unique opportunity to discuss on rice blast with scientists in the West, in where it is achieving the positions of a model pathogen and a model system to study plant microbe interactions. However, I and probably many of Japanese blast researchers have felt an accumulated frustration that, in world plant pathological conferences, the information presented from Asia is only a small fraction although they have much more information that may be valuable for world plant pathologists and resistance researchers. I have noticed also some concept gaps between the researchers in East and West, particularly in the field resistance and the multiline system. There seems to be few who study field resistance and multilines in the West, while in Japan they are expected to become mainstream strategies to suppress the disease. Surely there is a great value in having 3rd IRBC in Japan for cultivating wider communication between the scientists of the East and West.

Having this meeting in Japan in 2002 has another meaning, because it has developed 30 years before the unique non-fungicidal resistance-inducing blast control reagent, probenazole, indicating the future direction of plant disease control, and because of its initiative in the Rice Genome Project. 2002 was a commemorative year with the release of the genome data of both rice and blast fungus. In April the rapid data of rice genome were released from China and Syngenta, and then came the high precision data of the Rice Genome Consortium, in November. The genome data of rice blast fungus also came from Whitehead Institute and NC University. Correspondingly, several researches of functional genomics have started both in rice and blast fungus. These are really epoch making advances in the research of plant microbe interactions.

In these proceedings, the analyses of the mechanism and utilization of plant disease resistance is the keynote theme. Apparently, plant disease resistance is one of the most biologically interesting topics and this rice blast system provides the best-established model to analyze it in crops. Application of rice resistance against blast also provides one of the furthest advanced practical plant protection systems, which recently made a big advance in Japan, but was not sufficiently acknowledged in the international community.

I appreciate greatly for the researchers who have consented opening the 3rd IRBC in Japan and colleagues who have collaborated to realize it. It is my great pleasure that so many recent works on the rice blast resistance are summarized here. I hope this book may stimulate further advances in the researches of plant microbe interactions and can contribute to a wider field of biological sciences and practical plant protection. I also hope the next meeting may be held in shorter interval.

Shinji Kawasaki

OVERVIEW ON THE CHEMICAL CONTROL OF RICE BLAST DISEASE

*Isamu Yamaguchi

RIKEN Plant Science Center, 1-7-22 Suehiro, Tsurumi, Yokohama 230-0045, Japan.

**E-mail: yamaism@postman.riken.go.jp*

Abstract: Rice blast disease is the most serious and damaging of all the plant diseases in rice production in such temperate and humid climate like that of Japan, which resulted in the development of many excellent chemicals for this particular disease, e.g., blasticidin S, kasugamycin, iprobenphos (IBP), edifenphos (EDDP), isoprothiolane, ferimzone, and metominostrobin. These chemicals mostly act on the growth of *M. oryzae* by exhibiting specific fungicidal effects on the pathogen. The wide use of these site specific chemicals sometimes causes development of resistance in the pathogens. In fact, resistance against kasugamycin and organophosphorus thiolate fungicides were recognized in the field where they were too intensively used. Solutions to the problem were contrived for the effective use of the chemicals at hand, such as application of mixtures with different action mechanisms or their use in rotation. While all modern fungicides are developed through extensive safety evaluation, there is a deep public concern about their side effects on non-target organisms and on the environment.

Thus there is a great deal of interest in non-fungicidal disease controlling agents since they are supposedly specific to target organisms and are less likely to lead to resistance problems. Actually, two groups of non-fungicidal rice blast chemicals are currently on the market; melanin biosynthesis inhibitors (MBIs), e.g., fthalide, tricyclazole, pyroquilon, carpropamid, diclocymet and fenoxanil, and the so called priming effectors or plant defense activators such as probenazole, acibenzolar-S-methyl and tiadinil, which induce host resistance against the pathogen's attack.

S. Kawasaki (ed.), Rice Blast: Interaction with Rice and Control, 1–13.
© 2004 *Kluwer Academic Publishers. Printed in the Netherlands.*

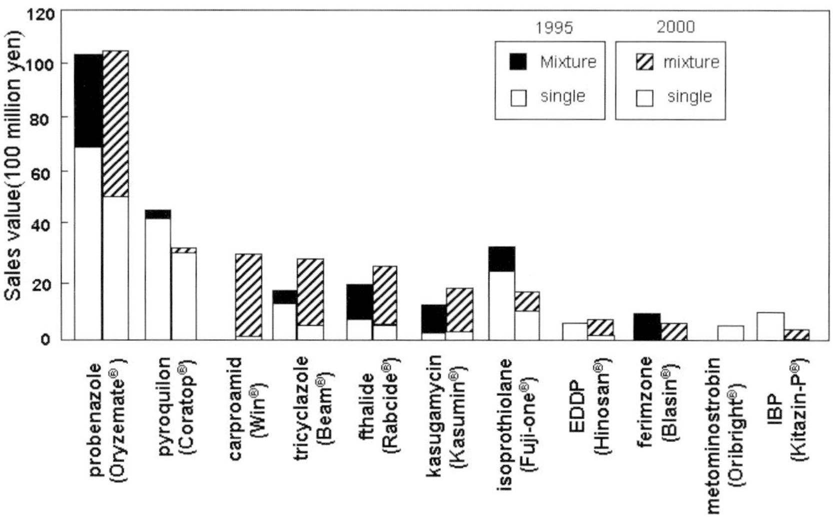

***Figure 1**. Sales of anti-blast fungicides in Japan.*

Figure 1 shows "sales of blast fungicides in Japan", which indicates that non-fungicidal plant activator, probenazole, and melanin biosynthesis inhibitors like pyroquilon, carpropamid, tricyclazole and fthalide are main fungicides for the blast control.

The exploitation and development of new anti-blast fungicides have been becoming more and more difficult in recent years; cost effective studies on the toxicity of the chemicals and residue analyses in crops, animals and the environment are highly required to secure human health and to preserve the ecosystems. Studies on molecular action mechanism of new MBIs and plant activators are currently attracting researchers' interest for the unique features of these chemicals. In this article, recent topics of the blast-controllers used in practice were overviewed mainly in terms of their action mechanisms, which may hopefully suggest the qualities that ideal anti-blast fungicides should possess.

1. INTRODUCTION

Rice blast disease, caused by the fungus Magnaporthe oryzae, has been observed in almost every rice growing country, and it has brought about significant yield losses in the production of rice which is the principal axis of agriculture in most Asian countries. Therefore, the greatest precautions have long been taken for the control of the disease. In Japan, organomercuric compounds were initially introduced for seed treatment and then applied in fields for the control of rice blast since this practice contributed greatly to the protection of rice plant from the disease and resulted in increased rice production. In 1956, however, mercuric poisoning was reported as "Minamata disease". Though the disease was proved not to be caused by agricultural mercuric chemicals but by industrial wastes, due to the dreadful mercuric poisoning, use of organomercuric chemicals for agriculture was prohibited and, as a result of earnest research, excellent non-mercuric blast-controllers were developed for practical use.

In this article, characteristics of the anti-blast chemicals used in practical control will be overviewed, mainly in terms of their modes of action. While advanced molecular biology of M. oryzae and the genetic breeding of blast-resistant rice varieties will be reviewed in this Proceeding, it should be noted that all the research on the rice blast was historically initiated by and based on "how to control the rice blast disease".

There are 16 excellent chemicals against the blast disease at present, as shown in Table 1 and 2. It is quite an exceptional case in crop protection to have so many chemicals for a single disease but it certainly shows the importance of rice blast control in rice production.

2. FUNGICIDAL CHEMICALS TO CONTROL RICE BLAST DISEASE

Conventional anti-blast chemicals such as blasticidins S, kasugamycin, iprobenphos, edifenphos and isoprothiolane have curative effects on the disease caused by M. oryzae, acting on primary metabolism of the pathogen, e.g., protein biosynthesis and phospholipid biosynthesis in the membrane. Some of the modern anti-blast chemicals such as ferimzone and metominostrobin seem to act on the primary metabolism of the pathogen, but they show novel specific anti-blast activity with less phytotoxicity and mammalian toxicity.

2.1. Pesticides of microbial origin

Blasticidin S was the first successful agricultural antibiotic developed in Japan by aiming only at protection of plant disease (Table 1). It was isolated from the culture filtrate of Streptomyces griseochromogenes (Takeuchi et al., 1958). The curative effect of blasticidin S on rice blast is quite high (10-20 ppm; 1-3g/10a) when compared with other synthetic chemicals developed in the 1960's. The antibiotic occasionally caused phytotoxic effects on tobacco plant, eggplant, and beans. Benzylaminobenzene sulfonate of blasticidin S

Table 1. Anti-blast chemicals with fungicidal activities

Chemicals	Chemical structure	Physico-chemicalproperty	Toxicity		
			LD50 (mg/kg)		TLm(ppm)
Blasticidin S	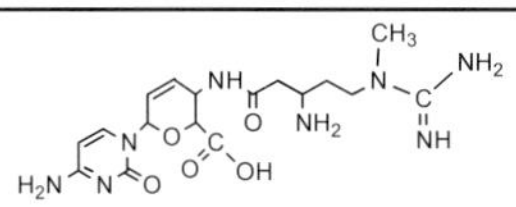	Mp 235-236 ℃,Sol. in water, ethanol,Unstable in alkaline water and to UV	M (O) M (D)	53.3 220	8.7 (carp)
Kasugamycin	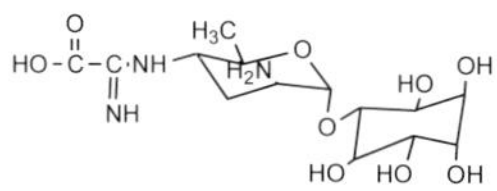	Dp 202-204 ℃, Sol. in water, Insol. in most organic solvents	M (O)	>2,000	>40 (carp)
Iprobenphos(IBP)	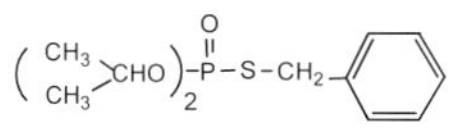	Bp 126 ℃/0.04 mmHg, Sparingly sol. in water, Sol. in ethanol, ether and Xylene, Unstable in alkaline soln.	M (O)	1,830	>10 (carp)
Edifenphos (EDDP)	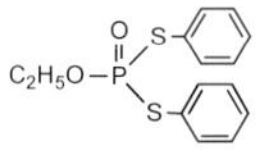	Bp 154 ℃/0.01 mmHg, Insol. in water, Sol. in acetone and xylene, Gradually decomposed in alkaline soln.	M (O) R (O)	214 212	2.5 (carp)
Isoprothiolane	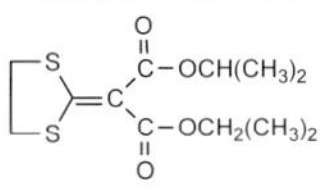	Mp 54-54.5 ℃ Sparingly sol. in water, Sol. in ethanol, acetone and benzene	M (O) R (O)	1,520 1,340	6.8 (carp)
Ferimzone	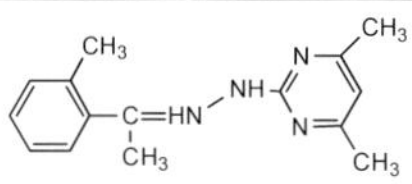	Mp 175-176 ℃, Sparingly sol. in water, Sol. in alcohol and most solvents log Pow 2.9	M (O) R (O)	542 642	27 (carp)
Metominostrobin	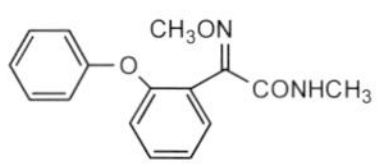	Mp 89 ℃, Sparingly sol. in water, Sol. in acetone and most organic solvents log Pow 2.3	M (O) R (O)	1,413 708	17.5 (carp)

was found to be less phytotoxic to plants without reducing control efficacy and has been industrially produced for practical use.

Biosynthesis of blasticidin S revealed that the pyrimidine ring came from cytosine and the sugar from glucose; the amino acid portion originated from arginine and methionine (Seto et al., 1961). The antibiotic markedly inhibits the protein synthesis of M. oryzae (Misato et al., 1959) and it was shown to act in the step of peptidyl transfer on the ribosomes (Yukioka et al., 1975).

Application of blasticidin S by dusting had occasionally caused conjunctivitis if the dust accidentally came in contact with the eyes, thus an improved dust formulation which contains calcium acetate has been used to alleviate eye trouble, together with harmless wettable powder or emulsion.

The fate of blasticidin S in the environment was investigated to indicate that the antibiotic was easily broken down by sunlight on plants and by microorganisms in soil (Yamaguchi et al., 1972), and no residue was detected in rice grains. In the course of study, several microbes capable of metabolizing and detoxifying blasticidin S were found (Yamaguchi et al., 1975), and two genes to code blasticidin S deaminase were isolated; BSD gene from Aspergillus sp. (Kimura et al., 1994) and another BSD gene from Bacillus sp. (Kamakura et al., 1987). The enzyme BSD belongs to the cytidine deaminase superfamily and is considered to be a good model to see the role of Zn in the enzyme (Kimura et al., 2000). Further, since BS is a potent protein synthesis inhibitor, it proved to function as an excellent selection marker for transformation of Magnaporthe as well as mammal and plant cells (Kimura &Yamaguchi, 1996).

Another anti-blast antibiotic, kasugamycin, is an aminoglycoside isolated from Streptomyces kasugaensis (Umezawa et al., 1965). The antifungal activity of kasugamycin is not significant in vitro, but it indicated marked curative effect against rice blast in a pot test in vivo. This is because kasugamycin exhibits inhibition of fungal growth specifically at pH 4.5~5.0, which roughly coincides with the hydrogen ion concentrations in the tissue of rice plants.

Mammalian toxicity of kasugamycin is quite low and no phytototoxicity was observed in most crops. After several years of intensive practical application of kasugamycin, however, reduction of its efficacy was recognized. Since then, kasugamycin combined with chemicals of different action mechanisms has been used in practice. Fortunately, once the single use of kasugamycin was discontinued, the proportion of resistant strains in the field rapidly declined in a few years.

Aminoglycoside antibiotics generally inhibit protein synthesis by causing codon misreading, whereas kasugamycin was proved to specifically interfere with the formation of the initiation complex, i.e., the binding of aminoacyl-tRNA to the mRNA-30S ribosomal subunit (Yamaguchi & Tanaka, 1966). The remarkable difference in sensitivity between microbes and mammals can be explained by the diversity of the ribosomal nature.

2.2. Anti-blast chemicals to inhibit phospholipid biosynthesis in the membrane

Since the discovery of remarkable antifungal activity of phosphorothiolate compounds, iprobenphos (IBP) and edifenphos (EDDP) were widely used for practical control of the blast disease. IBP is systemic to the rice plant, favoring its application to the paddy water, and EDDP is not systemic but has a potent antifungal activity against the blast fungus, being effective by foliar application. Isoprothiolane is a systemic fungicide primarily effective in preventing the invasion process of the fungus into the host plant (Taninaka et al., 1976; Araki et al., 1976). The chemical structure of isophothiolane (diisoprophyl 1,3-dithiolan-2-yliden- malonate) is seemingly different from those of phosphorothiolate (PTL) fungicides, however, cross-resistance between PTL fungicides and isoprothiolane suggested a similarity in their mode of action (Katagiri & Uesugi, 1977).

IBP, EDDP and isoprothiolane were reported to specifically inhibit the conversion of phosphatidylethanolamine to phosphatidylcholine (Kodama et al., 1979). In as much as phospholipid is an essential component of the fungal membrane, the inhibition of phosphatidylcholine synthesis induces deterioration of membrane permeability and/or enzyme activities associated with membrane, leading to the potent antifungal activities of PTL and isoprothiolane. In addition, the secondary effects to control spindly growth of rice seedlings and plant hoppers were reported for isoprothiolane, and retardant activity on rice plants was known for IBP.

2.3. Novel anti-blast chemicals to disturb membrane permeability and respiration

Ferimzone and metominostrobin are novel systemic fungicides developed for the control of rice blast. The antifungal activity of ferimzone is fungistatic and it does not give fatal damage to the pathogen in vitro, however, ferimzone causes specific leakage of acidic electrolytes from mycelia of M. oryzae. The results obtained so far suggest that ferimzone disturbs membrane permeability for specific molecules or ions in influx as well as in efflux. Thus it has been used as a synergistic component of mixtures with other anti-blast chemicals (Okuno et al., 1989). Metominostrobin belongs to the methoxyacrylate fungicides and it inhibits the fungal respiration by blockage of electron flux through the cytochrome bc1 segment in the mitochondrial respiratory chain. The mycelial cells of M. oryzae, however, induce cyanide-resistant respiration to recover respiratory activity when the cytochrome-mediated pathway is blocked by metominostrobin. Superoxide anion is supposed to be involved in metominostrobin-dependent induction mechanism of the cyanide-resistant respiration. Flavonoids found in plants have an ability to scavenge superoxide anion generated by blockage of electron flux through the cytochrome bc1 segment and thus inhibit metominostrobin-dependent induction of cyanide-resistant respiration. Consequently, metominostrobin controls rice blast in conjunction with rice plant components (Mizutani et al., 1996).

3. NON-FUNGICIDAL CHEMICALS TO CONTROL RICE BLAST DISEASE

Most modern anti-blast chemicals are non-fungicidal and act on the secondary metabolism of the pathogen, such as melanin biosynthesis, or they induce plant defense mechanism, so called systemic acquired resistance (SAR).

3.1 Melanin biosynthesis inhibitors

Anti-blast chemicals such as fthalide, tricyclazole and pyroquilon are primarily not toxic to the vegetative growth of M. oryzae, but perform outstanding control of blast disease in vivo.

The pathogen has a specific infection mechanism to overcome the self-defense in rice plants; when the pathogen recognizes the surface character of the host plants such as hardness and lipophilicity, the germinating hypha starts differentiating an appressorium to infect through the hard surface of rice plants with silicate. For the formation of appressoria, signal transduction is involved and many genes are expressed. Then appressorial cells are matured by forming a melanin layer in between the plasma membrane and cell wall, which can generate a high osmotic pressure inside the cell to penetrate the cuticle layer of host plants. Thus melanin biosynthesis inhibitors can exert their remarkable effects as blast controllers (Yamaguchi & Kubo, 1992).

Actually there are two types of melanin biosynthesis inhibitors: hydroxynaphthalene inhibitors (fthalide, tricyclazole and pyroquilon) and scytalone dehydratase inhibitors (carpropamid, diclocymet and fenoxanil).

Fthalide has a potent protective activity against rice blast with long residual effectiveness. The metabolic fate and behavior of fthalide were examined in detail and no toxic metabolites to mammals were detected, and the phytotoxicity of main metabolites was indicated to be negligible (Tokuda et al., 1976). On the host plant, fthalide interferes with the invasion process of M. oryzae through inhibition of the melanin biosynthesis in the appressoria of fungus (Chida & Sisler, 1987).

Tricyclazole is a highly potent systemic protectant for the control of rice blast. It allows spore germination, appressorial formation and mycelial growth of M. oryzae in vitro but specifically prevents penetration by the fungus into rice tissue by inhibiting the melanin biosynthesis. The synthesis of vermelone, a late intermediate in allomelanin synthesis in M. oryzae, was totally inhibited by the chemical at concentrations as low as 0.01 μg/ml (Tokousbalides & Sisler, 1978; Woloshuk et al., 1980). Pyroquilon has similar modes of action. As a result of this inhibition, predominant accumulation of scytalone and 2-hydroxyjuglone (2-HJ) was observed (Yamaguchi et al., 1982). Flaviolin accumulation was also induced at higher concentrations of tricyclazole and pyroquilon, indicating the presence of another step of inhibition (but catalyzed by the same enzyme) in the pathway.

Table 2. Non-fungicidal anti-blast chemicals

Chemicals	Chemical structure	Physico-chemicalproperty	Toxicity		
			LD_{50}(mg/kg)		TLm(ppm)
Fthalide	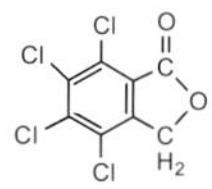	Mp 209 ℃, Sparingly sol. in water, Sol. in most organic solvents, Stable except in strong alkaline soln	M (O)	>10,000	320 (carp)
Tricyclazole	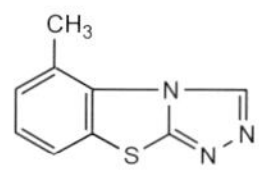	Mp 187 ℃, Partly sol. in water, Sol. in chloroform, acetone and ethanol.	M (O)	250	13.5 (gold fish)
Pyroquilon	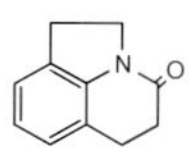	Mp. 112℃ Sol, in acetone, dichlormethane and methanol log Pow 1.6	M (O) R (O)	740 850	25 (carp)
Carpropamid	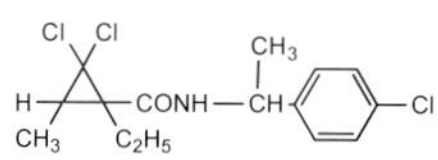	Mp 147-149℃, Insol. in water, Sol. in acetone and most organic solvents log Pow 4.2	M (O) R (O)	> 5,000 > 5,000	5.6 (carp)
Diclocymet	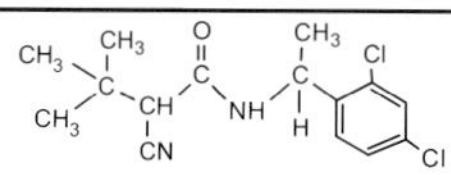	Mp. 155-156℃ Insol. in water, Sol, in acetone, methanol and xylene	R (O)	> 5,000	
Fenoxanil	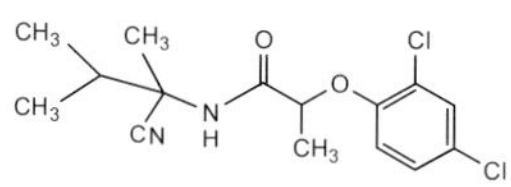	Mp 153 ℃, Insol. in water and hexane Sol. in acetone and methanol. Unstable to UV. log Pow 3.5	M (O) R (O)	5,000 5,000	10.1 (carp)
Probenazole (PBZ)	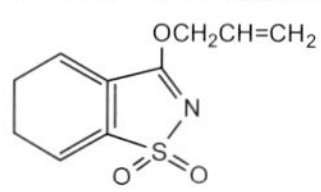	Mp 138 ℃, Sparingly sol. in water, Sol. in acetone and chloroform	M (O) R (O)	2,750 2,030	6.3 (carp)
Acibenzolar-S-methyl	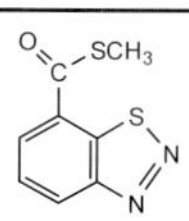	Mp 133 ℃, Insol. in water, Sol. in acetone, toluene and dichlormethane Unstable to UV and alkaline soln.log Pow 3.1	M (O) R (O)	>5,000 >5,000	0.8 (carp)
Tiadinil	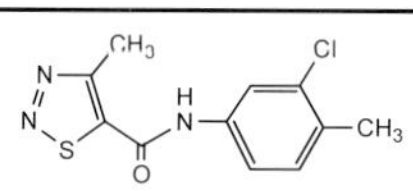	Mp 112℃, Insol. in water log Pow 3.7	R (O)	> 6,147	7.0 (carp)

The chemical also inhibits secondary infection under field conditions by reducing the sporulation of M. oryzae. Field studies indicated long-term control activity of tricyclazole and pyroquilon either by foliar spray, soil drench in the transplant flat, or submerged application.

The buff mutants of M. oryzae were shown to phenotypically resemble the tricyclazole-treated wild-type strains and were essentially nonpathogenic to rice plant, which suggests that a melanin biosynthetic pathway is indispensable for the appressoria to mature and to obtain infestiosity (Yamaguchi & Kubo, 1992).

Carpropamid was developed as a potent controlling agent against the rice blast (Kagabu & Kurahashi, 1998). Enzyme kinetics data showed that carpropamid is a tight-binding competitive inhibitor of scytalone dehydratase (STD), an enzyme in the melanin biosynthesis. The X-ray crystal structure analysis of STD complexed with carpropamid identified interactions that determine the tight-binding (Nakasako et al, 1998) as shown in Fig. 2.

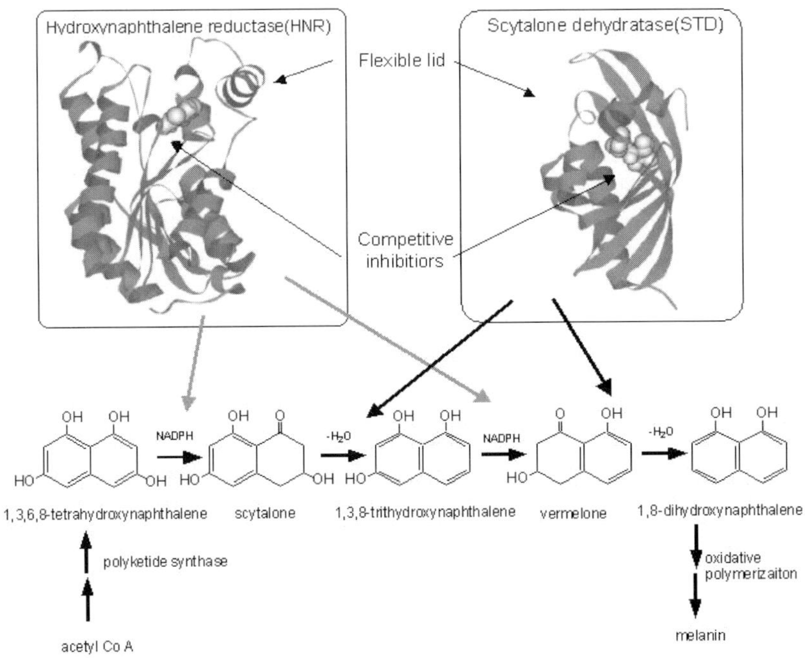

Figure 2. *Binding of inhibitors with the key enzymes in the melanin biosynthesis*

Three types of interactions were considered to be important in the binding. First, a strong hydrogen bond was observed between Asn 131 and chlorine. Second, two hydrogen bonds mediated by two water molecules were found. Third, the phenyl group of carpropamid interacted with aromatic amino acid residues of STD. Structural data also inferred that the C-terminal region (154G-172K) is important for the enzyme function. In fact, enzyme kinetics for the STD-mutants with or without carpropamid suggested involvement of the C-terminal region in enzyme function and in the inhibitor binding. Of the four isomers of carpropamid, only KTU3616B (1S, 3R, 1'R) is active in melanin biosynthesis inhibition in blast control (Motoyama *et al.*, 2002). Inhibition of STD by KTU3616B was observed at very low concentrations of the inhibitor close to the enzyme concentrations, suggesting that KTU3616B is a competitive tight-binding inhibitor of STD. The calculated *K*i value was 140 pM, which is more than 10^5 times smaller than the *K*m for scytalone.

The C-terminal region is supposed to act as a flexible lid which controls access of substrates and inhibitors to the active site pocket; deletion of amino acids in the C-terminal region showed a complete loss of enzyme activity, indicating that the C-terminal portion of STD was important for catalysis and/or structural integrity. Some amino acid residues involved in the inhibitor binding were also exchanged by site-directed mutagenesis. The Y30F, V75A, and F169A mutants still retained enzyme activity (more than 5% of original STD activity) but the *K*i values for carpropamid significantly increased in V75A and F169A, suggesting the specific involvement of V75 and F169 of STD in carpropamid binding. While the binding between mutated STD and carpropamid appears not so weak, the results implied that V75 and F169 point mutations might produce carpropamid-resistant strains. In fact, field resistant mutants with V75M STD have emerged in Western Japan in 2001. Detailed analyses of the function of the catalytic region of STD may help both understanding the resistant mechanism against STD-inhibitors and designing more efficient chemicals.

Recently, novel STD-inhibitors such as diclocymet (Manabe *et al.*, 2002) and fenoxanyl (Nishimura & Hino, 2002) have been developed and registered as anti-blast chemicals in Japan. They are considered to have action mechanisms similar to carpropamid.

3.2 Anti-blast chemicals to induce systemic acquired resistance to M. oryzae

Another attractive type of non-fungicidal blast controllers is called plant activators or priming effectors to induce systemic acquired resistance (SAR) in plants.

Probenazole (PBZ) is a systemic agent that is quite effective against rice blast disease when applied to the root system of rice plant (Watanabe et al., 1977). The application of PBZ to rice plants was shown to augment enzyme activities related to resistance at the invaded cells. The long lasting activity of PBZ was attributed to this host mediated defense action. PBZ as a plant defense activator is described more in detail by Iwata et al. in this Proceeding.

Acibenzolar-S-methyl (BTH) was also reported to induce SAR in rice plants (Yamaguchi, 1998). The chemical structure of BTH looks similar to PBZ but it was found that they have different sites of action; PBZ and its active metabolite (1,2-benzisothiazole-1,1- dioxide; BIT) act in the step prior to the salicylic acid but BTH induces SAR by acting downstream of salicylic acid as shown in Fig. 3 (Yoshioka *et al.*, 2001; Nakashita *et al.*, 2002). Recently, tiadinil, another priming effector, was developed and registered as an anti-blast chemical (Umetani *et al.*, 2003) but its detailed action mechanism still remains to be elucidated.

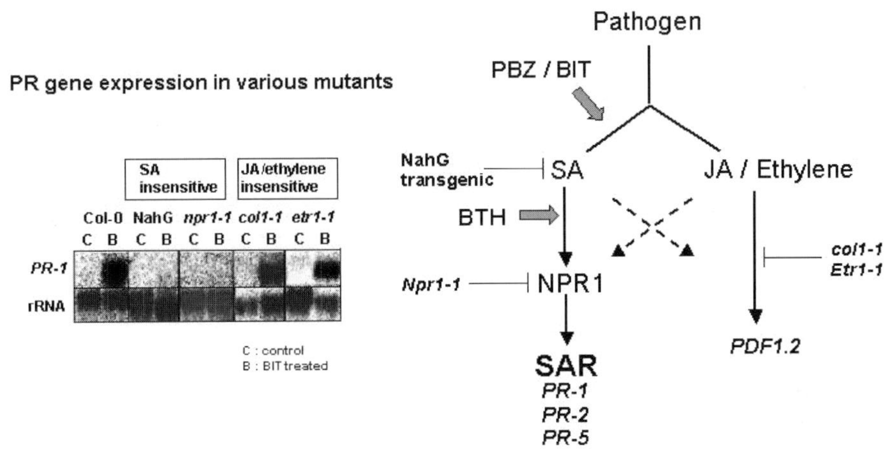

***Figure 3**. Probenazole and BIT induces systemic acquired resistance (SAR)*

REFERENCES

Araki, F., & Miyagi, Y. (1976). Effect of isoprothiolane on the infection process of *Pyricularia oryzae*. *Ann. Phytopathol. Soc. Jpn*. 42, 401-406.

Chida, T., & Sisler, H.D., (1987). Effect of inhibitors of melanin biosynthesis on appressorial penetration and reductive reactions in *Pyricularia oryzae* and *Pyricularia grisea*. *Pestic. Biochem. Physiol*. 29, 244-251

Kagabu, S., & Kurahashi, Y. (1998). Carpropamid, a new fungicide for rice blast control: its selection based on stereoisomer-activity relationships. *J. Pesticice Sci.* 23, 145-147.

Kamakura, T., Kobayashi, K., Tanaka, T., Yamaguchi, I., & Endo, T. (1987). Cloning and expression of a new structural gene for blasticidin S deaminase, a nucleoside aminohydrolase. *Agric. Biol. Chem.* 51, 3165-3168

Katagiri, M., & Uesugi, Y. (1977). Similarities between fungicidal action of isoprothiolane and organophosphorus thiolate fungicides. *Phytopathol.* 67, 1415-1417.

Kimura, M., Kamakura, T., Tao, Q-Z., Kaneko, I., & Yamaguchi, I. (1994). Cloning of the blasticidin S deaminase gene (BSD) from *Aspergillus terreus* and its use as a selectable marker for *Schizosacchromyces pombe* and *Pyricularia oryzae*. *Mol. Gen. Genet.* 242, 121-129.

Kimura, M., Sekido, S., Isogai, Y., & Yamaguchi, I. (2000). Expression, purification and characterization of blasticidin S deaminase (BSD) from *Aspergillus terreus*: The role of catalystic zinc in enzyme structure. *J. Biochem.* 127, 955-963.

Kimura, M. & Yamaguchi, I.(1996). Recent development in the use of Blasticidin S, a microbial fungicide, as a useful reagent in molecular biology. *Pestic. Biochem. Physiol.* 56, 243-248.

Kodama, O., Yamada, H., & Akatsuka, T. (1979). Kitazin P, inhibitor of phosphatidylcholine biosynthesis in *Pyricularia oryzae*. *Agric. Biol. Chem.* 43, 1719-1725.

Manabe, A., Maeda, K., Enomoto, M., Takano, H., Katoh, T., Yamada, Y., & Oguri, Y. (2002). Synthesis and fungicidal activity of α-cyanoacetamide derivatives: Discovery of a new rice blast fungicide, diclocymet (S-2900). *J. Pesticide Sci.* 27, 257-266.

Misato, T., Ishii, I., Asakawa, M., Okimoto, Y., & Fukunaga, K. (1959). Antibiotics as protectant fungicides against rice blast. II. The therapeutic action of blasticidin S. *Ann. Phytopathol. Soc. Jpn.* 24, 302-306.

Mizutani, A., Miki, N., Yukioka, H., Tamura, H., & Masuko, M. (1996). A possible mechanism of control of rice blast disease by a novel alkoxyiminoacetamide fungide, SSF126. *Phytopathol.* 86, 295-300.

Motoyama, T., Kagabu, S., Matsuoka, M., & Yamaguchi, I. (2002). Differential inhibition of a melanin biosynthetic enzyme scytalone dehydratase by carpropamid, a fungicide for rice blast control, and its isomers. *J. Pesticide Sci.* 27, 141-144.

Nakasako, M., Motoyama, T., Kurahashi, Y., & Yamaguchi, I. (1998). Cryogenic X-ray crystal structure analysis for the complex of scytalone dehydratase of a rice blast fungus and its tight-binding inhibitor, Carpropamid: The structural basis of tight-binding inhibition. J. Biochem. 37, 9931-9939.

Nakashita, H., Yoshioka, K., Yasuda, M., Nitta, T., Arai, Y., Yoshida, S., & Yamaguchi, I. (2002). Probenazole induces systemic acquired resistance in tobacco thorough salicylic acid accumulation. *Physiol. Mol. Plant Pathol* .61, 197-203.

Nishimura, A., & Hino, I. (2002). Fenoxanil (ACHI-BU)-*Anew systemic fungicide for the control of rice plan-agrochemicals Japan*. 13-15.

Okuno, T., Furusawa, I., Matsuura, K., & Shishiyama, J. (1989). Mode of action of ferimzone (TF-164), a novel systemic fungicide for rice diseases: Effect on the general metabolism of *Pyricularia oryzae*. *Ann. Phytopath. Soc. Japan.* 55, 281-289.

Seto, H., Yamaguchi, I., Otake, N., & Yonehara, H. (1961). Studies on the biosynthesis of blasticidin S Part I. *Agric. Biol. Chem.* 32, 1292-1298.

Takeuchi, S., Hirayama, K., Ueda, K., Sakai, H., & Yonehara, H. (1958). Blasticidin S, a new antibiotic. *J. Antibiot., Ser.* A 11, 1-5.

Taninaka, K., Kurono, H., Hara, T., & Murata, K. (1976). Rice blast controlling activities of bis (alkoxycarbonyl) ketene dithioacetals and their related compounds. *J. Pesticide Sci*, 1, 115-122.

Tokuda, T., Nishiki, M., Hoshi, H., Shinoda, K., Ishida, M., & Misato, T. (1976). Metabolic fate of fthalide (4,5,6,7-tetrachlorophthalide) in compost. *J. Pesticide Sci.* 1, 283-294.

Tokousbalides, M. C., & Sisler, H. D. (1978). Effect of tricyclazole on growth and secondary metabolism in *Pyricularia oryzae. Pestic. Biochem. Physiol.* 8, 26-32.

Umetani, K., Hakuno, F., Nishimura, A., Yamamoto, Y., & Hirooka, T. (2003). Studies on NNF-9850 (Tiadinil, V-GET), a novel fungicide. *Abst. PSJ Ann. Meeting.2003 (in Japanese).*

Umezawa, H., Okami, Y., Hashimoto, T., Suhara, Y., Hamada, M., & Takeuchi, T. (1965). A new antibiotic, kasugamycin. *J. Antibiot., Ser. A* 18, 101-103.

Watanabe, T., Igarashi, H., Matsumoto, K., Seki, S., Mase, S., & Sekizawa Y., (1977). The characteristics of probenazole (Oryzemate) for the control of rice blast. *J. Pesticide Sci.* 2, 291-296.

Woloshuk, C. P., Sisler, H. D., Tokousbalides, M. R., & Dutky, S. R. (1980). Melanin biosynthesis in *Pyricularia oryzae* – site of tricyclazole inhibition and pathogenicity of melanin-deficient mutants. *Pestic. Biochem. Physiol.* 14, 256-264.

Yamaguchi, H. & Tanaka, N. (1966). Inhibition of protein synthesis by blasticidin S. *J. Biochem.* 60, 632.

Yamaguchi, I. (1998). Activators for systemic acquired resistance. *Fungicidal Activity-Chemical and Biological Approaches to Plant Protection.* 193- 219.

Yamaguchi, I., & Kubo, Y. (1992). Target sites of melanin biosynthesis inhibitors. *Target sites of fungicide action.* 101-114.

Yamaguchi, I., Shibata, H., Seto, H., & Misato, T. (1975). Isolation and purification of blasticidin S deaminase from *Aspergillus terreus. J. Antibiot.* 28, 7-14.

Yamaguchi, I., Takagi, K., & Misato, T. (1972). The fate of blasticidin S, an agricultural antibiotic, in the environment (Part1) The site for degradation of blasticidin S. *Agric. Biol. Chem.* 36, 1719- 1727.

Yamaguchi, I., Sekido, S., & Misato, T. (1982). The effect of non-fungicidal anti-blast chemicals on the melanin biosynthesis and infection by Pyricularia oryzae. *J. Pesticide Sci.* 7, 523-529.

Yoshioka, K., Nakashita, H., Klessig, D., & Yamaguchi, I. (2001). Probenazole induces systemic acquired resistance in *Arabidopsis* with a novel type of action. *Plant J.* 25, 149-157.

Yukioka, M., Hatayama, T., & Morisawa, S. (1975). Affinity labeling of the ribonucleic acid component adjacent to the peptidyl recognition center of peptidyl transferase in *Escherichia coli* ribosomes. *Biochem. Biophys. Acta.* 390, 192-208.

RNA SILENCING IN THE BLAST FUNGUS *MAGNAPORTHE GRISEA*

*Hitoshi Nakayshiki, Naoki Kadotani, Yukio Tosa and Shigeyuki Mayama

Faculty of Agriculture, Kobe University, Kobe, Japan

**Email: hnakaya@kobe-u.ac.jp*

Abstract. We have examined RNA silencing in the blast fungus *Magnaporthe grisea* using the green fluorescence protein (GFP) gene as a model. The fluorescence in the GFP transformants was efficiently silenced by introducing a plasmid expressing double-stranded GFP RNA. Consequently, GFP mRNA accumulation was drastically reduced in the silenced transformants. In addition, small interfering RNAs of GFP were observed only in the silenced transformants. These results indicate that RNA silencing operates in *M. grisea* through processes very similar to those reported in other eukaryotes.

1. INTRODUCTION

RNA-mediated gene silencing is a novel gene regulatory mechanism that operates at the RNA level, thus as a post-transcriptional process. This unique mechanism has been independently discovered and studied in different classes of eukaryotes as co-suppression or post-transcriptional gene silencing (PTGS) in plants (Napoli *et al*., 1990; van der Krol *et al*., 1990), quelling in *Neurospora crassa* (Romano and Macino, 1992) and also as RNA interference (RNAi) in animals (Fire *et al*, 1998). Several lines of evidence suggest that common mechanisms underlie in those phenomena (Hammond *et al*., 2001). Therefore, here we use RNA silencing as a general term to describe them. RNA silencing is now regarded as an intrinsic defence mechanism to counter invading viruses and endogenous transposable elements in a wide range of eukaryotes (Waterhouse *et al*., 2000).

S. Kawasaki (ed.), Rice Blast: Interaction with Rice and Control, 15–21.
© 2004 *Kluwer Academic Publishers. Printed in the Netherlands.*

In several model organisms (eg. *Caenorhabditis elegans*), RNA silencing has been shown to provide a powerful reverse genetic tool to convert accumulating genomic information into biological function by silencing a gene of interest using dsRNA homologous to the target (Hammond *et al.*, 2001). This novel approach has several advantages over conventional methods such as gene disruption. First, this technology provides rapid and simple methods to attenuate gene expression *in vivo*. For example, in nematode, RNA silencing is induced by the injection of, or even by feeding, the homologous dsRNA to the worm (Fire *et al*, 1998). In plants, VIGS (virus-induced gene silencing), in which a target sequence carried by a virus vector is introduced into plants simply by inoculation, is widely used (Angell and Baulcombe, 1999). Those methods do not require a time-consuming transformation process. Second, RNA silencing would be applicable to genes that are difficult to study with the gene disruption method (eg. multigene families and lethal genes) since it even functions against homologous RNA species with some sequence diversity, and usually results in various, but not always total, degrees of silencing.

Rice blast disease, caused by the ascomycete *Magnaporthe grisea*, is one of the most devastating threats to rice production worldwide. This fungus represents a model fungal pathogen for elucidating plant-microbe interactions. The rice blast genome project was launched in 1998 by the international rice blast genome (IRBG) consortium, and this year (2002) the almost-complete genome sequence of this fungus has been published and become available on a web database (Martin *et al.*, 2002). Effective means to exploit the wealth of the genome information are now anticipated. In filamentous fungi, RNA silencing has been experimentally identified only in two species, *N. crassa* and *Cryptococcus neoformans* to date (Liu *et al.*,

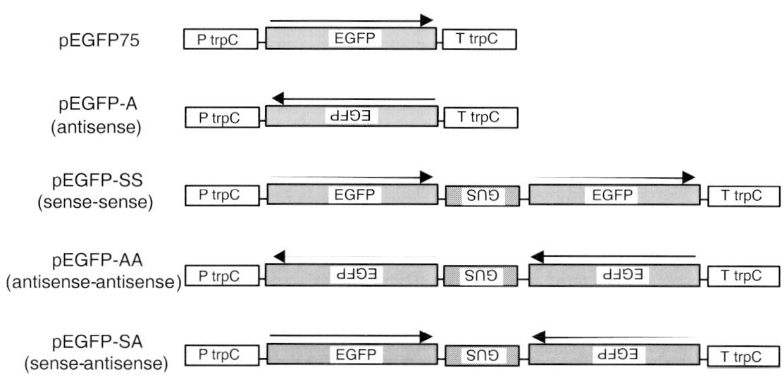

***Figure 1**. Schematic representation of transformation constructs used in this study. PtrpC, Aspergillus nidulans trpC promoter; T trpC, A. nidulans trpC terminater; EGFP, enhanced green fluorescence protein; GUS, beta-glucuronidase*

2002), and is seldom used as a tool for gene analysis. In this study, we show the operation of RNA silencing in the rice blast fungus using the green fluorescence protein (GFP) gene as a model, which leads the way in making use of RNA silencing in this fungus for genome-wide gene analysis.

2. HAIRPIN RNA EFFICIENTLY INDUCES GENETIC INTERFERENCE IN *M. GRISEA*

Plasmids used in this study are shown in Fig. 1. First we introduced pEGFP75, a plasmid carrying the EGFP (enhanced green fluorescence protein) gene under the control of the *Aspergillus nidulans* trpC promoter, into a wheat isolate of *M. grisea*. One of the transformants exhibiting strong GFP fluorescence and possessing about ten copies of pEGFP75 was designated as GFP10 and used in subsequent transformations as a recipient.

Next, we made double transformants with various silencing constructs expressing a single or duplicated GFP-RNA(s) in different orientations (Fig. 1). One of the silencing constructs pEGFP-A expresses single anti-sense GFP RNA, and the others (pEGFP-AA, pEGFP-SS and pEGFP-SA) produce double GFP RNAs in the following three orientations: sense-sense, antisense-antisense and sense-antisense, all having a partial sequence of the GUS gene as an internal spacer. Each of the constructs was introduced into GFP10 together with pII99, carrying the geneticin resistance gene. For controls, we also made double transformants with only pII99.

Eighty double transformants were randomly picked out from each

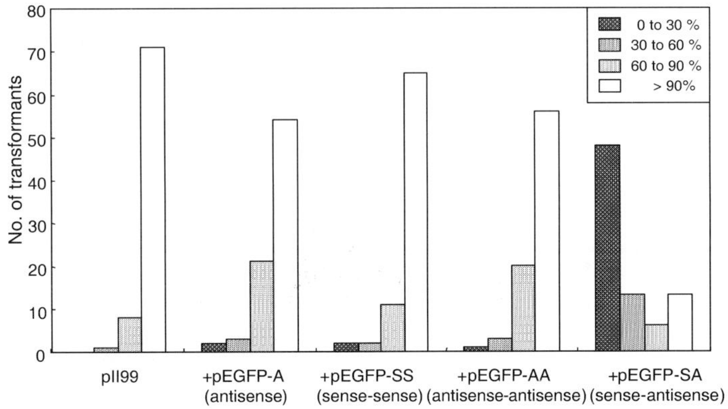

Figure 2. *The double transformants with constructs expressing GFP RNA(s) in various orientations were classified into four categories (0 to 30, 30 to 60, 60 to 90, and above 90 %) based on relative fluorescence to the original single-transformant GFP10. Number of the double transformants in each category (out of 80) was plotted in the graph.*

transformation and their GFP fluorescence was measured with an image analyser. The double transformants were then classified into four categories based on relative fluorescence to the original single-transformant GFP10, and number of the double transformants in each category was plotted in Fig. 2. The loss of GFP fluorescence was most frequently observed in the double transformants with pEGFP-SA expressing a hairpin GFP RNA with a double-stranded structure at the stem. It is noteworthy that various degrees of silencing were observed in the pEGFP-SA double transformants. In contrast, a few silenced double transformants were recognized with other constructs expressing antisense (pEGFP-A), antisense-antisense (pEGFP-AA) and sense-sense (pEGFP-SS) GFP RNAs, but the number of silenced transformants among them was low. Genomic Southern analysis of pEGFP-SA transformants revealed that almost all the pEGFP75 (the original GFP expressing plasmid) copies retained intact, eliminating the possibility that GFP silencing in the transformants was caused by genetic changes in the pEGFP75 loci (data not shown). Those results indicated that hairpin RNA (dsRNA) was a potent trigger of gene silencing in M. grisea.

3. GFP SILENCING IS CAUSED BY LOW LEVEL ACCUMULATION OF GFP mRNA AND NO METHYLATION IS INVOLVED

Since it has been described that RNA silencing in plants and animals is due to post-transcriptional degradation of targeted mRNA, we tested whether the loss of GFP fluorescence correlates with a lower level of GFP mRNA accumulation. Total RNA was isolated from GFP10 and the pEGFP-SA double transformants with or without GFP fluorescence, followed by a northern blot analysis. High levels of GFP mRNA were detected in the single transformant GFP10 in addition to the double transformants exhibiting strong GFP fluorescence (Fig. 3). In contrast, significantly reduced levels of

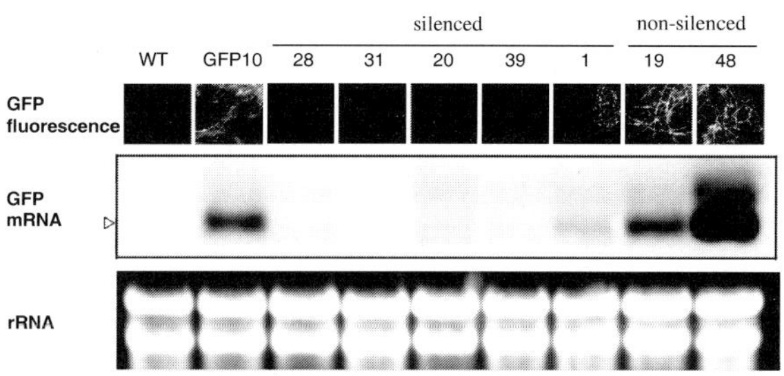

Figure 3. *Northern blots and GFP fluorescence of silenced and non-silenced double transformants with pEGFP-SA. rRNAs stained with ethidium bromide is shown below the blot.*

GFP mRNA accumulation were detected in the silenced transformants with no or low GFP fluorescence. In some of the silenced transformants, such as No. 20 and 1, mRNA accumulation was detected at very low levels; however, it was below levels detectable by northern analysis in the other silenced transformants. The different levels of mRNA accumulation in the silenced transformants may be the cause of various degrees of silencing.

DNA methylation is sometimes reported to be associated with RNA silencing, especially in plants. On the other hand, quelling in *N. crassa* is known to be a methylation-independent process (Cogoni *et al.*, 1996). Thus, we examined involvement of DNA methylation in GFP silencing of *M. grisea* by isochizomer analysis. No methylation was observed on the promoter nor on the coding sequences of the GFP gene in any of the silenced and non-silenced double transformants tested (data not shown), suggesting that DNA methylation was not involved in the GFP silencing in *M. grisea.*

4. SMALL INTERFERING RNAS ARE INVOLVED IN GFP SILENCING IN *M. GRISEA*

Small interfering RNAs (siRNAs) are believed to be key molecules in the RNA silencing process, thus recognized as its hallmark. Finally, we examined the involvement of siRNA in GFP silencing of *M. grisea* to gain further insight into its molecular mechanisms. Total RNA was extracted from the silenced and non-silenced double transformants, run on a polyacrylamide gel, and probed with FITC-labeled sense GFP RNA. Small hybridizing RNA molecules were detected with sizes of 19 to 24 nt in the silenced transformants but not in the non-silenced ones or the single transformant GFP10 (Fig. 4). Therefore, the presence of siRNA correlated with GFP silencing. This result strongly supports the idea that GFP silencing in *M. grisea* is processed by a similar mechanism to RNA silencing reported in other eukaryotes.

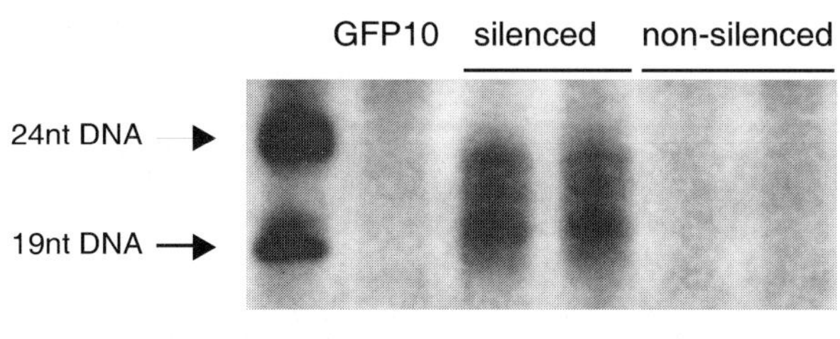

Figure 4. *Detection of small interfering RNAs in the GFP silenced double transformants. Total RNA from silenced and non-silenced double transformants as well as one from GFP10 were fractionated on a 15% polyacrylamide gel and probed with GFP.*

5. DISCUSSION

It has been shown that there are several ways to trigger RNA silencing in various eukaryotes. Co-suppression in plants, which is now regarded to be caused via the RNA silencing pathway, was originally discovered in Petunia plants over-expressing the chalcone synthase mRNA (Napoli *et al.*, 1990; van der Krol *et al.*, 1990); therefore induced by sense RNA. Quelling found in the fungus *N. crassa* occurred in transformants with multiple transgenes having no promoter (Romano and Macino, 1992). Injection of dsRNA into nematode (RNAi) results in silencing of a homologous endogenous gene (Fire *et al*, 1998). The anti-sense strategy, which has been used to attenuate the expression of a gene of interest in a wide range of organisms, is also believed to function, at least partly, through the processes of RNA silencing.

To establish RNA silencing as a gene analysis tool in a given organism, it is important to know which trigger induces RNA silencing most efficiently. Consequently, we examined various RNA species with different orientations to find a potent trigger of RNA silencing in *M. grisea*. The results indicated that dsRNA (hairpin RNA) worked best for the purpose in *M. grisea*, which is also the case with most of the plants and animals examined previously (Fire *et al.*, 1998; Wesley *et al.*, 2001). Surprisingly, there was not much difference between sense and anti-sense RNAs in the efficacy of silencing induction in *M. grisea*, since antisense RNA was expected to readily form dsRNA by annealing with the target mRNA *in vivo*. In a current model of RNA silencing, occurrence or synthesis of dsRNA is believed to be the first and key step in initiating the following processes. Our results implied that most of the dsRNA molecules induced by both sense and antisense RNAs may be produced by a host-encoded RdRP (RNA-dependent RNA polymerase)-mediated process with some recognition system for "aberrant RNA". Direct interaction of RNA molecules to target mRNA, on the other hand, may rarely be involved.

The first discovery of the biochemical agent that is directly involved in RNA silencing was reported by Hamilton and Baulcombe (1999) in plants. That was siRNA, an RNA species about 23 to 25 nt long that includes both sense and antisense target sequences, and is present only in silenced plants. Not only are siRNAs supposed to work as guide molecules to target cognate mRNA for degradation, but also to be a primer for multiple dsRNA synthesis cycles by the action of RdRP, leading to amplification of degradation signals (Lipardi *et al.*, 2001). Here, we show that an accumulation of small RNAs ~24 nt long was associated with RNA silencing in *M. grisea*. This finding strongly suggests the presence of the RNA silencing mechanism in *M. grisea*. Actually, counterparts of *qde-1*, *qde-2* and *qde-3*, genes required for quelling in *N. crassa* (Cogoni and Macino, 1997), are present in the genome database of *M. grisea* (data not shown). This information is also consistent with the conclusion. It is worthwhile examining if those homologs play a role in RNA silencing of this fungus. This issue will be addressed in the future.

REFERENCES

Angell S.M., Baulcombe D.C. (1999) Technical advance: potato virus X amplicon-mediated silencing of nuclear genes. Plant J. 20: 357-362.

Cogoni C., Irelan J.T., Schumacher M., Schmidhauser T.J., Selker E.U., Macino G. (1996) Transgene silencing of the al-1 gene in vegetative cells of Neurospora is mediated by a cytoplasmic effector and does not depend on DNA-DNA interactions or DNA methylation. EMBO J. 15: 3153-3163.

Cogoni C., Macino G. (1997) Isolation of quelling-defective (qde) mutants impaired in posttranscriptional transgene-induced gene silencing in *Neurospora crassa.* Proc. Natl. Acad. Sci. U.S.A. 94: 10233-10238

Fire A., Xu S., Montgomery M.K., Kostas S.A., Driver S.E., Mello C.C. (1998) Potent and specific genetic interference by double-stranded RNA in *Caenorhabditis elegans*. Nature 391: 806-811.

Hamilton A.J., Baulcombe D.C. (1999) A species of small antisense RNA in posttranscriptional gene silencing in plants. Science 286: 950-952.

Hammond S.M., Caudy A.A., Hannon G.J. (2001) Post-transcriptional gene silencing by double-stranded RNA. Nat. Rev. Genet. 2: 110-119.

Martin S.L., Blackmon B.P., Rajagopalan R., Houfek T.D., Sceeles R.G., Denn S.O., Mitchell T.K., Brown D.E., Wing R.A., Dean R.A. (2002) *Magnaporthe*DB: a federated solution for integrating physical and genetic map data with BAC end derived sequences for the rice blast fungus *Magnaporthe grisea*. Nucleic Acids Res. 30: 121-124.

Napoli C., Lemieux C., Jorgensen R. (1990) Introduction of a chimeric chalcone synthase gene into petunia results in reversible co-suppression of homologous genes *in trans.* Plant Cell 2: 279-289.

Lipardi C., Wei Q., Paterson B.M. (2001) RNAi as random degradative PCR: siRNA primers convert mRNA into dsRNAs that are degraded to generate new siRNAs. Cell 107: 297-307.

Liu H., Cottrell T.R., Pierini L.M., Goldman W.E., Doering T.L. (2002) RNA interference in the pathogenic fungus *Cryptococcus neoformans*. Genetics 160: 463-470.

Romano N., Macino G. (1992) Quelling: transient inactivation of gene expression in *Neurospora crassa* by transformation with homologous sequences. Mol. Microbiol. 6: 3343-3353.

Van der Krol A.R., Mur L.A., Beld M., Mol J.N., Stuitje A.R. (1990) Flavonoid genes in petunia: addition of a limited number of gene copies may lead to a suppression of gene expression. Plant Cell 2: 291-299.

Wesley S.V., Helliwell C.A., Smith N.A., Wang M.B., Rouse D.T., Liu Q., Gooding P.S., Singh S.P., Abbott D., Stoutjesdijk P.A., Robinson S.P., Gleave A.P., Green A.G., Waterhouse P.M. (2001) Construct design for efficient, effective and high-throughput gene silencing in plants. Plant J. 27: 581-590.

Waterhouse P.M., Wang M.B., Lough T. (2001) Gene silencing as an adaptive defence against viruses. Nature 411: 834-842.

SAGE (SERIAL ANALYSIS OF GENE EXPRESSION) IN *MAGNAPORTHE GRISEA*

Profiling of cAMP-inducible genes involved in appressorium formation

*Toshikazu Irie[1], Hideo Matsumura[2], Ryohei Terauchi[2] and Hiromasa Saitoh[2, 3]

[1]*Department of Environmental Science, The university of Shiga Prefecture.*
[2]*Iwate Biotechnology Research Center.*
[3]*Japan Society for the Promotion of Science*

**E-mail: tirie@ses.usp.ac.jp*

Abstract. SAGE (Serial Analysis of Gene Expression) is a highly accurate quantitative method for profiling gene expression. We exploited SAGE to study the global gene expression of *Magnaporthe grisea* during appressorium development induced by cAMP treatment of conidia. More than 3000 tags each were extracted and compared between the cAMP-treated and non-treated (control) conidia. Sequences of the tags for which frequency was considerably more or less in the treatment with cAMP as compared to the control were used for homology search with EST data registered in Genbank and *M. grisea* genome v.1 (Fungal Genomics Laboratory - North Carolina State University and Whitehead Institute Center for Genome Research). cAMP-induced genes were categorized according to their putative function. Among the cAMP-inducible genes with known functions, those involved in sugar metabolism (respiration) and nucleic acid metabolism were represented as the top two categories. In view of the established EST database and the availability of whole genome sequence in *M. grisea*, the utility of SAGE for dissecting its pathogenicity will expand in future.

S. Kawasaki (ed.), Rice Blast: Interaction with Rice and Control, 23–28.
© 2004 *Kluwer Academic Publishers. Printed in the Netherlands.*

1. INTRODUCTION

In recent years, considerable progress has been made to understand molecular mechanisms of *Magnaporthe grisea* infection to rice plants. A number of genes involved in infectious processes have been cloned and characterized (Talbot, 1995; Hamer & Talbot, 1998; Balhadére & Talbot, 2000; Xu, 2000; Idnurm & Howlett, 2001). However, available information on molecular events in blast infection is still fragmentary. Aiming at identification of more genes involved in host infection, EST analysis has been carried out using cDNAs obtained from libraries constructed from blast-infected rice leaves (Kim et al. 2001; Rauyaree et al. 2001). However, the number of ESTs analyzed in these studies is far from satisfactory to obtain an overview of gene expression changes in blast fungal infection. To improve the situation, we are applying the SAGE (Serial Analysis of Gene Expression) to obtain the global gene expression profiles of *M. grisea* in infection processes.

2. SAGE METHODOLOGY

Serial analysis of gene expression (SAGE), developed by Velculescu, Zhang, Vogelstein, & Kinzler (1995), is a highly accurate, quantitative analysis of expression of thousands of genes at a time, and recommends itself as an efficient tool for gene expression studies in the post-genomic era (Velculescu et al. 1995). Briefly, a 9-11 bp DNA fragment (tag) adjacent to the *Nla* III site located closest to the 3' end of each cDNA is extracted by a series of linker-ligations and restriction digestions. Tags are once amplified by PCR, concatenated, cloned into a plasmid vector and sequenced. The frequency of each tag in the sequence directly reflects the abundance of each mRNA in the sample. The size of tags (10 bp) in most cases contains sufficient information to uniquely identify the genes, provided that it is extracted from defined positions of the transcripts. The possible maximum number of different tag sequences (4^{10}=1,048,456) is considered to be far more than the number of genes of organisms ensuring the unique identification of a gene by the tag sequence. By using the tag sequence as query, one can perform BLAST search in EST database to identify the EST corresponding to the tag. It is also possible to extract longer (26 bp) tag sequence by altering the restriction enzyme used in the procedure (Matsumura et al. 2003). With the longer tags, BLAST search can be carried out not only against EST database but also against the whole genome sequence, increasing the probability of successful annotation of tags. In contrast to the analog data obtained by the hybridization-based techniques such as microarray analysis, SAGE describes the abundance of gene expression by counting the tags in the sample, so that the nature of data is by definition digital. This favorable property of SAGE data has been successfully exploited for constructing human SAGE database (e.g. NCBI see http://www.ncbi.nlm.nih.gov/SAGE/). We have successfully applied this technique to profile gene expression in rice seedlings (Matsumura et al. 1999), and to show the global gene expression change in rice cells after treatment with *M. grisea* cell wall elicitor (Matsumura et al. 2003). In the latter study, a rice gene encoding Bax-inhibitor (OsBI1) was shown to be dramatically down-regulated after the elicitor treatment, and further analysis suggested *OsBI1* gene expression change as a regulatory point of rice cell death caused by the elicitor.

3. APPRESSORIUM INDUCTION WITH cAMP

Key processes of *M. grisea* infection to host plants include the appressorium formation leading to invasive growth of hyphae in host tissues. In natural condition, conidia of *M. grisea* are known to form appressorium on hydrophobic surface. On the basis of this observation, gene expression of *M. grisea* on hydrophobic surface has been studied (Beckerman & Ebbole, 1996; Fang & Dean, 2000; Kamakura et al. 2002). This system, although reflecting the natural condition, poses difficulty in obtaining sufficient amount of fungal samples to be analysed. Lee & Dean (1993) reported that cAMP could induce appressorium formation even on hydrophilic membranes. This is an alternative and less technically demanding system to induce appressoria. Thus, we are comparing the gene expression profiles between cAMP-treated and non-treated *M. grisea* conidia on water-saturated dialysis membranes to identify the genes that are involved in appresorium development and the following infection processes.

As described in Lee & Dean (1993), the application of cAMP to *M. grisea* conidia induced appressorium development (Figure 1). More than 75% of conidia formed appressoria in 50 mM cAMP 20 h after the start of treatment, but no appressorium formation was detected in the control (a sample treated with 0 mM cAMP). We isolated mRNA from germinating conidia of *M. grisea* 6 h after the treatment with 50 mM or 0 mM cAMP.

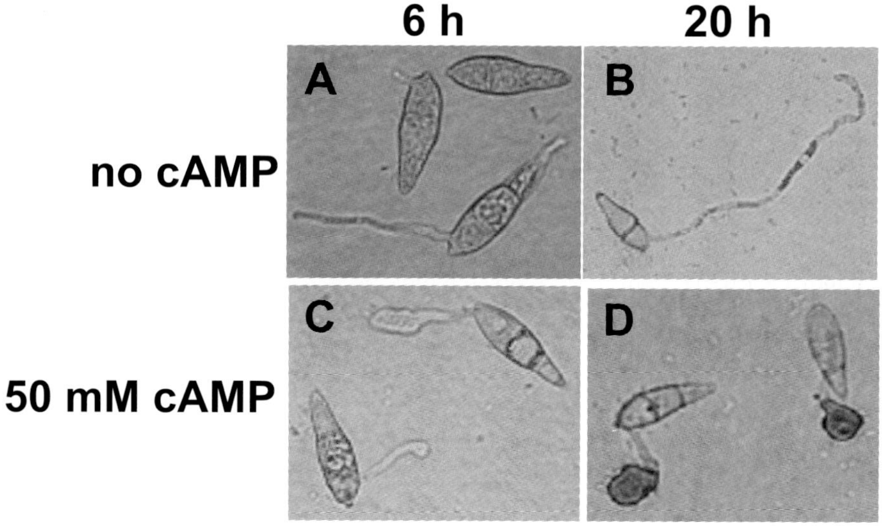

Figure 1. *Effect of cAMP. A; Conidia were incubated for 6 h at 25°C in 0 mM cAMP. They germinated, but did not form appressorium. B; The conidia were incubated for 20 h at 25°C in 0 mM cAMP. The conidia developed only hypha. C; The conidia were incubated for 6 h at 25°C in 50 mM cAMP. Germ tubes started to show apical expansion which transformed into appressoria. D; The conidia incubated for 20 h at 25°C in 50 mM cAMP. They formed melanized appressoria.*

4. GENE EXPRESSION CHANGES ACCOMPANYING APPRESSORIUM DEVELOPMENT

The original protocol of SAGE was described by Velculescu, Zhang, Vogelstein, & Kinzler (1995). We performed SAGE according to the procedure described by Matsumura, Nirasawa, & Terauchi (1999).

Messenger RNAs were isolated from the conidia incubated on the membrane for 6 hours with or without cAMP, and were used to create SAGE libraries (Irie et al. 2003). More than 3000 tags were extracted from cAMP-treated and non-treated SAGE libraries. There were about 100 different tags showing statistically significant difference in the frequencies between cAMP-treated and non-treated samples. Half of them were cAMP-induced and another half cAMP-repressed. About half of the tags corresponding to cAMP-induced genes could be annotated by using EST database (TIGR database), and the rest could be annotated by using genomic sequences available as *M. grisea* genome v.1 (Fungal Genomics Laboratory North

Carolina State University and Whitehead Institute Center for Genome Research).

5. CATEGORY OF c AMP-INDUCIBLE GENES

Functional categories of cAMP-inducible genes so far analysed are summarized in Figure 2. About half of the cAMP-inducible genes are with unknown functions. Among the genes with known functions, those related to sugar metabolism (respiration), nucleic acid metabolism share the top two categories with high abundancy, followed by the genes for transcription factors and protein, amino acid metabolisms. As expected, several genes for melanization presumably involved in appressorium melanization are also found among cAMP-inducible genes.

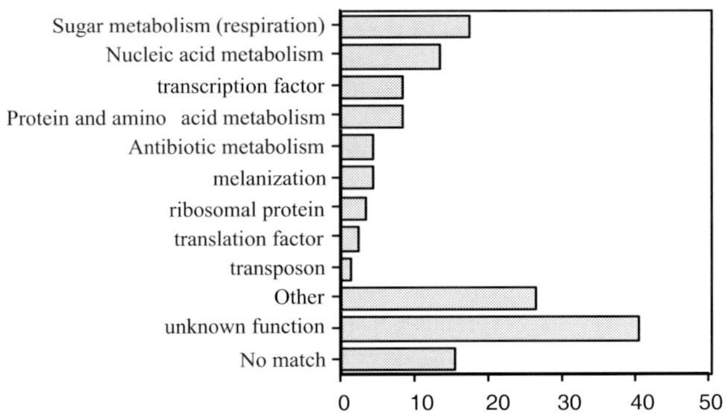

Figure 2. *Categories of the cAMP-inducible genes. No match indicates the gene that did not match DNA sequences in the available databases. Genes categorized to be with unknown function didn't have any significant homology to functional protein genes.*

Many genes with unknown functions were also identified among the cAMP-inducible genes. After confirmation of expression changes by RT-PCR or Northern analysis, selected genes will be used for gene disruption study to see whether they are indispensable for its pathogenicity. SAGE is an ideal technique for gene expression analysis of *M. grisea* which has now an established infrastructure for genomic research.

REFERENCES

Balhadére P. V. & Talbot, N. J. (2000). Fungal pathogenicity-establishing infection. In M. Dickinson & J. Beynon (Eds.), *Molecular Plant Pathology-Annual Plant Reviews, Volume 4* (pp. 107-123). Sheffield: Sheffield Academic Press.

Beckerman, J. L., & Ebbole, D. J. (1996). *MPG1*, a gene encoding a fungal hydrophobin of *Magnaporthe grisea*, is involved in surface recognition. *Mol Plant Microbe Interact, 9*(6), 450-456.

Fang, E. G., & Dean, R. A. (2000). Site-directed mutagenesis of the *magB* gene affects growth and development in *Magnaporthe grisea. Mol Plant Microbe Interact, 13*(11), 1214-1227.

Hamer, J. E., & Talbot, N. J. (1998). Infection-related development in the rice blast fungus *Magnaporthe grisea. Curr Opin Microbiol, 1*(6), 693-697.

Idnurm, A. & Howlett, B. J. (2001). Pathogenicity genes of phytopathogenic fungi. *Mol Plant Pathol, 2*(4), 241-255.

Irie, T., Matsumura, H., Terauchi,R., & Saitoh, H. (2003) Serial analysis of gene expression (SAGE) of Magnaporthe grisea: genes involved in appressorium formation. Mol Gen Genomics, 270(2) 181-189.

Kamakura, T., Yamaguchi, S., Saitoh, K., Teraoka, T., & Yamaguchi, I. (2002). A novel gene, *CBP1*, encoding a putative extracellular chitin-binding protein, may play an important role in the hydrophobic surface sensing of *Magnaporthe grisea* during appressorium differentiation. *Mol Plant Microbe Interact, 15*(5), 437-444.

Kim, S., Ahn, I. P., & Lee, Y. H. (2001). Analysis of genes expressed during rice-*Magnaporthe grisea* interactions. *Mol Plant Microbe Interact, 14*(11), 1340-1346.

Lee, Y. H., & Dean, R. A. (1993). cAMP regulates infection structure formation in the plant pathogenic fungus *Magnaporthe grisea. Plant Cell, 5*(6), 693-700.

Matsumura, H., Nirasawa, S., & Terauchi, R. (1999). Transcript profiling in rice (*Oryza sativa* L.) seedlings using serial analysis of gene expression (SAGE). *Plant J, 20*(6), 719-726.

Matsumura, H., Nirasawa, S., Kiba, A., Urasaki, N., Saitoh, H., Ito, M., Kawai-Ymamada, M., Uchimiya, H. & Terauchi, R. (2003). Overexpression of Bax-inhibitor suppresses the fungal elicitor-induced cell death in rice (*Oryza sativa* L.) cells. *Plant J,* 33(3) 425-434.

Matsumura, H., Reich, S., Ito, A., Saitoh, H., Kamoun, S., Winter, P., Kahl, G., Reuter, M., Kruger, D. H., & Terauchi, R. (2003) Gene expression analysis of plant host-pathogen interactions by SuperSAGE. Proc Natl Acad Sci USA, 100(26) 15718-15723.

Rauyaree, P., Choi, W., Fang, E., Blackmon, B., & Dean, R. A. (2001). Genes expressed during early stages of rice infection with the rice blast fungus *Magnaporthe grisea. Mol Plant Pathol, 2*(6), 347-354.

Talbot, N. J. (1995). Having a blast: exploring the pathogenicity of *Magnaporthe grisea. Trends Microbiol, 3*(1), 9-16.

Velculescu, V. E., Zhang, L., Vogelstein, B., & Kinzler, K. W. (1995). Serial analysis of gene expression. *Science, 270*(5235), 484-487.

Xu, J. R. (2000). Map kinases in fungal pathogens. *Fungal Genet Biol, 31*(3), 137-15.

APPROACH TO UNDERSTAND METABOLIC NETWORKS INVOLVED IN APPRESSORIUM FUNCTION OF *COLLETOTRICHUM LAGENARIUM*

*Gento Tsuji, Satoshi Fujii, Naoki Fujihara, Seiji Tsuge and Yasuyuki Kubo

Laboratory of Plant Pathology, Graduate School of Agriculture, Kyoto Prefectural University, Kyoto 606-8522, Japan

**Email: a9730602@kpu.ac.jp*

1. INTRODUCTION

In most plant pathogenic fungi, cellular differentiation of infection-specific structures is required for successful infection to host plant. The infection process is generally composed of several key steps; conidial attachment on host plants, germination; appressorium formation, penetration, and subsequent invasive growth in host plants. *Colletotrichum lagenarium* (Pass.) Ellis and Halsted is the causal agent of cucumber anthracnose disease. In our laboratory, extensive studies have been done for clarification of the molecular mechanism on infection-related morphogenesis and metabolism of this fungus.

This fungus produces darkly melanized appressoria. Studies with melanin deficient mutants and inhibitors demonstrated that melanin is essential for appressorium-mediated penetration (Kubo and Furusawa, 1991). These studies also clarified the melanin biosynthesis pathway of this fungus (Figure 1). We previously characterized three structural genes of *C. lagenarium*, *PKS1* encoding a polyketide synthase (Takano et al., 1995), *THR1* encoding a T3HN reductase (Perpetua et al., 1996), *SCD1* encoding a scytalone dehydratase (Kubo et al., 1996), and our attention was focused on regulation of melanin biosynthesis. Recently, we have successfully isolated

S. Kawasaki (ed.), Rice Blast: Interaction with Rice and Control, 29–37.
© 2004 *Kluwer Academic Publishers. Printed in the Netherlands.*

a regulatory gene, *CMR1* (Tsuji et al., 2000). Functional analysis of *CMR1* is presented in section 2.1.1.

Recent molecular genetic studies have demonstrated that the mitogen-activated protein (MAP) kinase signal transduction pathways play pivotal roles for fungal pathogenesis (Xu, 2000). In *Magnaporthe grisea*, two MAP kinase genes *PMK1* and *MPS1,* homologues of the yeast MAP kinase gene *FUS3/KSS1* and *SLT2,* were isolated and characterized for the involvement of pathogenicity, respectively (Xu and Hamer 1996; Xu et al. 1998). In *C. lagenarium,* the MAP kinase gene *CMK1* encoding the PMK1 MAP kinase homologue has been shown to regulate diverse steps of infection-related morphogenesis (Takano et al., 2000). Our advanced studies have been directed toward the characterization of downstream components. *Saccharomyces cerevisiae* Ste12p is known to be a downstream transcription factor regulated by Fus3p/Kss1p MAP kinases (Cook et al., 1997; Lengeler et al., 1998; Madhani and Fink, 1998). We have isolated a regulatory gene, *CST1,* coding for yeast *STE12* homologue from *C. lagenarium.* Functional analysis of *CST1* is presented in section 2.1.2.

Our efforts have also been focused on cloning novel genes involved in pathogenicity of *C. lagenarium.* Random insertional mutagenesis using a marker DNA fragment is one of the effective methods for extensive gene-hunting. Restriction enzyme mediated integration (REMI) is known to be an efficient tool for random insertional mutagenesis in several fungi (Balhadere *et al.*, 1999; Bolker *et al.*, 1995; Linnemannstons *et al.*, 1999; Lu *et al.*, 1994; Sweigard *et al.*, 1998). We have previously applied REMI to *C. lagenarium*; however, unpredictable deletion of the transformed plasmid or rearrangement of a chromosomal DNA has restricted the molecular genetic analysis of obtained mutants. Thus, we have sought an alternative approach for further effective gene tagging of this fungus. Recent study has indicated that *Agrobacterium tumefaciens* could transfer T-DNA not only to plant but also fungal cells (Bundock *et al.*, 1995 ; 1999, Gouka et al., 1999, Mullins et al., 2001, Rho et al., 2001). Therefore, we applied *A. tumefaciens*-mediated transformation (AtMT) to *C. lagenarium* as a novel tool for random insertional mutagenesis. Evaluation of AtMT will be described in section 2.2.

Here, we review recent work on regulation of infection-related morphogenesis and metabolism in *C. lagenarium.* We also evaluate the application of AtMT for isolation of novel genes involved in fungal pathogenicity.

2. RESULTS AND DISCUSSION

2.1. Characterization of regulatory genes required for melanin biosynthesis and infectious morphogenesis of C. lagenarium

2.1.1 CMR1 (Colletotrichum melanin regulator)

CMR1 is a regulatory gene involved in melanin biosynthesis of *C. lagenarium* (Tsuji et al., 2000). Deduced amino acid sequence of the *CMR1* showed characteristics of a transcriptional factor with two putative DNA binding motifs, Cys2His2 zinc fingers and a Zn(II)2Cys6 binuclear cluster. The *cmr1*△ mutants formed melanized appressoria indistinguishable from those of wild type 104-T. However, the mutants were defective in melanin production in vegetative hyphae, and formed reddish-brown colonies on PDA medium. The expression of melanin biosynthesis genes in the *cmr1*△ mutants was examined by northern blotting. During the development of appressorium formation, transcripts of the melanin biosynthesis genes, *PKS1*, *THR1* and *SCD1* accumulated in the *cmr1*△ mutants at the same level as the wild type 104T. In contrast, accumulation of *THR1* and *SCD1* transcripts was hardly detected during vegetative hyphal growth of the *cmr1* △ mutants.

Accumulation of *PKS1* transcripts was detected in the *cmr1*△ mutants, but significantly lower than that in the wild type 104-T. This result indicated that Cmr1p regulates the transcription of *SCD1*, *THR1*, and partially *PKS1*, during the development of mycelial melanization (Figure 1). However,

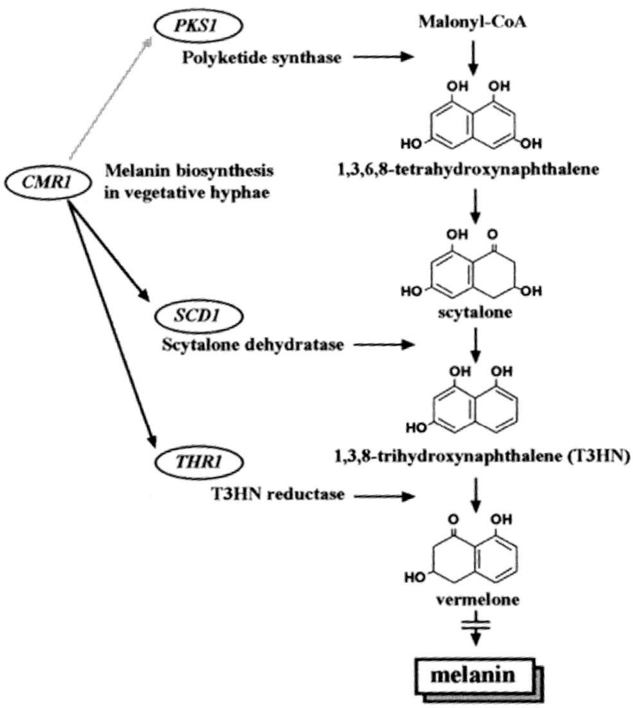

Figure 1. *Melanin biosynthesis pathway of* C. lagenarium *with cloned genes*

Cmr1p hardly contribute to the transcription of these genes during appressorium melanization, suggesting that a transcriptional factor(s) other than Cmr1p is involved in this process. Further experiments have been directed to identify novel regulators involved in appressorium melanization.

2.1.2. *CST1* (*Colletotrichum STE12 homologue*)

Recent studies have revealed that several plant pathogenic fungi require components of extensively conserved signal transduction pathways for pathogenicity (Xu, 2000). In *C. lagenarium*, the yeast *FUS3/KSS1* MAP kinase gene homologue *CMK1* has been shown to regulate diverse aspects of infectious morphogenesis, conidial germination, appressorium formation and invasive growth in host plants (Takano et al. 2000). Subsequently, we have cloned and characterized *CST1* gene homologous to the yeast *STE12* gene coding for a downstream transcriptional factor of *FUS3/KSS1* MAP kinase.

CST1 was isolated from *C. lagenarium* by polymerase chain reaction (PCR)-based screening. The deduced amino acid sequence of the *CST1* contained two putative DNA-binding motifs, a homeodomain and Cys2His2 zinc fingers. The *cst1*Δ mutants completely lacked the ability to form lesions when inoculated on intact host leaves, but formed yellow-brown lesions from wounded sites of leaves. To determine the steps leading to loss of pathogenicity, infection-related morphogenesis was investigated on cellulose membrane. The *cst1*Δ mutants produced darkly melanized appressoria as well as the wild type strain. However, appressoria of the *cst1*Δ mutants completely lacked ability to produce infectious hyphae into cellulose membrane, suggesting that the loss of pathogenicity of the mutants is due to a defect in appressorium penetration. These results indicate that *CST1* plays an essential role for appressorium penetration in *C. lagenarium* (Figure 2).

The *cst1*Δ mutants showed no critical defect in conidial germination, appressorium formation, and invasive growth. However, the involvement of *CST1* in these steps could not absolutely be excluded. Especially, it is possible to postulate that *CST1* is involved partially in appressorium formation and infectious growth. When incubated in the nutrient solution, conidia of the *cst1*Δ strains tended to produce incomplete appressoria, showing an intermediate phenotypic appearance between the wild-type strain 104-T and the *cmk1*Δ strain. In addition, the *cst1*Δ strain retained infectious growth ability in host leaves, but the lesions formed on wounded leaves were relatively smaller than those formed by the control strain. Thus, *CST1* may regulate diverse aspects of fungal infectious morphogenesis by means of co-regulation with different factors at different stage of cellular differentiation (Figure 2).

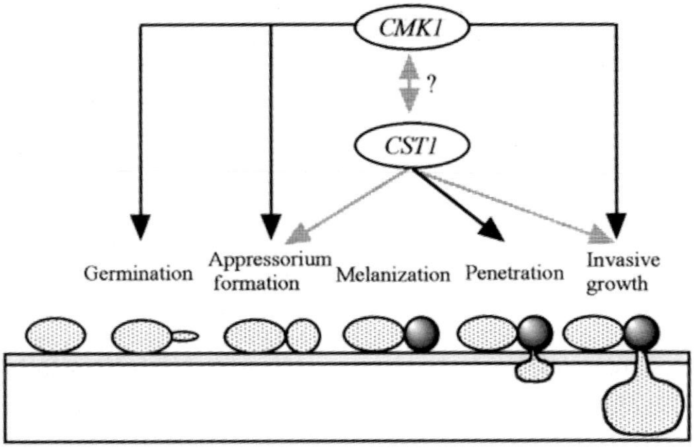

Figure 2. *Involvement of* CST1 *in infection-related morphogenesis of* C. lagenarium

2.2. Evaluation of random insertional mutagenesis by AtMT for isolation of novel genes from C. lagenarium

Random insertional mutagenesis using a marker DNA fragment is one of the effective methods to identify the genes relevant to pathogenicity, morphogenesis, metabolisms, and so on. We estimate the application of AtMT for random insertional mutagenesis of *C. lagenarium*.

We constructed a binary vector pBIG2RHPH carrying a hygromycin-resistant (*hph*) gene cassette between the right and left borders of T-DNA. Co-cultivation of *C. lagenarium* wild-type 104-T with pBIG2RHPH-introduced *A. tumefaciens* C58C1 brought about the production of 200-300 hygromycin-resistant transformants per 10^6 conidia in optimized condition. Subsequently, we analyzed the transformants for the mode of T-DNA insertion by Southern blotting. Eighteen of 21 transformants carried a single insert of T-DNA, and 21 of 24 inserted T-DNA showed no significant deletion in the genome of analyzed transformants. The mode of T-DNA insertion is to facilitate the recovery of tagged genes from the genomic DNA of the transformants.

A total of 5096 transformants were obtained and assayed for pathogenicity to host cucumber leaves. Nine mutants displayed reduced or loss of pathogenicity. We also screened transformants for defects in melanin biosynthesis and fatty acid metabolism, resulting in isolation of 8 and 5 mutants, respectively. Melanin biosynthesis is known to be required for appressorium penetration (Kubo and Furusawa, 1991). The melanin-deficient mutants were visually selected from the transformants cultured on PDA medium. Recently, peroxisomal metabolic function, probably fatty acid ß-oxidation in peroxisome, is shown to be required for appressorium-mediated plant infection (Kimura et al., 2001). The ability for fatty acids utilization was assessed with colonial growth assay on fatty acid medium.

The thermal asymmetric interlaced PCR (TAIL-PCR) was first reported in *A. thaliana* as an efficient tool for isolation of tagged genes (Liu et al., 1995). We applied TAIL-PCR to isolate genomic DNA segments adjacent to an inserted T-DNA from the transformants of *C. lagenarium*. Target regions flanking the T-DNA insertion were efficiently amplified from the genome of mutants, and subsequently sequenced (Table 1). Although the vector 'backbone' sequences were found at moderate frequency in fungal genome linked either to the left or right T-DNA border, the fungal genomic DNA segments flanking T-DNA were identified from at least 8 of 13 analyzed mutants. Seven of 8 identified fungal DNA segments contained a putative reading frame, and 5 of them showed significant homology to amino acid sequences identified from other organisms. Subsequently, sequenced fragment was used as a probe to screen a genomic cosmid library of *C. lagenarium*. Isolated cosmids were tested for complementation of deficient phenotype in mutants. Characteristics of each mutant of which rescued gene was responsible for the deficient phenotype are described bellow.

pd1. The mutant pd1 showed delayed melanin production in vegetative hyphae cultured on PDA medium. Appressoria of the mutant showed weak melanization, and loss of penetration ability. Sequence from the cosmid that could complement the defective phenotype of the mutation contained a reading frame with homology to *S. cerevisiae CCC2* gene encoding a Cu^{2+}-transporting P-type ATPase.

pd6. The mutant pd6 displayed complete loss of melanin production in both appressoria and vegetative hyphae. Sequence from the left border of T-DNA was identical with polyketide synthase gene *PKS1* of *C. lagenarium* (Takano et al., 1995). The mutant pd6 showed similar phenotypic appearance with the *pks1* knockout mutant that we isolated before. Thus, the deficient phenotype of this mutant is closely related to the insertion of T-DNA.

fa1624. The mutant fa1624 showed a deficient in vegetative hyphal growth on fatty acid medium. This mutant also lacked the ability for melanin biosynthesis only in appressoria. Sequence from the cosmid that could complement the defective phenotype of the fa1624 mutant displayed a significant homology with *S. cerevisiae PEX13*.

fa3822. The mutant fa3822 showed a defect in vegetative hyphal growth on neither fatty acid medium nor minimal medium. The growth could be restored by addition of arginine, suggesting that this mutant lacks ability for arginine biosynthesis. Sequence from both borders of inserted T-DNA showed significant homology with *S. cerevisiae ARG5* coding for acetyl glutamate kinase involved in arginine biosynthesis.

Table 1. Recovery of the DNA segments flanking the T-DNA from mutants of C. lagenarium
[a]Tentative gene name

Mutants	Gene[a]	Pathogenicity	LB[b]	RB	Homology	Complementation
pd1	*CPD1*	-	CR	VB	*S. cerevisiae CCC2*	+
pd2	*CPD 2*	-	VB	VB		
pd3	*CPD 3*	-	CR	CR	*A. pseudotamarii AFLR*	N.D.
pd4	*CPD 4*	+/-	CR	CR	*S. cerevisiae MAC10*	N.D.
pd5	*CPD 5*	-	VB	VB		
pd6	*CPD 6*	-	N.D.[c]	CR	*C. lagenarium PKS1*	+
pd7	*CPD 7*	-	N.D.	VB		
pd8	*CPD 8*	-	N.D.	nCR	no homology	N.D.
Fa1624	*CFA1*	-	CR	N.D.	*S. cerevisiae PEX13*	+
Fa2637	*CFA2*	-	VB	VB		
Fa2678	*CFA3*	+/-	N.D.	CR	no homology	N.D.
Fa3812	*CFA4*	+/-	N.D.	VB		
Fa3822	*CPM1*	-	CR	CR	*S. cerevisiae ARG6*	+

[b]Sequence showing a vector backbone (VB), predictable coding region (CR) or not (nCR)
[c]Not determined yet

3. CONCLUDING REMARKS

To understand the molecular mechanism of fungal pathogenicity, our efforts have been focused on cloning genes involved in infection-related morphogenesis and metabolism. We have successfully isolated two regulatory genes, *CMR1* and *CST1* involved in infection-related metabolism and morphogenesis, respectively. Functional analysis of these genes revealed that *CMR1* positively regulates the transcription of melanin biosynthesis structural genes in a development specific manner, and that *CST1* plays an essential role for appressorium penetration. These findings also indicated that additional regulatory factors must be involved in melanin biosynthesis and infection-related morphogenesis of *C. lagenarium.*

To obtain extensive genes involved in the mechanism, we developed AtMT to *C. lagenarium*. This method allowed us to carry out the isolation of mutants and subsequent gene rescue systematically. A high efficiency of fungal transformation enabled us to obtain a large population of T-DNA-integrated mutants in a short period. Southern blotting also displayed that T-DNA was mainly integrated at a single site in the genome of the transformants, and hardly deleted at the integrated site. This resulted in a high percentage of tagged gene recovery from the transformants by TAIL-PCR. Although vector 'backbone' sequences were detected in several mutants, the benefit of AtMT described above will overcome this disadvantage. Actually, we have confirmed in at least 4 of 13 mutants that the tagged gene accounts for the deficient phenotype. Thus, we propose that AtMT is a highly efficient tool for gene tagging in *C. lagenarium*. High throughput analysis of tagged genes by AtMT would accelerate the development of our research for the understanding of molecular regulatory network associated with infectious morphogenesis and metabolism of this fungus.

ACKNOWLEDGEMENTS

This work was supported in part by research grant from the Ministry of Education, Science, Sports and Culture of Japan (No.12460024 and 12052219), and Research Fellowships for Young Scientists from the Japan Society for the Promotion of Science (No.05878).

REFERENCES

Balhadere, P.V., Foster, A.J., and Talbot, N.J. Identification of pathogenicity mutants of the rice blast fungus *Magnaporthe grisea* by insertional mutagenesis. Mol Plant-Microbe Interact 1999; 12: 129-142.

Bolker, M., Bohnert, H.U., Braun, K.H., Gorl, J., and Kahmann, R. Tagging pathogenicity genes in *Ustilago maydis* by restriction enzyme mediated integration (REMI). Mol Gen Genet 1995; 248. 547-552.

Bundock, P., Dulk-Ras, A., Beijersbergen, A., and Hooykaas, P.J. Trans-kingdom T-DNA transfer from *Agrobacterium tumefaciens* to *Saccharomyces cerevisiae*. The EMBO J. 1995; 14: 3206-3214.

Bundock, P., Mroczek, K., Winker, A.A., Steensma, H.Y., and Hooykas, P.J. 1999. T-DNA from *Agrobacterium tumefacins* as an efficient tool for gene targeting in *Kluyveromyces lactis*. Mol Gen Genet 261: 115-121.

Cook, J.G., Bardwell, L., and Thorner, J. Inhibitory and activating functions for MAPK Kss1 in the *Saccharomyces cerevisiae* filamentous growth signaling pathway. Nature 1997; 390:85-88.

Gouka, R.J., Gerk, C., Hooykaas, P.J.J., Bundock, P., Musters, W., Verrips, C.T., and Groot, M.J.A. Transformation of *Aspergillus awamori* by *Agrobacterium tumefaciens*-mediated homologous recombination. Nature Biotecnol 1999; 17: 598-601.

Kimura, A., Takano, Y., Furusawa, I., and Okuno, T. Peroxisomal metabolic function is required for appressorium-mediated plant infection by *Colletotrichum lagenarium*. The Plant Cell 2001; 13: 1945-1957.

Kubo, Y., and Furusawa, I. "Melanin biosynthesis: Prerequisite for successful invasion of the plant host by appressoria of *Colletotrichum* and *Pyricularia*." In *The Fungal Spore and Disease Initiation in Plants and Animals*. Cole G.T. and Hoch H.C., eds. New York, Plenum Publishing, 1991.

Kubo, Y., Takano, Y., Endo, N., Yasuda, N., Tajima, S., and Furusawa, I. Cloning and structural analysis of the melanin biosynthesis gene *SCD1* encoding scytalone dehydratase in *Colletotrichum lagenarium*. Appl Environ Microbiol 1996; 62:4340-4344.

Lengeler, K.B., Davidson, R.C., D'souza, C., Harashima, T., Shen, W., Wang, P., Pan, X., Waugh, M., and Heitman, J. Signal transduction cascades regulating fungal development and virulence. Microbiol Mol Biol Rev 2000; 64: 746-785.

Linnemannstons, P., Vob, T., Hedden, P., Gaskin, P., and Tudzynki, B. Deletions in the gibberellin biosynthesis gene cluster of *Gibberella fujikuroi* by restriction enzyme-mediated mutagenesis. Appl Environ Microbiol 1999; 65: 2558-2564.

Liu, Y.G., Mitsukawa, N., Teruo, O., and Whittier, R.F. Efficient isolation and mapping of *Arabidopsis thaliana* T-DNA insert junctions by thermal asymmetric interlaced PCR. The Plant J 1995; 8: 457-463.

Lu, S., Lyngholm, L., Yang, G., Bronson, C., and Yorder, O.C. Tagged mutations at the Tox1 locus of *Cochliobolus heterostrophus* by restriction enzyme-mediated integration. Proc Natl Acad Sci USA 1994; 91: 12649-12653.

Madhani, H.D, and Fink, G.R. Combinatorial control required for the specificity of yeast MAPK signaling. Science 1997; 275:1314-1317.

Mullins, E.D., Chen, X., Romaine, P., Raina, R., Geiser, D.M., and Kang, S. *Agrobacterium*-mediated transformation of Fusarium oxysporum: An efficient tool for insertional mutagenesis and gene transfer. Phytopathology 1001; 91: 173-180.

Perpetua, N.S., Kubo, Y., Yasuda, N., Takano, Y., and Furusawa, I. Cloning and characterization of a melanin biosynthesis *THR1* reductase gene essential for appressorial penetration of *Colletotrichum lagenarium*. Mol Plant-Microbe Interact 1996; 9:323-329.

Rho, H.S., Kang, S., and Lee, Y. H. *Agrobacterium tumefaciens*-mediated transformation of the plant pathogenic fungus, *Magnaporthe grisea*. Mol Cells 2001; 12: 407-411.

Sweigard, J. A., Carroll, A., Farrall, L., Chumley, F., and Valent, B. *Magnaporthe grisea* pathogenicity genes obtained through insertional mutagenesis. Mol Plant-Microbe interact 1998; 11: 404-412.

Takano, Y., Kikuchi, T., Kubo, Y., Hamer, J. E., Mise, K., and Furusawa, I. The *Colletotrichum lagenarium* MAP kinase gene *CMK1* regulates diverse aspects of fungal pathogenesis. Mol Plant-Microbe Interact 2000; 13:374-383.

Takano, Y., Kubo, Y., Shimizu, K., Mise, K., Okuno, T., and Furusawa, I. Structural analysis of *PKS1*, a polyketide synthase gene involved in melanin biosynthesis in *Colletotrichum lagenarium*. Mol Gen Genet 1995; 249:162-167.

Tsuji, G., Kenmochi, Y., Takano, Y., Sweigard, J., Farrall, L., Furusawa, I., Horino, O., and Kubo, Y., 2000. Novel fugal transcriptional activators, Cmr1p of *Colletotrichum lagenarium* and Pig1p of *Magnaporthe grisea*, contain Cys2His2 zinc finger and Zn(II)2Cys6 binuclear cluster DNA-binding motifs and regulate transcription of melanin biosynthesis genes in a developmentally specific manner. Mol Microbiol 2000; 38:940-954.

Xu, J.R. MAP kinases in fungal pathogens. Fung Genet Biol 2000; 31:137-152.

Xu, J.R., and Hamer, J.E. MAP kinase and cAMP signaling regulate infection structure formation and pathogenic growth in the rice blast fungus Magnaporthe grisea. Genes Dev 1996; 10:2696-2706.

Xu, J.R., Staiger , C.J., and Hamer, J.E. Inactivation of the mitogen-activated protein kinase Mps1 from the rice blast fungus prevents penetration of host cells but allows activation of plant defense responses. Proc Natl Acad Sci USA. 1998; 95:12713-12718.

PROTEOMICS OF MAGNAPORTHE GRISEA: LIQUID CHROMATOGRAPHY MASS SPECTROMETRY FOR THE IDENTIFICATION OF EXTRACELLULAR PROTEINS

*Sheng-Cheng Wu, Jeremi Johnson, Alan G. Darvill, Peter Albersheim and Ron Orlando

Complex Carbohydrate Research Center, University of Georgia, 220 Riverbend Road, Athens, Georgia 30602-4712, USA

**E-mail: wusc@ccrc.uga.edu*

1. INTRODUCTION

The recent wealth of sequence information from a broad spectrum of prokaryotic and eukaryotic genomes has brought rapid technological innovations towards high-throughput functional analysis of whole genomes. Functional genomics addresses expression and function of all genes with regard to cell type, organ and physiological state [11]. One branch of functional genomics is the study of proteins, using proteomics technology [3, 6,7,13]. Proteomics is the study of the complete protein component, or a proteome, of an organism. The proteome is highly dynamic. For example, the presence of proteins, their abundance, state of modification and sub-cellular location, are all affected by the physiological state of the cell or tissue. It is estimated that, depending on the species, as many as a few thousand to ten thousand proteins may be present in a cell at a given moment [3]. Proteomics was developed with the realization of this dynamic complexity of a proteome, as well as the tremendous discrepancy among genes predicted from a whole-genome sequence, transcripts detected using

S. Kawasaki (ed.), Rice Blast: Interaction with Rice and Control, 39–46.
© 2004 *Kluwer Academic Publishers. Printed in the Netherlands.*

DNA microarray assays [11], and proteins actually present in an organism's particular state, organs or cells.

Traditional proteomics involves identification by mass spectrometry of proteins separated by two-dimensional gel electrophoresis (2DE) [6]. Using this technology, gel plugs containing protein 'spots' from 2DE are treated with an endoprotease (e. g. trypsin), and the released peptides eluted from each of the gel plugs analyzed by Matrix-Assisted Laser Desorption/Ionization mass spectrometry (MALDI MS), or electrospray mass spectrometry (ES-MS). The mass spectra of the peptides, or MS fingerprints, are then queried against a non-redundant sequence database that contains lists of molecular weights of peptides predicted from proteolysis of all known or theoretical protein sequences. To increase query specificity, an aliquot of the digested peptide sample can also be subjected to tandem mass spectrometry (MS/MS) on a Quadrupole/Time-of-Flight (Q-Tof) mass spectrometer system to obtain specific amino acid sequences from a few of the peptides present in the sample [2,4,16]. The amino acid sequence and the mass of a peptide in the sample can then be combined into what is commonly called a 'Peptide Sequence Tag' or PST, which can be used to query against a sequence database resulting in great specificity.

Two-dimensional gel electrophoresis is a 40-year-old technique that retains a high resolving power for protein separation [6]. It has been used to demonstrate the dynamic proteome of an organism and display gene expression at the protein level. However, using 2DE for proteomics has some documented deficiencies. For instance, it is labor-intensive, it requires great technical skills to achieve good resolution and reproducibility, and in reality it does not represent the 'whole proteome' due to limited capacity and resolution of the gel. It is often hard to detect by 2DE low abundant proteins and also proteins with poor water-solubility, extreme isoelectric points and extreme low or high molecular weight. Most importantly, 2DE cannot be directly coupled with mass spectrometric analysis, thus limiting its use as a high-throughput technique [4].

Consequently, new technologies are being developed to either complement or replace the traditional 2DE technology [5], with liquid chromatography (LC) being the most promising technique for proteomic analyses. While 2DE separates proteins solely based on pI and molecular weight there exist a variety of liquid chromatographic methods with differing separation principles. In a multidimensional liquid chromatography system (MDLC), different LC methods, such as ion-exchange, affinity, and reverse-phase chromatography, are used in combination to achieve separation of many hundreds and thousands of proteins and/or peptides. In addition, on-line capillary and nano-based LC systems can be coupled with electrospray mass spectrometers for high-throughput protein identification [16].

Extracellular proteins (ECPs) are defined as proteins secreted outside the plasma membrane of a cell. In bacteria, fungi and plants, ECPs also include proteins associated with the cell walls (structural and apoplastic proteins). ECPs serve diverse and essential biological functions in various cell types exposed to differing physiological states. Some fungal ECPs are required for uptake of molecules for nutrition, most notable are hydrolytic enzymes that breakdown macromolecules such as cell wall polysaccharides [19]. Some are for growth, cell sensing and communication [5,9,10,14]. The most important ECPs, however, are those associated with pathogenicity and pathogen-host associations because of their critical implications in the quest for disease control [14,17]. For instance, in fungal plant pathogens, most of the isolated avirulence factors are ECPs [17].

The study on extracellular proteins secreted by *M. grisea*, conducted in our laboratory [19 and unpublished data], is limited mostly to cell wall degrading enzymes (CWDE). For example, we have identified six *endo*-β-1,4-xylanases secreted by *M. grisea*, three of which are required for pathogenicity [manuscript in preparation]. Other *M. grisea* ECPs identified are mostly deduced from cloned genes that encode a typical signal peptide [1], with the most comprehensive database of putative ECPs presented at http://plpa2linux.tamu.edu/.

Despite the importance of ECPs, there have been few reports on systemic profiling of ECPs from fungi. As fungal genomics enters a new era with the completion, or near completion, of *Saccharomycete cerevisiae*, *Neurospora crassa*, and *M. grisea* genomes [18; http://www-genome.wi.mit.edu/annotation/fungi/], it has become apparent that development of high-throughput proteomics technologies for identification and profiling of fungal ECPs and other proteins is required.

In this paper, we present a preliminary report describing the use of liquid chromatography coupled to electrospray ionization time-of-flight mass spectrometry (LC MS/MS) as an alternative to 2DE for the identification of extracellular proteins (ECPs) secreted into the culture medium of *Magnaporthe grisea*.

2. EXPERIMENTAL AND DISCUSSION

2.1. Preparation and gel analysis of ECPs

To obtain sufficient amounts of extracellular proteins for method development and optimization, we targeted ECPs present in the culture media. *M. grisea* strain CP987 was grown in either 1 liter of Vogel's minimal medium with 5 g of sucrose (SUC), or Vogel's minimal medium with 5 g of purified rice cell walls (RCW), or a complete medium (CM) as described previously [19]. The culture media were harvested, by filtration through double layers of glass microfiber filters (type GF/F; Whatman International Ltd., Maidstone, England), after 8 days of growth at 22°C in the dark with constant shaking (150 rpm). The media were concentrated by freeze-drying then re-dissolved in water to 50 mL, followed by dialysis at 4°C against a 10 mM sodium acetate buffer, pH 5.0, in a dialysis bag with a molecular weight cut-off of 1000 (Spectra/Por® 6; Spectrum Laboratories, Inc., Rancho Dominguez, California, USA). The dialyzed samples were stored at -20°C for further analysis.

An aliquot (100μg) of each ECP sample was analyzed in parallel by 2DE. Preliminary results revealed 70-150 Coomassie Brilliant Blue stained proteins spots, with number, density and distribution of the spots varying among the three different samples (data not shown). In each case, however, the majority of the protein spots were displayed in an area of pI 4-6 and molecular mass of 5,000-50,000, indicating that *M. grisea* secretes mostly acidic and small proteins under the specified growth conditions and as analyzed by 2DE.

2.2. *Fractionation of ECPs*

In addition to the dynamic nature and complexity of a proteome within an organism, there are significant challenges dealing with non-volatile salts, detergents, chaotropes, and other agents often present as a result of protein extraction or biological preparation. Therefore, sample cleanup prior to mass spectrometry is essential since many of the aforementioned agents can interfere with ionization of protein samples. Another benefit of one or more fractionation steps is that it allows separation of proteins into sub-groups based on their physical properties, thus reducing the number of proteins present in a sample and the complexity of LC-MS spectra. The initial attempt to purify the above dialyzed ECP samples was by cation-exchange chromatography (CEC). Each ECP sample, in 10 mM sodium acetate buffer, pH 5.0, was applied onto a 1-mL Hitrap SP XL column (Amersham Pharmacia Biotech AB, Uppsala, Sweden), and the column washed with 10 mL of 10 mM sodium acetate buffer, pH 5.0. Unbound proteins in the flow-through and the wash fluid were collected for the next round of separation. The proteins bound to the column were eluted using 3 mL of 1 M NaCl in 10 mM sodium acetate buffer, pH 5.0. The unbound proteins from the CEC at pH 5.0 were dialyzed in 10 mM sodium acetate buffer, pH 4.0, and then subjected to CEC at pH 4.0. Similarly, The unbound proteins from the CEC at pH 4.0 were dialyzed in 10 mM sodium acetate buffer, pH 3.6, and then subjected to CEC at pH 3.6. Protein analysis data of the batch-separated

ECP samples are presented in Table 1. Based on these data, *M. grisea* grown in sucrose-supplemented minimal medium produces the largest amount of ECPs, with most of them binding to the cation exchanger at pH 4.0. The total recovery of proteins bound to the cation exchanger at the three different pH values investigated is 73-89%, indicating some protein loss during fractionation.

Table 1. *Summary of ECP fractionation and LC MS/MS*

Sample	**CM**	**SUC**	**RCW**	
Yield:				
Initial amount (mg)	2.1	8.6	1.5	
Amount bound at pH 5.0 (mg)		0.63	0.43	
Amount bound at pH 4.0 (mg)		0.73	7.02	
Amount bound at pH 3.6 (mg)		0.26	0.21	1.1
Recovery (%)	80.5	89.1	73.3	
Total MS/MS spectra produced:				
ECP at pH 5.0		11		
ECP at pH 4.0		22		
ECP at pH 3.6		132	49	
No. of Spectra matching GenBank sequence entries:				
ECP at pH 5.0		0		
ECP at pH 4.0		0		
ECP at pH 3.6		7	5	

2.3. LC tandem mass spectrometry

All experiments were carried out using a Micromass Q-TOF^{II} (Manchester, UK) fitted with a Waters Cap-LC. Capillary effluent was directed into the instrument via a microspray source. A sample (12.5 μg in 50 μL of water) of proteins was mixed with 50 μL of 100 mM ammonium bicarbonate and 2 μg (2 μL) of a sequencing grade trypsin (Promega, Madison, WI). The mixture was incubated overnight at 37°C with constant shaking (200 rpm). An aliquot (5 μL) of this digest was injected onto a Waters Capillary LC fitted with a 75 μM x 15 cm pepmap C-18 column. The elution started at 10% solvent B (solvent A = 1% formic acid in water; solvent B = 1% formic acid in acetonitrile) for 50 min, ramped to 40% B over a period of 50 min, held at 40% B for 50 min, and ramped again to 80% over a period of 30 min. The flow rate was maintained at 250 nL/min. Column eluent was sampled under positive ion mode utilizing the following

MS conditions: capillary voltage 2700 V, cone voltage 37 V, collision energy 10 V, cone gas 100 l/h, gas cell 15 psi. Auto-switching was utilized to automatically obtain MS/MS data on peptide ions meeting criteria specified within the software system. The number of MS/MS spectra from each of the four analyzed ECP samples varies greatly (Table 1). Based on the estimated total number of ECPs (about 150 ECPs for the SUC sample) and the assumption that each ECP is cleaved four times by trypsin, the yield of quality MS/MS spectra is around 25%. The main reason for this yield of quality MS/MS spectra may be the presence of non-peptidic contaminants such as secondary metabolites secreted by *M. grisea* in the original sample. In addition, improvement of trypsin digestion and optimization of the capillary LC and mass spectrometry may also increase the yield of quality MS/MS spectra.

2.4. Database search

Each output file that contains the uninterpreted MS/MS data for a mixture of tryptic peptides from an ECP sample was used as a query for identity search against translated protein sequences of fungal ESTs in GenBank (http://www.ncbi.nlm.nih.gov/). The search engine employed was the online version of Mascot at http://www.matrixscience.com [12]. Parameters selected for the queries were: Enzyme = Trypsin, Peptide tolerance = 1 Dalton, MS/MS tolerance = 0.8 Dalton, Data format = Micromass (.PKL), Instrument = ESI-QUAD-TOF, and Monoisotopic = Average mass. Results are summarized in Table 1 and Table 2.

There are five significant hits that identify three RCW ECPs and seven that identify four SUC ECPs (see footnote 3 of Table 2 for definition of "significance"). The other queries did not produce any significant matches. The ratio of 7 hits versus a total of 214 MS/MS spectra from all MS/MS assays is small. Although the quality of the MS/MS spectrum queries could be part of the cause, it is more likely that the small hit to spectra ratio is a result of the limited number of *M. grisea* sequence entries in GenBank. Fortunately, the annotation of the *M. grisea* genome database is near completion (http://www-genome.wi.mit.edu/annotation/fungi/magnaporthe/). A complete, MS-spectra-searchable protein database will tremendously enhance the speed of protein identification by proteomics.

Among the sequences identified here, mgecp0002 corresponds to RCW58p recently characterized by our laboratory (manuscript in preparation). RCW58p is similar to a probable phytotoxin isolated from fungal plant pathogens (GenBank Nos: AF074941 and P81702), as well as an antigen from fungal human pathogens [GenBank Nos: O60022 and Q00398]. Mgecp0004 is similar to a serine peptidase, and mgecp0007 is a hydrophobin named MPG1p [9]. The other four ECPs identified in Table 2 are not homologous to any known proteins.

Table 2. *Summary of MS/MS and database searching*

Spectrum[1]	Peptide Mass		Score[2]	Peptide[3]	Hit[4]	ECP ID
	Expected	Calculated				
Sample RCW at pH 3.6:						
675.38(2)	1348.75	1348.66	67	TVDNTPTNVNFK	BM872369	mgecp0001
739.43(2)	1476.84	1476.76	85	KTVDNTPTNVNFK	BM872369	mgecp0001
618.85(2)	1235.69	1235.60	40	YGWQTQGQIR	AA415101	mgecp0002
1099.96(2)	3296.85	3296.66	110	SINVLAIDHASGFNIGLA AMNDLTNGQAGSLGR	AA415101	mgecp0002
822.4(2)3	1642.85	1642.72	50	HEVSWSSVDTDPER	BU640342	mgecp0003
Sample SUC at pH 3.6:						
703.89(2)	1405.77	1405.64	93	APFDSAEGATSPEK	BM862870	mgecp0004
1023.08(2)	2044.15	2043.94	42	YWGESSPYQELTTENLK	BM862870	mgecp0004
853.51(1)	852.50	852.42	54	DITETFK	BU637471	mgecp0005
1092.61(1)	1091.60	1091.49	56	YFSADFQSK	BU637471	mgecp0005
529.43(1)	528.42	528.36	26	GLLVK	BG810348	mgecp0006
638.39(2)	1274.76	1274.63	55	GPFTIDADGNLR	BG810348	mgecp0006
404.74(2)	807.47	807.43	43	GRPHWR	L20685	mgecp0007

[1] Peptide of the specified molecular weight, single (1) or double (2) charged, from the first MS. This peptide is the subject of the second MS (see footnote 3).

[2] Score is -10*Log(P), where P is the probability that the observed match is a random event. Individual ions scores larger than 41 indicate identity or extensive homology ($p<0.05$).

[3] Peptide sequence deduced from the ion masses of randomly fragmented peptides from the second MS.

[4] GenBank accession number for the sequence encoding a protein that matches with a MS/MS spectrum.

This report and the work of others [4,7,13,17] demonstrate that LC-MS/MS is a potentially important method for high throughput proteomics research. Besides deciphering protein populations that dictate cellular and biological functions, LC-MS/MS may also be valuable in the dissection of the dynamic nature of a proteome, the detection of post-translational modifications, and the annotation of a genome.

ACKNOWLEDGEMENTS

This research was supported by U.S. Department of Energy (DOE) grant DE-FG05-93ER20221, the DOE-funded (DE-FG02-93ER20097) Center for Plant and Microbial Complex Carbohydrates, The National Science Foundation (NSF grant number 9626835), and the National Institutes of Health (NIH grant number P41RRR05351).

REFERENCES

1. Antelmann H, Tjalsma H, Voigt B, Ohlmeier S, Bron S, van Dijl JM, Hecker M. 2001. *Genome Res.* 11: 1484-1502.

2. Brockman AH, Orlando R, Tarleton RL. 1999. *Rapid Commun. Mass Spectrom.* 13: 1024-1030.

3. Corthals G, Wasinger B, Hochstrasser D, Sanchez J. 2000. *Electrophoresis* 21: 1104.

4. Griffin TJ, Aebersold R. 2001. *J. Biol. Chem.* 276: 45497-454500.

5. Hahn M, Mendgen K. 2001. *Curr. Opin. Plant Biol.* 4: 322-327.

6. Hirose I, Sano K, Shioda I, Kumano M, Nakamura K, Yamane K. 2000. *Microbiol.* 146: 65-75.

7. Ho Y et al. 2002. *Nature* 415: 180-183.

8. Konishi H., Ishiguro K. and Komatsu S. 2001. *Proteomics* 1: 1162–1171.

9. Kershaw MJ, Wakley G, Talbot NJ. 1998. *EMBO J.* 17: 3838-3849.

10. Lee IH, Walline RG, Plamann M. 1998. *Mol Microbiol.* 29: 209-218.

11. Noordewier MO, Warren PV. 2001. *Tends Biotechnol.* 19: 412-415.

12. Perkins DN, Pappin DJ, Creasy DM, Cottrell JS. 1999. *Electrophoresis.* 20: 3551-3567.

13. Smolka MB, Zhou H, Purkayastha S, Aebersold R. 2001. *Anal. Biochem.* 297: 25-31.

14. Templeton MD, Rikkerink EH, Beever RE. 1994. *Mol. Plant Microbe Interact.* 7: 320-325.

15. Tunlid A, Talbot N J. 2002. *Curr. Opin. Microbiol.* 5: 513-519.

16. Wagner K, Miliotis T, Marko-Varga G, Rainer B, Unger K. 2002. *Anal. Chem.* 74: 809.

17. White FF, Yang B, Johnson LB. 2001. *Curr. Opin. Plant Biol.* 3: 291-298.

18. Wood V et al. 2002. *Nature* 415: 871-880.

19. Wu S-C, Ham K, Darvill AG, Albersheim P. 1997. *Mol. plant-Microbe Interac.* 10: 700-708.

IDENTIFICATION AND CHARACTERIZATION OF SECRETED PROTEINS FROM *MAGNAPORTHE GRISEA*

*Guodong Lu[1, 2], Cristina Filippi[1], Dan Li[1] and Daniel Ebbole[1]

[1] *Department of Plant Pathology & Microbiology, Texas A & M University, College Station, USA 77843*

[2] *College of Plant Protection, Fujian Agricultural and Forestry University, Fuzhou, China 350002*

**E-mail: guodonglu@yahoo.com*

1. INTRODUCTION

Rice blast disease (*Magnaporthe grisea*) causes significant economic damage every year either directly through crop loss or indirectly in the cost of fungicides. A key to developing effective resistance is gaining a better understanding of the host-pathogen interaction, which requires a thorough understanding of both the host and pathogen. Recognition of the rice blast fungus via major disease resistance genes generally involves detection of a fungal-specific molecule by the host plant. These recognition molecules are primarily thought to be proteins secreted by the fungus. Secreted proteins examined to date have been plant cell wall-degrading enzymes, such as two xylanases of *M. grisea* XYN22 and XYN33 (Wu et al., 1995) and polygalacturonase and pectate lyase of *Aspergillus nidulans* (Dean and Timberlake, 1989); proteins involved in virulence, such as ECP1 and ECP2 of *Cladosporium fulvum*, and GAS1 and GAS2 of *M. grisea* are the virulence factors to tomato and rice, respectively (Lauge et al., 1997; Xue et al., 2002); MPG1 of *M. grisea* involved in pathogenecity (Talbot et al., 1993); or avirulence gene products, such as Avr-Pita and Pwl2 gene of *M. grisea* (Orbach et al., 2000 and Sweigard et al., 1995); NIP1 of barley scald

S. Kawasaki (ed.), Rice Blast: Interaction with Rice and Control, 47–55.
© 2004 *Kluwer Academic Publishers. Printed in the Netherlands.*

pathogen *Rynchosporium secalis* (Rohe et al., 1995) and AVR4, AVR9, ECP1, ECP2, ECP3, ECP4 and ECP5 of *C. fulvum* (Joosten and de Wit, 1999 and Lauge et al., 2000). A more systematic examination of secreted proteins from the rice blast fungus would help to identify the fungal factors involved in the interaction with the plant host.

Systematic approaches to analyze virulence factors from fungal pathogens have been developed using vectors to express proteins in plants for analysis of plant responses. These approaches have been used to test specific proteins or to screen libraries of cDNAs that may contain proteins involved in plant interactions (Lauge et al., 2000; Takken et al., 2000 and Kamoun et al., 2003). An alternate approach is to directly apply fungal proteins to test for plant responses. This approach is conceptually similar to testing of phytotoxic compounds (De Wit et al., 1982 and Lauge et al., 2000). As part of a larger functional genomics project we have proposed to take a high throughput approach to functionally analyze 300 putative secreted proteins of *M. grisea* to determine if the fungal expressed proteins are capable of evoking a response on rice plants.

2. MATERIALS AND METHODS

2.1 Identification of candidate secreted protein genes

In order to identify candidate secreted protein genes we employed several computer algorithms. SIGNALP-2.0 (http://www.cbs.dtu.dk/services/SignalP-2.0) predicts the presence and location of signal peptide cleavage sites in amino acid sequences from different organisms: Gram-positive prokaryotes, Gram-negative prokaryotes, and eukaryotes. The method incorporates a prediction of cleavage sites and a signal peptide/non-signal peptide prediction based on two types of algorithms, artificial neural networks and a hidden Markov model (Nielsen et al., 1997; Nielsen and Krogh, 1998).

A second type of algorithm is employed by the ProtComp program (www.softberry.com). This program predicts the sub-cellular localization of eukaryotic proteins and combines several methods of protein localization prediction - neural networks-based prediction, direct comparison with homologous proteins of known localization; prediction of certain functional peptide sequences, such as signal peptide, signal-anchor, transit peptides of mitochondria and chloroplasts and transmembrane segments; and searches for certain localization-specific motifs.

To identify the candidate secreted protein genes in *M. grisea*, the proteins predicted from the *M. grisea* genome sequence were subjected to SignalP-2.0 analysis. Proteins scoring positive for secretion based on all criteria for both the neural network and hidden Markov model prediction algorithms were then subjected to ProtSort analysis. Proteins predicted to be secreted by ProtSort were then manually annotated with BLAST search results and EST sequence data to support the ORF prediction and N-terminus of the predicted proteins. Additionally, we compared homologs of predicted secreted proteins of *M. grisea* to predicted ORFs of *Neurospora crassa.* When related proteins from both organisms were predicted to be secreted, we considered the overall prediction to be of higher reliability.

2.2 Over-expression of candidate secreted protein genes

Strains The *M. grisea* strain 70-15 was used as host for over-expression of candidate secreted protein genes. The *Saccharomyces cerevisiae* strain W303 was used as host strain for yeast recombination cloning. *Escherichia coli* strain DH10B was used as host for plasmid amplification.

Vectors construction: An oligonucleotide encoding the peptide sequence RGSHHHHHH ($RGSH_6$) tag was added to the pTE11 vector to construct the pDL1 vector (Figure no shown). Coding regions cloned in-frame with the start codon and $RGSH_6$ tag will be expressed from the *M. grisea* RP27 promoter and the expressed protein can be detected using $RGSH_6$ tag-specific antibodies and affinity purified with Ni-NTA affinity chromatography.

Gene cloning: The candidate secreted protein genes were amplified by PCR. *M. grisea* 70-15 genomic DNA was used as template in most cases. The 5' primer was designed to include 18 nucleotides homologous to promoter sequence in the vector including the ATG plus an additional 20 nucleotides matching the gene. The 3' primer contained 18 nucleotides matching the RGSHHH sequence in the vector with 20 nucleotides matching the gene sequence starting at the final codon of the coding region. PCR products were cloned into the expression vector by homologous recombination in yeast cells (restriction site independent cloning). Approximately 0.3ug of *Swa*I digested pDL1 DNA and 10 ul of the PCR reaction were co-transformed into lithium acetate competent yeast cells. Standard lithium acetate-mediated yeast transformation was used (Agatep et al., 1998) except that no non-specific DNA was added to the reaction. The transformation cells were spread on selection plates (SC-Trp, SC drop-out medium supplemented with adenine (20 mg/L), histidine (20 mg/L), leucine (100 mg/L), lysine(30 mg/L), and uracil (20 mg/L)). The yeast transformants were harvested together by washing the plate. Yeast cell walls were digested with zymolyase and protoplasts were disrupted by sodium dodecyl sulphate treatment prior to total DNA extraction. The DNA isolated from yeast transformants was used to transform *E. coli* cells by electroporation. Direct PCR of cells from *E. coli* colonies was used to screen for constructs with the correct insert. The constructed plasmid DNA was recovered with a miniprep procedure (Qiagen Inc.). Plasmid DNA was used to transform protoplasts of *M. grisea*.

2.3 Expression and detection of proteins

M. grisea transformants were selected for resistance to hygromycin B purified by spreading conidia to obtain pure cultures. The resulting strains were grown in complete medium (Talbot et al., 1993) for several days to obtain mycelia, and then transferred to fresh medium for an additional 6 to 24 hr. Culture filtrates were harvested by filtration and two volumes of acetone were added to precipitate proteins. The harvested mycelium was extracted separately to obtain cellular proteins. For protein purification, Ni-NTA affinity purification was used according to the manufacturer's instructions (Qiagen, Inc). SDS-PAGE and Western blotting were conducted using standard protocols (Sambrook et al., 1989; QIAexpress Detection and Assay handbook, Qiagen, Inc.).

3. RESULTS AND DISCUSSION

3.1 Candidate secreted protein genes and cloning

An initial set of candidate secreted proteins was identified from previously cloned genes with sequences deposited in GenBank. This initial set of 17 genes was the first to be subjected to PCR amplification and testing of the pipeline for high-throughput analysis. To functionally test secretion of candidates, the candidate genes are being expressed in *M. grisea* with an $RGSH_6$ tag for western blot detection and protein purification. Each PCR product was cloned into pDL1 ($RGSH_6$ tag) by recombination cloning in yeast. DNA isolated from yeast cells was used to transform *E. coli* cells. We routinely characterize three colonies to screen for the construct. Approximately 75% of *E. coli* colonies contained the correct clone. This high efficiency of cloning is necessary for high-throughput analysis of our 300 target genes.

Our initial analysis of EST sequences led to the identification of over 140 putative secreted protein encoding cDNAs. Following the release of the *M. grisea* genome sequence by the Whitehead Institute Center for Genome Research (http://www-genome.wi.mit.edu/annotation/fungi/magnaporthe/) analysis by SignalP-2.0 was used to identify a group of 1238 potentially secreted proteins that were positive for all SignalP criteria indicating secretion. ProtSort was used as a second method that further identified 758 of the 1238 candidates as being secreted. Criteria such as predicted size, representation in the *M. grisea* EST databases, and homology to known proteins were used in choosing these initial genes for further analysis.

3.2 Expression and detection of proteins

The constructed genes were transformed into strain 70-15 of *M. grisea* by PEG-mediated transformation. Transformants with intact transgenes were identified by screening with PCR using primers in the vector sequence. For several genes, GFP fusion proteins were constructed to visualize protein localization. For proteins known to be secreted, such as MPG1, fluorescence was most intense at hyphal apices and cell walls, similar to observations of secreted proteins in *Aspergillus niger* (Gordon et al., 2000). Fluorescence was also found in septa and vacuoles as previously reported in *A. niger* (Gordon et al., 2000; Khalaj et al., 2001) and yeast (Braspenning et al., 1998)(Fig. 1).

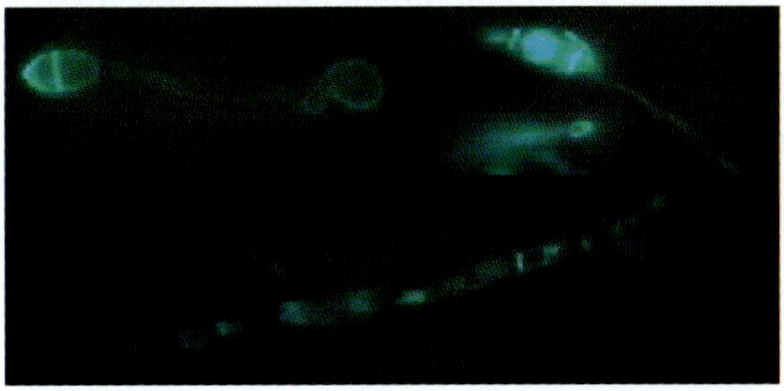

Figure 1. *Green fluorescence detected: Most intense at hyphal apices, cell walls, septa and vacuoles*

Next we tested the candidate proteins to determine if they were detected in the culture medium. Culture medium and mycelium were separated and the extracted proteins were subjected to SDS-PAGE and Western blotting analysis. The results showed that extracellular gene-His tag (anti-His) and gene::GFP (anti-GFP) proteins were detected in culture medium of young mycelia, which confirmed that they are secreted (Fig.2). To optimize the culture media for protein secretion, five media were tested, potato dextrose, complete medium, soymilk, malt extract and Vogels minimal medium. Complete medium produced the greatest yield of secreted protein as judged by comparing intensity of staining in Western blots (data no shown). We conducted experiments to determine optimal culture time for expression and to optimize the initial amount of mycelium to transfer to fresh medium to produce the greatest yield. Standardized conditions could be used to detect protein, however, the optimal medium, time and culture density for achieving the greatest protein yield differed for different proteins. Our initial

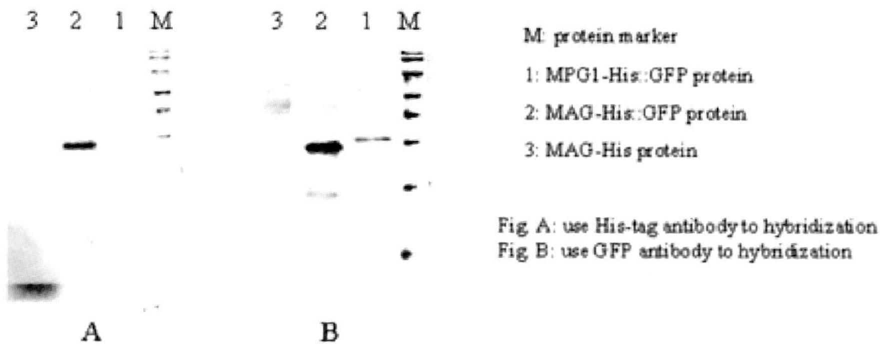

Figure 2. *Protein purification use Ni-NTA.*

tests used the magnaporin and xylanase genes to assess secretion. For magnaporin, the amount of secreted protein increased with increasing mycelial inoculum (1.2g/50 ml culture) and culture time (48hr), whereas for the xylanase-33 gene, the best protein yield was obtained with mycelial inoculum was 0.8g and 6-12hr of incubation (Fig. 3). These secreted proteins could be purified by the Ni-NTA affinity method, although the efficiency was different for different genes. In general, we found that purification was most efficient with more concentrated protein samples so that precipitation of proteins prior to Ni-NTA affinity purification yielded more protein than direct addition to Ni-NTA affinity material to culture filtrate (Fig. 2).

4. CONCLUSIONS

We conclude that the key steps of gene amplification, cloning and isolation of expression constructs, and transformation into *M. grisea* are efficient and amenable to high-throughput approaches. The characterization of *M. grisea* transformants, where multiple transformants require single-spore isolation of pure culture allows analysis of approximately 10 clones per person per week. Protein expression analysis is rapid but may require empirical optimization for some proteins. Our goal is to obtain sufficient quantities of protein (crude and affinity purified) for testing on rice plants. Proteins will be inoculated into wound sites on intact plants or on detached leaves to identify proteins that evoke a plant response. Our initial analysis suggests that the high-throughput pipeline for gene cloning and protein expression will allow us to rapidly test 300 proteins for secretion and to further test the secreted proteins in evoking plant responses.

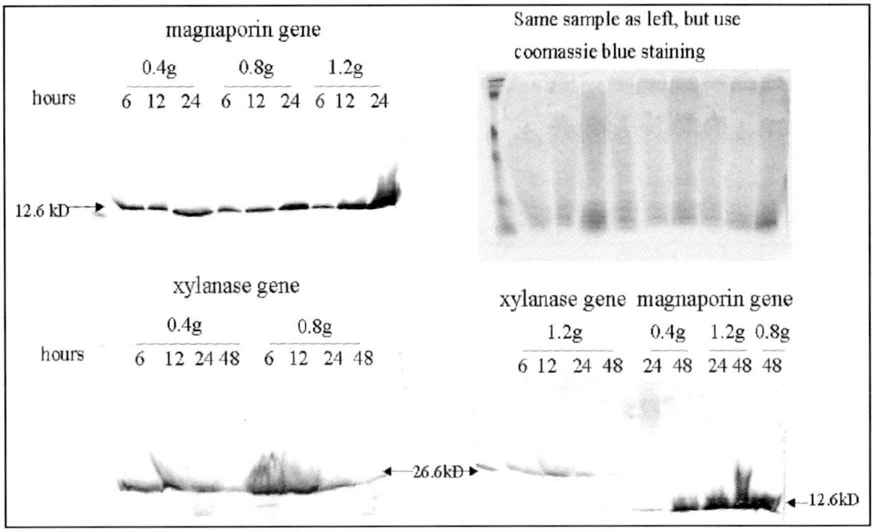

***Figure 3**. Effect of mycelial mass and culture growth time on yield of secreted proteins*

ACKNOWLEDGEMENTS

We thank Jim Sweigard for pTE11 vector. This work was supported by grants from the National Science Foundation DBI-0115642 and the United States Department of Agriculture 99-35303-8057.

REFERENCES

Agatep, R., Kirkpatrick, R. D., Parchaliuk, D. L., Woods, R. A., & Gietz, R. D. (1998). Transformation of *Saccharomyces cerevisiae* by the lithium acetate/single-stranded carrier DNA/polyethylene glycol protocol. Technical Tips Online, 1, 51: P01525

Braspenning, J., Meschede, W., Marchini, A., Muller, M., Gissmann, L., & Tommasino, M. (1998). Secretion of heterologous proteins from *Schizosaccharomyces pombe* using the homologous leader sequence of pho1+ acid phosphatase. Biochem. Biophys. Res. Commun., 7, 166-171.

De Wit, P. J. G. M., & Spikman, G. (1982). Evidence for the occurrence of race and cultivar-specific elicitors of necrosis in intercellular fluids of compatible interactions of *Cladosporium fulvum* and tomato. Physiol. Plant Pathol., 21, 1-11.

Dean, R.A., & Timberlake, W.E. (1989). Production of cell wall-degrading enzymes by *Aspergillus nidulans*: a model system for fungal pathogenesis of plants. Plant Cell, 1, 265-273.

Gordon, C. L., Archer, D. B., Jeenes, D. J., Doonan, J. H., Wells, B., Trinci, A. P., & Robson, G. D. (2000). A glucoamylase::GFP gene fusion to study protein secretion by individual hyphae of *Aspergillus niger*. J. Microbiol Methods, 42, 39-48.

Joosten, M. H. A. J., & de Wit, P. G. J. M. (1999). The tomato-*Cladosporium fulvum* interaction: a versatile experimental system to study plant-pathogen Interactions.
Annu. Rev. Phytopathol., 37, 335-367.

Kamoun, S., Hamada, W., & Huitema, E. (2003). Agrosuppression: a bioassay for the hypersensitive response suited to high-throughput screening. Mol. Plant Microbe Interact., 16, 7-13.

Khalaj, V., Brookman, J. L., & Robson, G. D. (2001). A study of the protein secretory pathway of *Aspergillus niger* using a glucoamylase-GFP fusion protein. Fungal Genet. Biol., 32, 55-65.

Lauge, R., Goodwin, P. H., de Wit, P. J. G. M., & Joosten, M. H. A. J. (2000). Specific HR-associated recognition of secreted proteins from *Cladosporium fulvum* occurs in both host and non-host plants. Plant J., 23, 735-745.

Laugé, R., Joosten, M. H. A. J., Van den Ackerveken, G. F. J. M., Van den Broek, H. W. J., & De Wit, P. J. G. M. (1997). The in planta–produced extracellular proteins ECP1 and ECP2 of *Cladosporium fulvum* are virulence factors. Mol. Plant Microbe Interact., 10, 725–734.

Nielsen, H., & Krogh, A. (1998). Prediction of signal peptides and signal anchors by a hidden Markov model., In *Proceedings of the Sixth International Conference on Intelligent Systems for Molecular Biology* (*ISMB* 6), AAAI Press, Menlo Park, California, pp. 122--130.

Nielsen, H., Engelbrecht, J., Brunak, S., & von Heijne, G. (1997). Identification of prokaryotic and eukaryotic signal peptides and prediction of their cleavage sites. Protein Engineering, 10, 1-6 .

Orbach, M. J., Farrall, L., Sweigard, J. A., Chumley, F. G., & Valent, B. (2000). A telomeric avirulence gene determines efficacy for the rice blast resistance gene Pi-ta. Plant Cell, 12, 2019-2032.

Rohe, M., Gierlich, A., Hermann, H., Hahn, M., Schmidt, B., Rosahl, S., & Knogge, W. (1995). The race-specific elicitor, NIP1, from the barley pathogen, *Rhynchosporium secalis*, determines avirulence on host plants of the Rrs1 resistance genotype. EMBO J., 14, 4168-4177.

Sambrook, J., Fritsch, E.F., & Maniatis, T. (1989). Molecular Cloning: A Laboratory Manual. 2nd ed Cold Spring Harbor, NY, Cold Spring Harbor Laboratory Press.

Sweigard, J. A., Carroll, A. M., Kang, S., Farrall, L., Chumley, F.G., & Valent, B. (1995). Identification, cloning, and characterization of PWL2, a gene for host species specificity in the rice blast fungus. Plant Cell, 7, 1221-1233.

Takken, F. L. W., Luderer, R., Gabriels, S. H. E. J., Westerink, N., Lu, R., de Wit, P. J. G. M., & Joosten, M. H. A. J. (2000). A functional cloning strategy, based on a binary PVX-expression vector, to isolate HR-inducing cDNAs of plant pathogens. Plant J., 24, 275-283.

Talbot, N. J., Ebbole, D. J., & Hamer J. E. (1993). Identification and characterization of MPG1, a gene involved in pathogenicity from the rice blast fungus *Magnaporthe grisea.* Plant Cell, 5, 1575-1590.

Wu, S.C., Kauffmann, S., Darvill, A.G., & Albersheim, P. (1995). Purification, cloning and characterization of two xylanases from *Magnaporthe grisea*, the rice blast fungus. Mol. Plant Microbe Interact., 8, 506-514.

Xue, C., Park, G., Choi, W., Zheng, L., Dean, R. A., & Xu, J.-R. (2002). Two novel fungal virulence genes specifically expressed in appressoria of the rice blast fungus. Plant Cell, 14, 2107-2119.

REPEAT-INDUCED POINT MUTATION (RIP) IN *MAGNAPORTHE GRISEA*: EVIDENCE FOR THE PRESENCE OF SEXUAL CYCLE IN NATURE

*Ken-Ichi Ikeda, Hitoshi Nakayashiki, Yukio Tosa and Shigeyuki Mayama

Faculty of Agriculture, Kobe University, Kobe, Japan

**E-mail: ikeken@affrc.go.jp*

Abstract: The RIP (repeat-induced point mutation) process introduces C:G to T:A transitions into duplicated sequences during a specific period in the sexual reproduction and has been reported in only two fungal species, *Neurospora crassa* and *Podospora anserina*. To investigate whether a RIP-like process operates in *Magnaporthe grisea*, the retrotransposon MAGGY and the hygromycin B phosphotransferase gene were introduced into the fungus as multiple transgenes and examined for sequence alterations after asexual growth and sexual crossing. Numerous C:G to T:A transitions were found in the transgenes, preferentially in (A/Tp)Cp(A/T) contexts, in the sexual progenies, suggesting that a RIP-like process functions in *M. grisea*. We also examined the sequence of another retrotransposon, Pyret, in field isolates of *M. grisea*. Even though the perfect stage rarely has been observed in *M. grisea* under field conditions, RIP-like transitions were found in all the field isolates tested and were correlated with the fertility of the isolates in Petri dishes. Analysis of RIP-like transitions in repetitive sequences may provide a tool to estimate the frequency of sexual recombination of *M. grisea* isolates.

1. INTRODUCTION

Possible beneficial roles of transposable elements in genome evolution have been proposed by many researchers (McDonald *et al.*, 1997; Hurst and Werren, 2001). We reported that the retrotransposon MAGGY was

S. Kawasaki (ed.), Rice Blast: Interaction with Rice and Control, 57–63.
© 2004 *Kluwer Academic Publishers. Printed in the Netherlands.*

transcriptionally activated under stress conditions (Ikeda *et al.*, 2001). Stress-responsive transposition may have some adaptive effects by creating new phenotypes, such as resistance to the stress. However, transpositions also pose a serious threat to the genome, exemplified by disruption and anomalous expression of essential genes of the host (Kumpatla *et al.*, 1998). Therefore, transposition of the elements is likely to be strictly regulated by host factors. In fungi, several gene inactivation mechanisms directed against repeated sequences have been reported, including RIP (repeat-induced point mutation), MIP (methylation induced premeiotically) and quelling (Faugeron, 2000). RIP causes genetic changes in duplicated sequences, characterized by C:G to T:A transitions in the target sequence during a specific period in the sexual cycle, with a significant preference for CpA contexts (Selker *et al.,* 1987; Cambareri *et al.*, 1989). RIP was originally identified in *Neurospora crassa* (Selker *et al.,* 1987) and also has been described recently in *Podospora anserina* (Hamann *et al.,* 2000). RIP-like C:G to T:A transitions are, however, also reported in the sequence of transposable elements in phylogenetically more distant filamentous fungi, including *Aspergillus fumigatus* (Neuveglise *et al.*, 1996), *A. nidulans* (Nielsen *et al.*, 2001), *Fusarium oxysporum* (Hua-Van *et al.*, 1998), and *Magnaporthe grisea* (Nakayashiki *et al.*, 1999). Interestingly, some of these fungi have no known sexual stage (e.g., *F. oxysporum* and *A. fumigatus*) or have it only under laboratory conditions (e.g., *M. grisea*). In addition, the

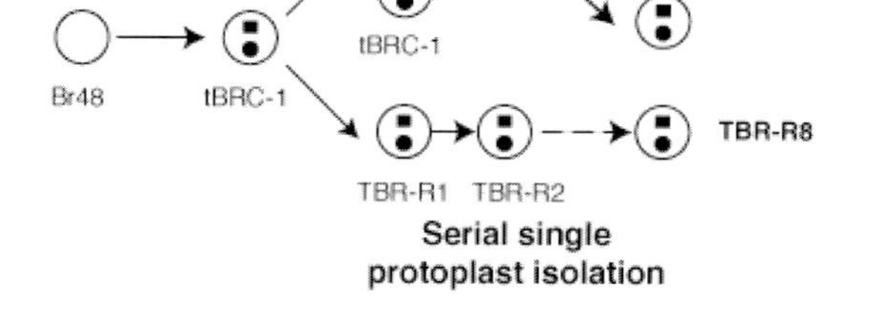

Figure 1. *The experimental scheme in this study. Multiple copies of the retrotransposon MAGGY (circles) and three copies of the hygromycin B phosphotransferase gene (boxes) were introduced into a wheat isolate of Magnaporthe grisea (Br48). F_1 progenies of the resulting transformant, tBRC-1, were analysed for nucleotide sequence changes relative to TBR-R8, which was obtained by serial single protoplast isolation from tBRC-1.*

target contexts for the C:G to T:A transitions in these fungi, for example, (A/Tp)Cp(A/T) in *M. grisea* (Nakayashiki *et al.*, 1999), Cp(A/G) in *F. oxysporum* (Hua-Van *et al.*, 1998), and CpG in *A. fumigatus* and *A. nidulans* (Neuveglise *et al.*, 1996; Nielsen *et al.*, 2001), differ significantly from the one for canonical RIP (CpA). Therefore, it is not clear whether the C:G to T:A transitions found in these fungi resulted from processes equivalent to RIP or not.

In this study, we introduced the retrotransposon MAGGY and multiple hygromycin B phosphotransferase (*hph*) genes into a fertile *M. grisea* isolate from wheat, and examined the transgenes for evidence of genetic alteration during the sexual cycle under laboratory conditions (Fig. 1). We also looked for RIP-like transitions in the sequence of another retrotransposon, Pyret (Nakayashiki *et al.*, 2001), in field isolates of *M. grisea*. Based on our results, we discuss the relationship between a RIP-like process in *M. grisea* and RIP in *N. crassa*, and the occurrence of the perfect stage of *M. grisea* in the field.

2. NUCLEOTIDE TRANSITION IN MULTIPLE TRANSGENES DURING SEXUAL REPRODUCTION

We amplified two PCR products, corresponding to partial fragments of MAGGY (5340bp) and *hph* (797bp), with a high fidelity Taq enzyme using genomic DNA of the F_1 progenies and TBR-R8 as templates. Three independent clones each of MAGGY and *hph* were analysed.

***Table 1**. Nucleotide changes in MAGGY and hygromycin B phosphotransferase gene sequences in TBR-R8 and F_1 progenies of Magnaporthe grisea*

Gene	*Isolate*	*Total substitutions (%)*	*C:G to T:A transitions (%)*	*C:G to T:A transitions /total substitutions (%)*
MAGGY	TBR-R8	0.10	0.06	56.25
	F_1-4	0.33	0.26	78.87
	F_1-6	0.84	0.72	85.47
hph	TBR-R8	0.00	0.00	-
	F_1-10	0.71	0.71	100
	F_1-12	0.50	0.50	100

Data are from Ikeda et al., 2002.

The clones derived from TBR-R8 and the F_1 progenies were all sequenced by the primer walking method. Sequence alterations on the MAGGY and *hph* sequences are summarized in Table1. In general, C:G to T:A transitions in MAGGY were detected in the F_1 progenies at a much higher frequency than in TBR-R8, which was obtained asexually. With the *hph* gene, the sequence alterations were restricted to C:G to T:A transitions and found only in the F_1 progenies, indicating that a RIP-like process operates in *M. grisea* during sexual reproduction.

The F_1-10 and F_1-12 progenies, from which the *hph* fragments were cloned, showed a hygromycin-sensitive phenotype. Actually, every *hph* gene contained at least one C:G to T:A transition giving rise to an amino acid substitution. These results demonstrated that the RIP-like process in *M. grisea* indeed acts as a gene silencing mechanism on duplicated sequences during the sexual phase of the life cycle.

3. TARGET SITE SPECIFICITY OF C:G TO T:A TRANSITION IN *MAGNAPORTHE GRISEA*

We further analysed target contexts for the RIP-like process in *M. grisea* and found that they were (A/Tp)Cp(A/T). Preferred contexts can be symmetrical (e.g., ApCpT or TpCpA) but in fact no preference for the symmetrical contexts could be detected. This conclusion was also reported previously for degenerated MAGGY element (Nakayashiki *et al.*, 1999).

4. RIP-LIKE TRANSITION IN FIELD ISOLATES OF *MAGNAPORTHE GRISEA*

We analysed sequences of a middle-repetitive retrotransposon, Pyret (Nakayashiki *et al.*, 2001), in four field isolates of *M. grisea* to investigate whether a RIP-like process had also operated in this fungus under natural conditions.

A Pyret fragment corresponding to the middle region of the *gag* domain was amplified by PCR with a high fidelity Taq enzyme. Three clones were isolated from each of the *M. grisea* isolates and sequenced. To estimate the frequency of RIP-like events in these clones, we used a "de-RIP" approach

Table 2. *Mating ability of the field isolates of* Magnaporthe grisea *and RIP-like transitions on Pyret sequences in the isolates.*

Isolate	*RIP-like transitions (%)*[a]	*AT contents (%)*	*Germination of ascospores (%)*[b]
Rice (1836-3)	7.34 ±1.17	47.13	-[c]
Rice (2012-1)	6.76 ±0.18	46.51	-[c]
Foxtail millet (GFSI1-7-2)	16.96±1.89	52.23	33.0
Wheat (Br48)	32.75±0.29	60.67	73.5

a, The percentage of C:G to T:A transitions with reference to the deRIP-Pyret sequence. b, Fertility was determined by crossing with the finger millet strains G10-1 (MAT1-1) and Z2-1 (MAT1-2). The percentage of germinated ascospores was calculated from at least 50 ascospores. c, No ascospores were obtained. Data are from Ikeda et al., 2002.

(Cambareri *et al.,* 1998) creating a hypothetical sequence *in silico*, deRIP-Pyret, to represent the gene in its unaltered state. Specifically, each thymine or adenine residue in deRIP-Pyret was replaced with cytosine or guanine if cytosine or guanine occurred at the corresponding position in any of the cloned Pyret sequences. We detected RIP-like transitions in every clone of Pyret from all isolates examined and consequently the AT contents of those elements had increased (Table 2).

The frequency of RIP-like transitions was correlated with the fertility of the isolates under laboratory conditions (Table 2). Since RIP-like transitions in *M. grisea* occurred only during the sexual stage of the life cycle in this study, it is likely that the numerous C:G to T:A transitions in Pyret in the field isolates resulted from the operation of the RIP-like process during the sexual cycle in the natural field context.

5. CONCLUSION

In fungi, RIP processes have been proposed as a potent defence mechanism against transposable elements, as typified in *Neurospora crassa* and *Podospora anserina* (Selker *et al*., 1987; Cambareri *et al*., 1989; Hamman *et al*., 2000). RIP-like processes found in other fungal species, however, did not always fulfill the conditions of canonical RIP. For example, *A. fumigatus* and *F. oxysporum* have no known sexual phase (Neuveglise *et al*., 1996; Hua-Van *et al*., 1998). Similarly, the perfect stage of *M. grisea* has been reported only under laboratory conditions. These other fungi also showed significant differences from *N. crassa* in the target contexts for C:G to T:A transitions. Thus, it was a matter of debate to what extent the RIP-like processes in other fungal species shared common features with RIP as described for *N. crassa* and *P. anserina*.

We, therefore, investigated the RIP-like process operating in *M. grisea* under laboratory conditions using fertile isolates of the fungus. The results indicated that the RIP-like process in *M. grisea* exhibited most of the main features originally described for RIP events. Thus, we conclude that the RIP-like transitions in *M. grisea* are generated by a process equivalent to RIP as in *N. crassa* and *P. anserina*.

The target contexts for the RIP-like transitions in *M. grisea* were (A/Tp)Cp(A/T), significantly different from the one for RIP in *N. crassa,* CpA (Cambareri *et al*., 1989). The target contexts for RIP-like transitions previously reported in several fungal species showed several variations depending on the phylogenetic relationship among the fungi. Therefore, the canonical target contexts for RIP, CpA, might be only one variation among several possible target sites for the RIP mechanism among a wider array of fungal species.

It is of great interest to know how widely RIP-like mechanisms are distributed among fungal species. All the above fungal species are classified into the subphylum *Pezizomycotina* in the phylum *Ascomycota*, supporting the idea that the fungi, in *Pezizomycotina* at least, may share the basis for the RIP mechanism. Actually, *Ascobolus immersus,* which is also a member of *Pezizomycotina,* possesses a gene silencing mechanism similar to RIP, termed MIP. This distinct mechanism also inactivates duplicated sequences during the sexual cycle, but does so by heavy methylation rather than sequence alterations (Goyon and Faugeron, 1989). It is very likely that some common mechanisms underlie RIP and MIP (Malagnac *et al*., 1997; Freitag *et al*., 2002).

M. grisea strains were clustered into several pathotypes based mainly on their corresponding gramineous host plants (Kato *et al*., 2000). The sequencing of the retrotransposon Pyret in four field isolates of *M. grisea* revealed that the Pyret elements had numerous RIP-like transitions and that the frequency of the transitions was correlated with the mating ability of the isolate in the Petri dish. These results strongly suggested that the *M. grisea* isolates have passed through the sexual cycle in nature. Since the sexual cross is the main source of genetic variation in eukaryotes, it is of interest for plant pathologists to know whether sexual crossing in nature gives rise to the pathogenic diversity of *M. grisea*. In fact, blast disease sometimes causes epidemics on plant species that previously were not regarded as hosts of *M. grisea*. Examples are the wheat blast that was first reported in Brazil in 1985 (Igarashi *et al.*, 1986) and the perennial ryegrass blast that was originally described in 1992 in the USA (Landschoot and Hoyland, 1992). Interestingly, in both cases the fertility of the newly appearing strains was high, at least under laboratory conditions (Urashima *et al*., 1993). Therefore, it is an attractive hypothesis that those new strains may have arisen from sexual crosses between fertile isolates of *M. grisea* in the field. Analysis of RIP-like transitions in repetitive sequences can provide a tool to estimate the frequency of sexual recombination of *M. grisea* isolates.

REFERENCES

Cambareri, E. B., Jensen, B. C., Schabtach, E., & Selker, E. U. (1989). Repeat-induced G-C to A-T mutations in *Neurospora*. *Science* **244:** 1571-1575.

Cambareri, E. B., Aisner, R., & Carbon, J. (1998). Structure of the chromosome VII centromere region in *Neurospora crassa*: degenerate transposons and simple repeats. *Mol Cell Biol* **18**: 5465-5477.

Faugeron, G. (2000). Diversity of homology-dependent gene silencing strategies in fungi. *Curr Opin Microbiol* **3**: 144-148.

Freitag, M., Welliams, R. L., Kothe, G. O., & Selker, E. U. (2002). A cytosine methyltransferase homologue is essential for repeat-induced point mutation in *Neurospora crassa*. *Proc Natl Aca Sci USA* **99:** 8802-8807.

Goyon, C., & Faugeron, G. (1989). Targeted transformation of *Ascobolus immersus* and *de novo* methylation of the resulting duplicated DNA sequences. *Mol Cell Biol* **9**: 2818-2827.

Hamann, A., Feller, F., & Osiewacz, H.D. (2000). The degenerate DNA transposon *Pat* and repeat-induced point mutation (RIP) in *Podospora anserina*. *Mol Gen Genet* **263**: 1061-1069.

Hua-van, A., Héricourt, F., Capy, P., Daboussi, M.J., & Langin, T. (1998). Three highly divergent subfamilies of the *impala* transposable element coexist in the genome of the fungus *Fusarium oxysporum*. *Mol Gen Genet* **259**: 354-362.

Hurst, G.D., & Werren, J. H. (2001). The role of selfish genetic elements in eukaryotic evolution. *Nat Rev Genet* **2**: 597-606.

Ikeda, K., Nakayashiki, H., Takagi, M., Tosa, Y., & Mayama, S. (2001). Heat shock, copper sulfate, and oxidative stress activate the retrotransposon MAGGY resident in the plant pathogenic fungus *Magnaporthe grisea*. *Mol Genet Genomics* **266:** 318-325.

Ikeda, K., Nakayashiki, H., Kataoka, T., Tamba, H., Hashimoto, Y., Tosa, Y., & Mayama, S. (2002). Repeat-induced point mutation (RIP) in *Magnaporthe grisea*: implications for its sexual cycle in the natural field context. *Mol Microbiol* **45:** 1355-1364.

Igarashi, S., Utiamada, C.M., Igarashi, L.C., Kazuma, A.H., & Lopes, R.S. (1986). *Pyricularia* em trigo. 1. Ocorrencia de *Pyricularia* sp. no estado do Parana. *Fitopatol Bras* **11**: 351-352.

Kato, H., Yamamoto, M., Yamaguchi-Ozaki, T., Kadouchi, H., Iwamoto, Y., Nakayashiki, H., Tosa, Y., Mayama, S., & Mori, N. (2000). Pathogenicity, mating ability and DNA restriction fragment length polymorphisms of *Pyricularia* populations isolated from Gramineae, Bambusideae and Zingiberaceae plants. *J Gen Plant Pathol* **66**: 30-47.

Kumpatla, S.P., Chandrasekharan, M.B., Iyer, L.M., Li, G., & Hall, T.C. (1998). Genome intruder scanning and modulation systems and transgene silencing. *Trend Plant Sci* **3**: 97-104.

Landschoot, P.J., & Hoyland, B.F. (1992). Grey leaf spot of perennial ryegrass turf in Pennsylvania. *Plant Dis* **76**: 1280-1282.

Malagnac, F., Wendel, B., Goyon, C., Faugeron, G., Zickler, D., Rossignol, J. L., Noyer-Weidner, M., Vollmayr, P., Trautner, T. A., & Walter, J. (1997). A gene essential for *de novo* methylation and development in *Ascobolus* reveals a novel type of eukaryotic DNA methyltransferase structure. *Cell* **91:** 281-290.

McDonald, J.F., Matyunina, L.V., Wilson, S., Jordan, I.K., Bowen, N.J., & Miller, W.J. (1997). LTR-retrotransposons and the evolution of eukaryotic enhancers. *Genetica* **100**: 3-13.

Nakayashiki, H., Nishimoto, N., Ikeda, K., Tosa, Y., & Mayama, S. (1999). Degenerate MAGGY elements in a subgroup of *Pyricularia grisea*: a possible example of successful capture of a genetic invader by a fungal genome. *Mol Gen Genet* **261**: 958-966.

Nakayashiki, H., Matsuo, H., Chuma, I., Ikeda, K., Betsuyaku, S., Kusaba, M., Tosa, Y., & Mayama, S. (2001). Pyret, a *Ty3/Gypsy* retrotransposon in *Magnaporthe grisea* contains an extra domain between the nucleocapsid and protease domains. *Nucleic Acids Res* **29**: 4106-4113.

Neuveglise, C., Sarfati, J., Latge, J.P., & Paris, S. (1996). *Afut1*, a retrotransposon-like element from *Aspergillus fumigatus*. *Nucleic Acids Res* **24**: 1428-1434.

Nielsen, M.L., Hermansen, T.D., & Aleksenko, A. (2001). A family of DNA repeats in *Aspergillus nidulans* has assimilated degenerated retrotransposons. *Mol Genet Genomics* **265**: 883-887.

Selker, E.U., Cambareri, E.B., Jensen, B.C., & Haack, K.R. (1987). Rearrangement of duplicated DNA in specialized cells of *Neurospora*. *Cell* **51**: 741-752.

Urashima, A.S., Igarashi, S., & Kato, H. (1993). Host Range, Mating Type And Fertility Of *Pyricularia Grisea* From Wheat In Brazil. *Plant Dis* **77:** 1211-1216.

RELATIONSHIP BETWEEN TWO AVIRULENCE GENES: *AVR-HATTAN 3* AND *AVR-PIKS*

*N. Yasuda[1], K. Hirayae[2], N. Hayashi[3], Y. Fujita[4], M. Tsujimoto and T. Nakajima[5]

[1]*National Agricultural Research Center, 3-1-1 Kannondai, Tsukuba, 305-8666, Japan*
[2]*National Agricultural Research Center for Kyusyu Okinawa Region, Suya 2421, Nishigoshi, 861-1192 Kumamoto, Japan*
[3]*National Institute of Agrobiological Sciences, 2-1-2 Kannon-dai, Tsukuba, 305-8602, Japan*
[4]*National Agricultural Research Center for Hokkaido Region,*
[5]*National Agricultural Research Center, Hokuriku Area, 1-2-1 Inada, Jyouetsu, 943-0193, Japan*

**E-mail: yasuda@affrc.go.jp*

1. INTRODUCTION

Race-cultivar specificity of the rice blast fungus complies with the 'gene-for-gene' system. Several avirulence genes, which determine cultivar specificity have been identified and characterized (Valent and Chumley, 1994, Notteghem *et al.*, 1994, Mandel *et al.*, 1997, Dioh *et al.*, 2000, Kang *et al.*, 2001, Fukiya *et al.*, 2001, Farman *et al.*, 2002). However, no common features of these genes other than recognition factors have been found. The molecular mechanisms behind race-cultivar interactions might vary with the host-pathogen combinations, while the mode and frequency of the avirulence gene mutations resulting in virulent races might vary with the genes. Studies on the molecular background and evolution of race-cultivar interactions will lead to the effective use of resistance genes. In this paper, we describe and identify new avirulence genes and corresponding resistance genes in the rice blast system.

S. Kawasaki (ed.), Rice Blast: Interaction with Rice and Control, 65–69.
© 2004 *Kluwer Academic Publishers. Printed in the Netherlands.*

2. GENETIC ANALYSIS OF AVIRULENCE IN *M. GRISEA*

We performed crosses between two blast isolates having different pathogenicity. A 1:1 segregation ratio of pathogenicity in the F_1 progenies suggested that pathogenicity was controlled by a single locus. This was shown by crossing two blast isolates, Y93-164a-1 and Y93-165g-1, which then produced 137 progenies. Y93-164a-1 was avirulent to rice cultivar Ishikarishiroke, while Y93-165g-1 was virulent. The ratio of avirulent to virulent progenies toward Ishikarishiroke was 68: 69, and it fitted the 1:1 ratio. This result indicates that an avirulence gene determines pathogenicity to Ishikarishiroke.

In the same way, we made three pairs of crosses and the resulting progenies were tested for avirulence/virulence on three rice cultivars. The ratio of virulent to avirulent progenies from each cross was close to 1:1, suggesting that three avirulence genes independently prevent infection in rice cultivars Aichiasahi, K 59 and Hattan 3.

Consequently, we identified four avirulence genes in four blast isolates namely: Y93-164a-1, Y93-165g-1, Y93-245c-2 and Y90-71 toward Ishikarishiroke, Aichiasahi, K59 and Hattan 3, respectively.

Rapid mapping of the three avirulence genes to Ishikarishiroke, Aichiasahi and Hattan 3 was made using RFLP or RAPD markers (Yasuda *et al.*, 2000). The RFLP markers were kindly provided by Dr. Leong at Wisconsin University and Dr. Teraoka at Tokyo University of Agriculture and Technology.

3. GENETIC ANALYSIS OF RESISTANCE IN RICE CULTIVARS

We examined the resistance genes corresponding to each of the four avirulence genes identified. Resistance genes to the three avirulence genes among them are from the three Japanese differentials Ishikarishiroke, Aichiasahi and K59 known to possess the *Pii*, *Pia* and *Pit*, respectively. We assumed that the avirulence gene toward Ishikarishiroke corresponds to *Pii*. This hypothesis was proven with F_3 lines from the cross between Ishikarishiroke having the resistance gene, *Pii* and a rice cultivar Norin 3, which does not. The isolate Ken54-20 known to have the *Avr-Pii* (Yamasaki and Kiyosawa, 1966) was used to identify the *Pii* gene. We compared reactions of each F_3 line to the isolates Ken54-20 and Y93-164a-1. All of the F_3 lines tested showed exactly the same reaction to Y93-164a-1 and Ken54-20. This indicates that the resistance gene in

Ishikarishiroke affecting resistance to Ken54-20 also controls resistance to Y93-164a-1. Hence, an avirulence gene toward Ishikarishiroke corresponds to *Pii*.

Furthermore, we revealed that an avirulence gene toward Aichiasahi corresponds to *Pia* and an avirulence gene toward K59 corresponds to *Pit*. However, we have not yet specified the resistance gene corresponding to *Avr-Hattan 3*.

4. RESISTANCE GENE CORRESPONDING TO *AVR-HATTAN 3*

The rice cultivar Hattan 3 belongs to the Shin 2 group wherein no major genes for resistance exist against most of the Japanese isolates. However, it was demonstrated that Shin 2 has the resistance gene *Piks* toward a Philippine isolate (Kiyosawa, 1969). Moreover, Dr. Hayashi and his coworkers identified the avirulence gene *Avr-Piks* in a Chinese blast isolate, CHNOS60-2-3 (unpublished). Both Hattan 3 and Shin 2 are resistant to the isolates Y90-71 and CHNOS60-2-3. The reactions of Hattan 3 and Shin 2 to these two isolates were very similar. Furthermore, the two cultivars showed strong resistance with no visible symptoms to both of the isolates. These facts led us to the hypothesis that *Avr-Hattan 3* corresponds to *Piks*.

Recently, a genetic marker closely linked to *Avr-Piks* was developed for PCR detection (unpublished). The PCR marker for *Avr-Piks* was found to co-segregate with *Avr-Piks* in a cross between CHNOS60-2-3 and 2145-R-57. With a set of PCR primers, a 1.2 kb fragment was amplified. That is if the progenies from the cross have *Avr-Piks,* while no fragment was amplified if the progenies did not hold *Avr-Piks*. We used the PCR marker for *Avr-Piks* to estimate whether *Avr-Hattan 3* is identical with *Avr-Piks* or not.

We investigated the *Avr-Piks* marker in the parental isolates, Y90-71 and 3514-R-2. The 1.2 kb fragment was amplified with the isolate Y90-71, which possesses *Avr-Hattan 3*. However, no fragment was detected with the other parent 3514-R-2, which does not posses *Avr-Hattan 3*. In the PCR analysis of progenies of the cross, all the isolates avirulent to Hattan 3 produced the 1.2 kb band, but all the virulent isolates did not generate the band (Figure 1). Segregation analysis of the 146 progenies revealed that the *Avr-Piks* marker completely co-segregated with *Avr-Hattan 3*.

Three spontaneous mutants pathogenic to Hattan 3 were obtained from the nonpathogenic parental isolate Y90-71. We investigated the marker in the three mutants of Y90-71. The 1.2 kb fragment was not

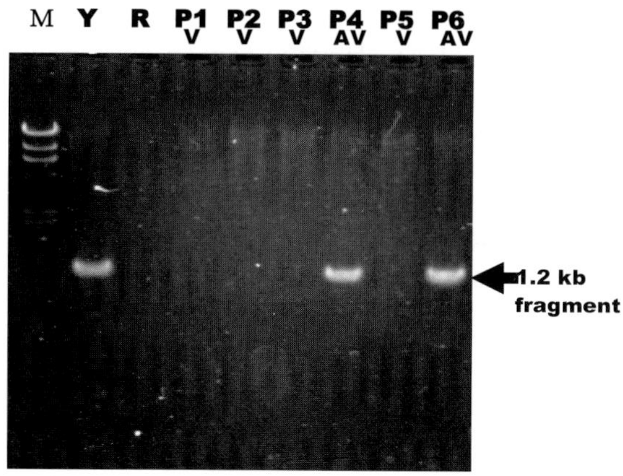

Figure 1. Amplification of the *Avr-Piks* marker in the parental isolates and progenies of their cross

amplified in these pathogenic mutants. Furthermore, in a cross between a mutant and an avirulent progeny, 10 progenies were avirulent and 10 of them were virulent to Hattan 3. The analysis with the *Avr-Piks* marker showed that all of the avirulent progenies had the 1.2 kb fragment while all virulent progenies did not produce the band.

These results convinced us that the *Avr-Piks* marker co-segregates with the pathogenicity to Hattan 3. Hence, we conclude that *Avr-Hattan 3* is identical with or closely linked to *Avr-Piks*.

We are now making BAC contigs of *Avr-Hattan 3* region using BAC library constructed by Nishimura and her co-workers (Nishimura *et al.*, 1998). We found that the *Avr-Piks* marker is located at the rightmost position of our contigs. Currently we are trying to isolate the *Avr-Hattan 3* gene by positional cloning.

REFERENCES

Dioh, W., Tharreau, D., Notteghem, J. L., Orbach, M., and Lebrun, M. H. (2000) Mapping of avirulence genes in the rice blast fungus, *Magnaporthe grisea*, with RFLP and RAPD markers. *Mol. Plant Microbe Interact.* 13: 217-27

Farman, M. L., Eto, Y., Nakao, T., Tosa, Y., Nakayashiki, H., Mayama, S., and Leong, S. A. (2002) Analysis of the structure of the *AVR1-CO39* avirulence locus in virulent rice-infecting isolates of *Magnaporthe grisea*. *Mol. Plant Microbe Interact.* 15(1): 6-16

Fukiya, S., Kodama, M., Kito, H., Sone, T., and Tomita, F. (2001) Establishment of a new cross of the rice blast fungus derived from Japanese differential strain Ina168 and hermaphroditic rice pathogen Guy11. *Biosci. Biotechnol. Biochem.* 65(7): 1464-73

Kang, S., Lebrun, M. H., Farrall, L., and Valent, B. (2001) Gain of virulence caused by insertion of a *Pot3* transposon in a *Magnaporthe grisea* avirulence gene. *Mol. Plant Microbe Interact.* 14: 671-4

Kiyosawa, S. (1969) Inheritance of resistance of varieties to a Philippine fungus strain of *Pyricularia oryzae*. *Jpn. J. Breed.* 19: 61-73.

Mandel, M. A., Crouch, V. W., Gunawardena, U. P., Harper, T. M., and Orbach. M. J., (1997) Physical Mapping of the *Magnaporthe grisea AVR1-MARA* Locus Reveals the Virulent Allele Contains Two Deletions. *Mol. Plant Microbe Interact.* 10: 1102-1105.

Nishimura, M., Nakamura, S., Hayashi, N., Asakawa, S., Shimizu, N., Kaku, H., Hasebe, A., and Kawasaki, S. (1998) Construction of a BAC library of the rice blast fungus *Magnaporthe grisea* and finding specific genome regions in which its transposons tend to cluster. *Biosci. Biotechnol. Biochem.* 62(8): 1515-21

Notteghem, J. L., Tharreau, D., Silue, D., and Roumen, E.(1994)Present knowledge of rice resistance genetics and strategies for *Magnaporthe grisea* pathogenicity and avirulence gene analysis. In: *Rice Blast Disease* 155-165 pp. Zeigler, R. S., Leong, S.A. and Teng, P. S. eds. CAB International, Wallingford

Valent, B. and Chumley, F. G. (1994) Avirulence genes and mechanisms of genetic instability in the rice blast fungus. In: *Rice Blast Disease* 111-134 pp. Zeigler, R. S., Leong, S.A. and Teng, P. S. eds. CAB International, Wallingford

Yamasaki, Y., and Kiyosawa, S. (1966) Studies on inheritance of resistance of rice varieties to blast. 1. Inheritance of Japanese varieties to several strains of the fungus. *Bull. Natl. Inst. Agr. Sci.* D14, 39-69.

Yasuda, N., Fujita, Y., and Tsujimoto, M. (2000) Identification of avirulence genes of *Magnaporthe grisea* and selection of molecular markers. In: *Advances in rice blast research* 303-307 pp. Tharreau, D., Lebrun, M. H., Talbot, N.J., and Notteghem, J. L. eds. Kluwer Academic Publishers, Dordrecht, The Netherlands.

GENOME EVOLUTION AND FUNCTION OF RESISTANCE GENES

Ko Hirano and *Shinji Kawasaki

National Institute of Agrobiological Sciences (NIAS), 2-1-2, Kannondai, Tsukuba, Ibaraki, 305-8602, Japan

**E-Mail: kawasa@nias.affrc.go.jp*

Abstract: In order to analyze the mechanism of R-genes diversification, as a model of a non-clustering isolated R gene, we have compared the base changes in the neighborhood region of blast resistance gene *Pi-b*, between resistant *japonica* and suceptible *indica* cultivars, and found the trace of dynamic mobilization of genome in this region. The mutation rate in this region, about 60-80 kb around *Pi-b*, was more than ten times higher than the average of the genome. It was noticeable that this high-variavility was not localized within the R-gene, of which specific part may have been subject to some diversification selection. Major types of the mobilization were accumulation of less than a few bases of small substitutions and insertion-and-deletions (indels). Base transitions were rather suppressed in this dynamic region. The characters of the indels were more complex around the *Pi-b* than the average. Density of small transposable elements like mites was about the same level as the other regions.

To confirm this tendency of mobilization, genome structures around some isolated R genes of *Arabidopsis* were analyzed between ecotypes Columbia and Landsberg and found to be higher in the mutation rate than the average in most cases. Analysis about 15 kb around these rather isolated R genes, revealed that the variability of the genome in this region is about 5 times higher than the average of the total genome, although there are gradations in the extent of this variability, from the normal level to more than ten times of that.

Key words: resistance gene, resistance gene analogue, hypervariability, genome mobilization, rice blast resistance

S. Kawasaki (ed.), Rice Blast: Interaction with Rice and Control, 71–77.
© 2004 *Kluwer Academic Publishers. Printed in the Netherlands.*

1. INTRODUCTION

The disease resistance (R) genes confer resistance to plants against pathogens with only a single gene, providing ability to somehow recognize products of the corresponding avirulence genes in the pathogens. Several domain swapping experiments of the R genes of most common NBS-LRR type (Dodds et al. 2001), and the less frequent LRR kinase type (He et al. 2000) and the membrane-anchored LRR type (Wulff et al. 2001), indicated that the race-specificites of the R genes are localized mainly in the LRR region although participation of other regions were also suggested (Luck et al. 2000). However, pathogens escape quickly from this host surveillance system by frequent mutation of their avilurence genes corresponding to the specific R genes, and the variety of plant R genes (analogues) in an individual plant is by-far limited than that of the immunoglobulin proteins of vertebrates with circulating system, having a more direct pathogen-recognizing function. Therefore, there must be some mechanism to maintain the variety of these R genes (analogues) also in plants, and wild plants may be keeping the needed variability with the outcrossing large population. The self-incompatibility system may be an important adaptation to keep this variety of R gene analogues as the protecting resouvor against the future emergence of new types of pathogens. It is natural that in the artificial monoculture condition, where there is the immense possibility of pathogen mutation in wide area, and there is very limited pool of the R gene (analogues) in the plant genome, the cultivated plants' R genes break down in rather short periods.

Although it was anticipated that plant must have some system to diversify their R genes to provide their new resources for their protection from future pathogens, hitherto, the main cause to produce these new varieties was supposed to be mutual recombinations between the clustered R genes and their analogues, because these diversification analyses have been done mostly on the R-gene clusteres (Chin et al., 2001, Meyers et al., 1998, Noel et al. 1999, Parniske and Jones 1999). However, R genes present not only as cluster but also in isolated environment in the genome, but the experiments analyzing the role of de-novo mutations as a possible important mechanism of R gene diversification in these rather isolated cases have been rare. In the course of identification of the rice blast resistant *Pi-b* gene through the complementation analysis of subclones covering the region, we have noticed a greatly enhanced variability around this R gene. This extensive comparative analysis was made possible because the *Pi-b* gene was a rather isolated R gene, and the critical single base level nucleotide comparison between the indica and japonica cvs. could be done. After confirming the elevated genome variability around Pi-b, we have extended the examination of the extent of this variability around the R gene in other rather isolated genes in *Arabidopsis* and confirmed the generality of this phenomenon.

2. HYPERVARIABLE REGION AROUND THE RICE BLAST GENE *PI-B*.

We have cloned a rice blast resistance gene *Pi-b* by a systematic genome complementation of the 30-50 kb of subcontigs constructed with a high capacity binary vector pBIGRZ (in submission). This subcontig covers about 100 kb of the *Pi-b* region of the resistant cultivar (cv.: with *Pi-b*) BL-1 (*indica* in this region) genome, and was constructed by collecting the clones from the tentative libraries transformed with the total genome, by carefully referencing the common marker ends of the subcontig of the a 180 kb of BAC clone, covering the region of a susceptible cv. (Shimokita: *japonica* in this region). The *Pi-b* gene was determined by successful complementation by a single clone within the subcontig, and had a structure of NBS and LRR in the center and last one third of the product protein.

We have experienced great difficulty to construct this parallel subcontig of the *indica* resistant cv. from the *japonica* subcontig. When the completed two subcontigs were compared, surprisingly, these restriction maps of the central *Pi-b* region of about 100 kb were totaly different between the them, although the outer regions the patterns were almost the same. There were two and one corresponding R gene analogues in the center of the respective regions of the resistant and susceptible cvs, respectively. The *Pi-b* in the BL-1 was duplicated and one RGA (resistance gene analogue: named as *Pi-b'*) was inserted with a large transposon, and a counterpart RGA in susceptible Shimokita (named as *Pi-b''*) was disrupted by two small deletions in the C terminal region. By comparison of the corresponding regions of these three RGAs it was reveealed that the base change between *Pi-b* and *Pi-b'* was 6% while that between *Pi-b* and *Pi-b''* was 9%. This differences in *Pi-b* region are more than 10 fold greater than that of the genome average, 0.6 %. Considering that the base difference between human and chimp is 1.2%, the extent of the variation between the corresponding RGAs is thought to be great.

Therefore, as we have got the two parallel subcontigs completely arranged with the common scattered markers, we compared the whole 100 kb sequences of the *Pi-b* region between the resistant indica and susceptible japonica. Although some parts of them was disrupted by insertion of large transposons, the several segements of BL-1 could be compared with the corresponding reference of the susceptible japonica cv. Shimokita. The region with highly elevated mutation rate around the *Pi-b* was revealed and this hypervariable (HV) region was divided into four domains, left HV, Pi-b', Pi-b, right HV. The average mutation rate reached about 7 %, which is more than 10-30 times higher than that of the genome average 0.2-0.6%, obtained from the randomly selected BAC end sequences. In this HV region, there were no other complete ORFs than the *Pi-b* and the derivatives but the incomplete ones that may have been disrupted through this high rate of mutations. Among these base changes, 75-85% of the mutations were substitutions, and this rate is not different from the flanking stable regions' 80%. Apparently these types of mutations cannot be explained only by the repeated recombination of two homologues. Insertions and deletions (indels) within the HV region constitute the remaining 15-25% of mobilization. A characteristic of the indels in this region is that percentage of indels of more than 50 bases was very high (about 20%) in the R gene and derivatives, compared to the outer regions, 0-3% in the flanking stable regions and 5-12% in the flanking HV regions. Although the influence of transposable elements was suspected at first, correlation of the genome variability with them was not apparent. Density of the small repetitive elements like MITE seems not to be different from the outside of the HV region.

As this HV island around *Pi-b* gene was very impressive and consistent with the condition required for the R gene, we started an extensive investigation for the variability of the R gene regions.

3. ARE OTHER R GENE REGIONS HYPERVARIABLE TOO?

In order to check for general applicability of the rule of hyper-variability around R gene (analogue)s, rather isolated R genes or RGAs in *Arabidopsis* were selected from the genome database of Columbia, and as a counterpart, BAC clones of ecotype Landsberg genome library were selected from the corresponding regions and their full sequence was analyzed to compare with those of Columbia. As the examples *RPP8* on chromosome 5, *RPP13* on chr. 3, *RPS5/RFL1* (at the same locus) on chr. 1, *RPM1* on chr. 3 were selected and their neighborho of 87kb, 120kb, 94kb, and 190 kb sequences were analyzed and compared between the ecotypes, respectively.

As a result, in the cases of *RPP8*, *RPP13*, and *RPS5/RFL1*, the average mutation rates around respective genes were 2.6%, 2.4%, and 1.9%. In comparison, the average mutation rate of the total genome determined from the BAC end sequences was 0.5% as above. Only in the case of RPM1, the mutation rate around the gene was lower than the average and was 0.3%.

All the data of 15 kb regions examined around these R genes were combined, and the distribution of the mutation rate frequencies was compared with those obtained from the BAC ends data randomly obtained troughout the genome. To make the tendency of distribution more apparent, the accumulated frequency of the data was also studied. Although the raw frequency distribution showed occurrence of rather high mutation rate around the R genes, accumuraated frequency distribution showed clear shift of the distribution to the higher mutation rate around the R genes. The median value of the mutation rate of the R gene region was 2.0%, while that of the random genome region was 0.3%, and it was clearly indicated that the average mutation rate was also significantly higher in the R gene regions than the average.

4. DISCUSSION

Rice blast is a most devastating disease of the first crop of which full genome sequence was clarified (Sasaki et al. 2002), and its pathogen's genome was also almost fully sequenced (Martin et al. 2002) as the first fungal plant pathogen. This system is suited for analyzing the resistant interaction of plants against pathogens. More than 20 rice resistance genes against blast fungus have been identified so far and for most of them a series of near isogenic lines have been constructed (Fukuta et al. in this report).

Basically true resistance genes of plants are recognizing systems of pathogen molecules by plants, which have a rather limited pool of R genes in the genome. We have developed a special system to scan the R genes in genome, and typical motifs of the LRR or NBS in the genome were identified and it was estimated that there are at least 1600 RGAs in rice genome (Shimizu et al. in preparation). This number may be definitely insufficient if their structure is rigid and fixed against the rapidly mutating hosts of possible pathogens. Although from the analyses of the clustered R genens, the groups of Michelmore (Chin et al., 2001, Meyers et al., 1998) and Jones (Noel et al. 1999, Parniske and Jones 1999) proposed that the variability of the plant R genes are mainly due to mutual repeated recombinations between the homologous R genes, we supposed that if the variability of R genes relys only on the recombination between them it could be a rather rigid system for minute mutations of the pathogen's avirulence genes. Fine-tuning of recognizing sites of pathogen-molecules may be more smoothly achieved through single or a few bases levels of mutations, although for the occurrence of novel pathogens, the dynamic recombinations of R-genes may be more efficient.

By a comparative analysis of the neighborhood region of the recently cloned rice blast resistance gene *Pi-b*, the gene was revealed to localize in the center of a highly mutating region. It was impressive that in the 180 kb BAC clone encompassing the *Pi-b*, only the very neighborhood of the *Pi-b* was very high in the mutation rate and exceeding 20 times of the total genome average.

The general applicability of this tendency was checked through similar comparative analyses between the two ecotypes of *Arabidopsis*, Columbia and Landsberg. By analyzing 4 loci of R genes, it was apparent that 2 R gene loci, RPP8 and RPP13, were highly mutating, and exceeding 5 times of the average total genome mutation rate, one loci RPS1 was medium in the mutation activity but still high in the limited region around the R gene. The last case of RPM1, the region was very serene and the mutation rate was less than the average of 0.5%. However, in total, the average mutation rate in the R genes loci seems to be about 5 times higher than the total genome average. Therefore, we believe that there may be some mechanism to increase the mutation rate of the genome around the R gene and its analogues. This was also supported by the insertional mutation analysis of the rice retro-transposon TOS 17 by Miyao and Hirochika (in this issue), which has the tendency to be inserted in a region close to the R genes. This fact may be reflecting some dynamic activities of the regions.

It is noteworthy that in the case of rice blast resistance gene *Pi-ta*, it was observed that the mutation rate in the gene was rather low between resistant *indica* and susceptible *japonica* cvs. (Bryan et al. 2000). There may be a gradational variation in the mutation rate of or around R genes. A temporal change or shift in the mutational activites in the genome may occur, as there are local variations in its activity. Once a very durable gene was established through these mutations, it will be benefitable for plants that the gene is to be maintained by lowering the mutation activity in a long range of period.

As cause of the higher mutation rate we are speculating about selective operation of miss-restoration of the double strand breaks of the chromosome, which may be happening persistently in the course of cell activities. Especially the miss-paring of the nucleotides by some error-prone DNA polymerase may be a candidate, although its presence in plants is not yet confirmed. The spatial difference of the mutation rate may be generated by the difference of the extent of DNA relaxation in the intermediate phase of the nucleus. If the region was relaxed and near to the thin free strand without the protection of nucleosomes, the chances of mutation will be elevated. We are currently outlining experiments to check for these hypotheses.

REFERENCES

Bryan GT, Wu KS, Farrall L, Jia Y, Hershey HP, McAdams SA, Faulk KN, Donaldson GK, Tarchini R, Valent B. (2000) A single amino acid difference distinguishes resistant and susceptible alleles of the rice blast resistance gene Pi-ta. Plant Cell. Nov;12(11):2033-46.

Chin DB, Arroyo-Garcia R, Ochoa OE, Kesseli RV, Lavelle DO, Michelmore RW. (2001) Recombination and spontaneous mutation at the major cluster of resistance genes in lettuce (*Lactuca sativa*). Genetics 157(2): 831-49.

Dodds PN, Lawrence GJ, Ellis JG. (2001) Six amino acid changes confined to the leucine-rich repeat beta-strand/beta-turn motif determine the difference between the P and P2 rust resistance specificities in flax. Plant Cell 13(1): 163-78.

He Z, Wang ZY, Li J, Zhu Q, Lamb C, Ronald P, Chory J. (2000) Perception of brassinosteroids by the extracellular domain of the receptor kinase BRI1. Science 288(5475): 2360-3.

Luck JE, Lawrence GJ, Dodds PN, Shepherd KW, Ellis JG. (2000) Regions outside of the leucine-rich repeats of flax rust resistance proteins play a role in specificity determination. Plant Cell 12(8): 1367-77.

Martin SL, Blackmon BP, Rajagopalan R, Houfek TD, Sceeles RG, Denn SO, Mitchell TK, Brown DE, Wing RA, Dean RA. (2002) MagnaportheDB: a federated solution for integrating physical and genetic map data with BAC end derived sequences for the rice blast fungus *Magnaporthe grisea*. Nucleic Acids Res. 30(1):121-4.

Meyers BC, Shen KA, Rohani P, Gaut BS, Michelmore RW. (1998) Receptor-like genes in the major resistance locus of lettuce are subject to divergent selection. Plant Cell 10(11):1833-46.

Miyao A, Hirochika H (2004) Transposon-insertion lines of rice for analysis of gene function. *In this issue.*

Noel L, Moores TL, van Der Biezen EA, Parniske M, Daniels MJ, Parker JE, Jones JD. (1999) Pronounced intraspecific haplotype divergence at the RPP5 complex disease resistance locus of *Arabidopsis*. Plant Cell 11(11): 2099-112.

Parniske M, Jones JD. (1999) Recombination between diverged clusters of the tomato *Cf-9* plant disease resistance gene family. Proc Natl Acad Sci U S A. 11;96(10):5850-5.

Sasaki T, Matsumoto T, Yamamoto K, Sakata K, Baba T, Katayose Y et al. (2002) **The Genome Sequence and Structure of Rice Chromosome 1.** Nature, 420: 312-316

Wulff BB, Thomas CM, Smoker M, Grant M, Jones JD. (2001) Domain Swapping and Gene Shuffling Identify Sequences Required for Induction of an Avr-Dependent Hypersensitive Response by the Tomato *Cf-4* and *Cf-9* Proteins. Plant Cell 13(2): 255-72.

BROAD-SPECTRUM RESISTANCE GENES *PI2*(T) AND *PI9*(T) ARE CLUSTERED ON CHROMOSOME 6

*Guifu Liu, Shaohong Qu, Bo Zhou, Lirong Zeng and Guo-liang Wang

Department of Plant Pathology, The Ohio State University, 201 Kottman Hall, 2021 Coffey Road, Columbus OH 43210 USA

**Email: WANG.620@osu.edu*

Abstract: *Pi9(t)* was introgressed from the wild rice *Oryza minuta* and *Pi2(t)* was transferred from a breeding line developed at CIAT. Both *Pi2(t)* and *Pi9(t)* were previously mapped on rice chromosome 6 and confer high level of resistance to diverse blast isolates. To further test their resistance spectrum, 43 blast isolates collected from 13 countries were used to inoculate the *Pi2(t)* and *Pi9(t)* plants. *Pi9(t)* was highly resistant to all isolates tested and *Pi2(t)* was resistant to 36 isolates, confirming the broad-spectrum resistance of these two genes to diverse blast isolates. To understand the molecular basis of broad-spectrum resistance, we conducted fine-mapping of the two blast resistance genes. A high-resolution genetic map was established using RAPD markers tightly linked to *Pi9(t)*. Using these markers, a bacterial artificial chromosome (BAC) contig was constructed covering about 100 kb at the *Pi9(t)* locus. Interestingly, all the *Pi9(t)* markers closely segregated with the *Pi2(t)* gene in an F_2 mapping population. Whether *Pi9(t)* and *Pi2(t)* are tightly linked or allelic is under investigation.

1. INTRODUCTION

Rice blast is the most devastating fungal disease of rice. Use of host resistance (R) genes is still the most effective and economical way to control

S. Kawasaki (ed.), Rice Blast: Interaction with Rice and Control, 79–86.
© 2004 *Kluwer Academic Publishers. Printed in the Netherlands.*

the disease. Genetic analysis of resistance to blast began in the early 1960s when Goto established the differential system for races of *Magnaporthe grisea*, the rice blast fungus, in Japan (Ou 1985). Over 20 loci for complete resistance have been mapped relative to molecular markers on the rice molecular map (McCouch et al. 1994). Numerous molecular markers linked to these resistance genes are being used in marker-aided selection to facilitate breeding of new cultivars with high level of resistance to rice blast. Recent cloning of the two resistance genes, *Pib* and *Pita*, has provided insights on the blast resistance gene structure. The structure of these two genes shows high similarity with known resistance genes in other plant species. For example, the deduced amino acid sequence of the *Pib* gene contains a nucleotide binding site (NBS) and leucine-rich repeats (LRRs) (Wang et al. 1999), a common feature of many cloned plant resistance genes (Bent 1996). Interestingly, *Pita* is similar to *Pib* with a centrally localized nucleotide-binding site and leucine-rich domain (LRD) at the C-terminus. The AVR-Pita (176) protein, which lacks the secretory and pro-protein sequences, is shown to bind specifically to the LRD of the *Pi-ta* protein, both in the yeast two-hybrid system and in an in vitro binding assay, indicating that the AVR-Pita (176) protein binds directly to the *Pi-ta* LRD region inside the plant cell to initiate a *Pi-ta*-mediated defense response (Bryan et al. 2000; Jia et al. 2000). Comparison of the sequences of 6 resistant and 5 susceptible alleles of *Pita* has revealed overall amino acid polymorphism with only one single amino acid determining specificity.

Pi2(t) and *Pi9(t)* are two major resistance genes mapped on rice chromosome 6 in an interval of approximately 9.8 cM distance. The *Pi2(t)* gene was introgressed from 5173, a highly resistant breeding line developed at the International Center for Tropical Agriculture (CIAT) into the susceptible cultivar CO39. The derived isogenic line was named C101A51 (Mackill and Bonman 1992). The *Pi9(t)* gene was transferred from the wild rice *Oryza minuta* into the elite breeding line IR31917 through wild hybridization and repeated backcrossing and a highly resistant introgression line, 75-1-127, was developed a the International Rice Research Institute (IRRI) (Amante-Bordeos et al. 1992).

The resistance spectrum of the *Pi2(t)* carrying line C101A51 was tested extensively in the last few years. For example, Chen et al. (1996) collected 455 blast isolates from different regions of the Philippines and found C101A51 was susceptible to only a few isolates which all belong to lineage 44. Extensive field tests in several countries also indicated that *Pi2(t)* is one of the rice R genes with the broadest resistance spectrum (Chen et al. 1996). The spectrum of the *Pi9(t)* gene was tested at IRRI from isolates collected in Philippines. No compatible isolates to *Pi9(t)* were identified (H. Leung, IRRI, personal communication). To further examine the resistance spectra of *Pi2(t)* and *Pi9(t)* to isolates from other rice growing areas, we tested their resistance to 43 isolates collected from 13 different countries. The inoculation experiments confirmed that these two resistance genes provided high level of resistance to diverse blast isolates. To understand the molecular basis of broad-spectrum resistance to rice blast, we initiated the cloning project of both resistance genes using map-based cloning strategies. High-resolution mapping indicated that these two genes are clustered on chromosome 6.

2. RESULTS

2.1. Identification of markers linked to Pi9(t)

To identify markers linked to the *Pi9(t)* gene, the RAPD technique (Williams et al. 1990) and bulk segregant analysis (Michelmore et al. 1991) were performed with an F_2 population generated from a cross between the *Pi9(t)* carrying cultivar 75-1-127 and the susceptible cultivar CO39. Among 79 plants inoculated with Philippine blast isolate PO6-6, 58 were resistant and 21 were susceptible, confirming *Pi9(t)* is a single dominant gene (3:1 segregation ratio). Ten resistant and 10 susceptible plants were selected from the 79 plants to make resistant and susceptible pools, respectively. Approximately 20 ng of the pool DNA was used for each RAPD reaction. Random primers were purchased from Operon Technologies (Alameda, CA). To detect more polymorphic bands and for better separation of the amplified fragments, especially those between 100 bp and 1000bp, ^{33}P labeled dCTP nucleotide was added to the PCR reaction and the reaction mix was run on a 4.5% polyacrylamide sequencing gel. After screening with over 900 Operon random primers, three primers were found with reproducible polymorphic bands between the resistant and susceptible pools. After gel purification, all three fragments were cloned into the pGEM-T vector (Promega) and the corresponding recombinant clones were named pB8, pBA14 and pBV14. After sequencing the insert, specific primers were designed for each marker based on their sequences. Two markers (pB8 and pBA14) were then successfully converted into SCAR markers since a single band (700 bp for pB8 and 400 bp for pBA14) was amplified in 75-1-127 but not in CO39 (data not shown). No polymorphic band was found between resistant plant 75-1-127 and the susceptible line CO39 using primer pairs based on the sequence of marker pBV14, thus, pBV14 was not used in the succeeding experiment.

To fine map *Pi9(t)* on chromosome 6, a large F_2 population of 1280 plants derived from a cross between C101A51 and CO39 was inoculated with isolate PO6-6. The disease reaction of each plant was recorded 7 days after inoculation. Five of the 1280 F_2 plants showed recombination between pBA14 and *Pi9(t)*, indicating that the pBA14 marker lies approximately 0.2

cM (5/2560=0.2%) from *Pi9(t)*. No recombinant plants between pB8 and *Pi9(t)* were identified. Southern hybridization showed that pB8 is a single copy fragment present only in the resistant plant 75-1-127 and in *O. minuta*, suggesting that pB8 is within the introgressed region and might be a part of or tightly linked to the *Pi9(t)* gene. pBA14 is a medium repetitive element and only a faint polymorphic band was observed between 75-1-127 and CO39, which prevented its use in Southern hybridization analysis of resistant and susceptible plants (data not shown).

2.2. Construction of a physical and high-resolution genetic map around the Pi9 region

To pinpoint *Pi9(t)* on a physical map, a BAC library was constructed using high molecular weight (HMW) DNA isolated from 75-1-127 according to the procedure described by Wang et al. (1995). To maximize the chance of getting a DNA fragment containing the *Pi9(t)* gene, over 200,000 clones were collected and stored in 100 pools (about 2000 individual clones per pool). The library equaled approximately 21 genome equivalents based on the rice genome size (430 Mb) and the average insert size of the BAC clones was 45 kb. BAC DNA was mini-prepared from each pool. Since the fine-mapping results indicated that pB8 might lie within the *Pi9(t)* gene or tightly linked to the gene, the pB8 primer pairs were used to screen the 100 BAC pools. A 700 bp fragment was amplified in 12 of the 100 pools. Individual BACs were isolated and confirmed with membrane hybridization using pB8 as the probe.

To construct a BAC contig spanning the *Pi9(t)* region, two BAC clones (BAC19 and BAC12) were selected because of their different *Hind*III-restriction digestion patterns. Hybridization of 4 BAC ends (12R, 12L, 19R and 19L) isolated from the two BACs verified that BAC19 and BAC12 overlap. To extend the BAC contig past the 12L end, 12L was used as probe in hybridization with the membranes containing all of the BAC pool DNA. Over 10 positive BAC pools were identified containing the 12L sequence. BAC ends were isolated from 4 BACs. Among them, BAC3 (a 40 kb BAC) was chosen for further characterization because of its minimum overlapping with BAC12 based on restriction enzyme digestion patterns (data not shown). Southern hybridization confirmed that 3R overlapped with BAC12 but not with BAC19. Both 3R and 3L ends hybridized with both 75-1-127 and IR31917 (data not shown). From this information, a 100 kb BAC contig comprised of BAC19, BAC12 and BAC3 was constructed.

Using the six BAC ends isolated above, a high-resolution map was constructed at the *Pi9(t)* locus with a total of 596 plants (79 F_2, 55 F_3 and 462 F_5 plants). Hybridization results indicated that no recombinants were found between *Pi9(t)* and either 12R, 19L, or pB8. For 3R, only one recombinant (plant R198-10R) was identified. Two recombinants (R198-

10R and R174-11R, resistant phenotype and susceptible genotype) were found when end sequences 12L and 3L were used as probes. Based on these mapping results, the *Pi9(t)* gene was mapped between the BAC ends 19R and 3R (Fig. 1).

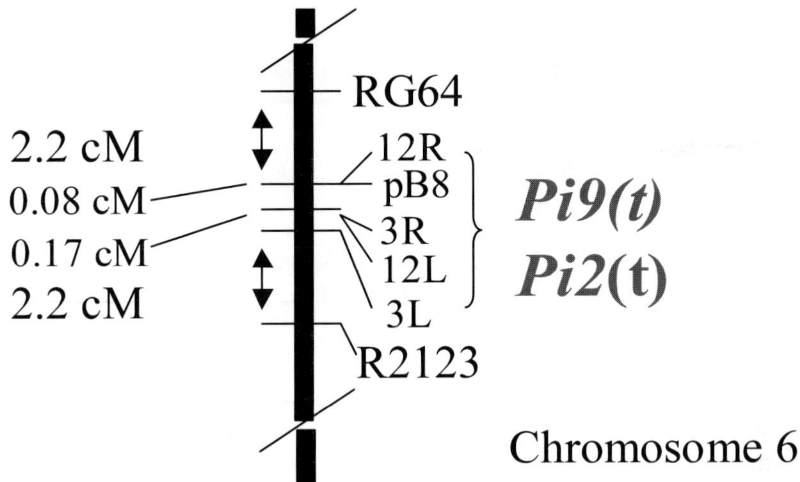

Fig. 1. *High-resolution map of the Pi9(t) region.* The genetic map was constructed using the mapping data from F2 populations of both *Pi9(t)* and *Pi2(t)*. The numbers in brackets are genetic distances in centi-Morgan.

2.3. Pi2(t) is tightly linked to Pi9(t) markers

The blast resistance gene *Pi2(t)* was previously mapped to rice chromosome 6 and is located 2.8 cM from the RFLP marker RG64 (Yu et al. 1991). Based on our mapping data, the *Pi9(t)* gene was also mapped to the same region between RFLP markers RG64 and R2131. To check whether *Pi2(t)* and *Pi9(t)* are tightly linked or allelic, a cross between the *Pi2(t)* near-isogenic line C101A51 (Mackill and Bonman 1992) and the susceptible cultivar CO39 was made. About 2000 F_2 plants were inoculated with blast isolate PO6-6 and DNA was extracted from 450 (50 resistant and 400 susceptible) F_2 plants exhibiting a clear resistant or susceptible reaction. Hybridization showed that 5 different recombinants were present for both markers R2123 and RG64, suggesting that *Pi2(t)* is 2.2 cM from both markers and lies in the same chromosomal interval as the *Pi9(t)* gene. To further confirm the location of the *Pi2(t)* gene, three BAC ends (12R, 12L and 3R) isolated from the *Pi9(t)* BAC contig were used to probe the F_2 population. No recombinants were identified between *Pi2(t)* and the three *Pi9(t)*-linked markers (Fig. 2), suggesting that *Pi2(t)* and *Pi9(t)* are tightly linked at the same region of chromosome 6 (Fig. 1).

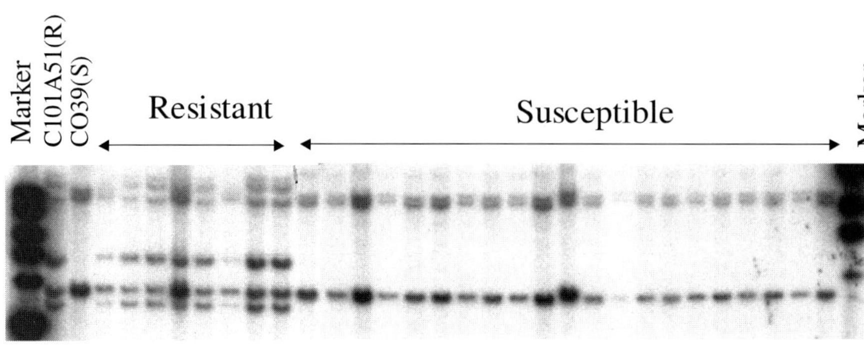

Fig. 2. *Pi9(t)*-linked marker 12L is tightly linked to *Pi2(t)*. DNA was digested with restriction enzyme *Bam*H1 and separated in 1% agarose gel. ^{32}P labeled BAC end 12L was used as a probe

3. DISCUSSION

Most of the known major resistance genes confer resistance to a few strains of a given pathogen and are short-lived in field conditions due to frequent changes of avirulence factors in the pathogen population. Although durable resistance is often associated with partial resistance (Wang et al. 1994), some cultivars with major resistance genes have been used in rice production for many years without completely losing resistance to blast (Bonman et al. 1992). For example, *Pi2(t)* has been used in rice breeding programs for over ten years and is still effective in many rice growing areas. One of the possible explanations for its durability to blast is its broad-spectrum resistance to diverse isolates (Chen et al. 1996). *Pi9(t)* was introgressed from the wild species *O. minuta* by Amante-Bordeos et al. (1992). Its resistance to Philippine isolates was tested at IRRI and no compatible isolates were identified (H. Leung, personal communication). Our inoculation test with 43 isolates collected from 13 countries showed *Pi9(t)* conferring high level of resistance to all of the isolates tested. The durability of its resistance in field conditions will be observed over time. However, it is tempting to speculate that *Pi9(t)* resistance may last a relatively long time since it has a broader spectrum and a higher level of resistance than that of the *Pi2(t)* gene. The wild rice origin of the resistance gene may also contribute to its durability, considering that it has been under the pressure of natural selection in the wild for many years.

Many plant resistance genes appear to be organized as complex clusters that contain multiple copies of closely-related gene sequences (Michelmore and Meyers 1998). For example, the *Xa21* resistance gene family in rice contains 8 highly homologous resistance gene-like sequences within a 270 kb region (Song et al., 1995). The *Dm3* locus in lettuce contains at least 22 sequences related to the *Dm3* resistance gene in a genomic region of 3.5 Mbp (Meyers et al., 1998). In most of these cases, one or two members at these loci confer resistance to the corresponding pathogens. *Pi2(t)* was introgressed into the susceptible indica cultivar CO39 from resistant breeding line 5173 from South America (Mackill and Bonman, 1992), while *Pi9(t)* was introgressed from the wild species *O. minuta* collected in the Philippines. As demonstrated in this study, the tight linkage of these two genes on chromosome 6 raises several intriguing questions on the relationship between these two genes. For example, why the *Pi9(t)* gene was integrated into the *Pi2(t)* genomic region and what level of DNA sequence diversity exists between these two genes. Recently, we have partially sequenced the *Pi9(t)* BAC contig and found at least 6 NBS/LRR candidate genes at the locus. Preliminary analysis of the *Pi2(t)* locus indicated that all these 6 candidate genes are also present at the *Pi2(t)* carrying plants. All the individual NBS/LRR genes in both *Pi9(t)* and *Pi2(t)* loci are being transformed to a susceptible cultivar to test the function of each candidate gene to rice blast. Hopefully, the above questions will be answered as the complete genomic sequence of this region becomes available and *Pi9(t)* and *Pi2(t)* are cloned in the near future.

ACKNOWLEDGEMENTS

We thank R. Nelson and H. Leung for providing the rice cultivars and blast isolates, N. Huang for providing the DH mapping population. Special thanks to D. Tharrenau, M. Farman and Y. H. Lee for providing rice blast isolates, M. Bonman and E. Mazur for critical reading of the manuscript. RFLP markers RG64 and R2123 were kindly provided by S. McCouch, Cornell University, and T. Sasaki, Japan Rice Genome Project, respectively. This project is, in part, supported by a grant from the Ohio Agricultural Research and Development Center (OARDC) and the Rockefeller Foundation to GLW.

REFERENCES:

Amante-Bordeos A, Sitch LA, Nelson R, Damacio RD, Oliva NP, Aswidinnoor H, Leung H (1992) Transfer of bacterial blight and blast resistance from the tetraploid wild rice *Oryza minuta* to cultivated rice, *Oryza sativa.* Theor. Appl. Genet. 84:345-354.

Bent AF (1996) Plant disease resistance genes: function meets structure. Plant Cell 8:1757-1771.

Bonman JM, Khush GS, Nelson R (1992) Breeding rice for resistance to pests. Annu. Rev. Phytopathol. 30:507-528.

Bryan GT, Wu KS, Farrall L, Jia Y, Hershey HP, McAdams SA, Faulk KN, Donalds GK, Tarchini R, Valent B (2000) A single amino acid difference distinguishes resistant and susceptible alleles of the rice blast resistance gene *Pi-ta.* Plant Cell. 12:2033-46.

Chen D, Zeigler RS, Ahn SW, Nelson RJ (1996) Phenotypic characterization of the rice blast resistance gene *Pi-2(t).* Rice. Plant Dis 80:52-56.

Jia Y, McAdams SA, Bryan GT, Hershey HP, Valent B (2000) Direct interaction of resistance gene and avirulence gene products confers rice blast resistance. EMBO J. 19:4004-14.

Mackill DJ, Bonman JM (1992) Inheritance of blast resistance in near-isogenic lines of rice. Phytopathology 82:746-749.

McCouch SR, Nelson RJ, Tohme J, Zeigler RS (1994) Mapping of blast resistance genes in rice. In Rice Blast Disease, eds Zeigler, R.S., Leong S.A. and Teng, P.S., CAB International. pp 167-186.

Meyers BC, Chin DB, Shen KA, Sivaramakrishnan S, Lavelle DO, Zhang Z, Michelmore RW (1998) The major resistance gene cluster in lettuce is highly duplicated and spans several megabases. Plant Cell 10:1817-32.

Michelmore RW, Meyers BC (1998) Clusters of resistance genes in plants evolve by divergent selection and a birth-and-death process. Genome Res 8:1113-30.

Michelmore RW, Paran I, Kesseli RV (1991) Identification of markers linked to disease-resistance genes by bulked segregant analysis: a rapid method to detect markers in specific genomic regions by using segregating populations. Proc Natl Acad Sci U S A. 88:9828-32.

Ou SH (1985) pp 109-201 in Rice Disease, Ed. 2. The Cambrian News Ltd., UK.

Song WY, Wang GL, Chen L, Zhai W, Kim HK, Holsten T, Zhu L, Ronald PC (1995) The rice disease resistance gene, *Xa21*, encodes a receptor-like protein kinase. Science: 270:1804-1806.

Williams JGK, Kubelik AR, Livak KJ, Rafalski JA, Tingey SV (1990) DNA polymorphisms amplified by arbitrary primers are useful as genetic markers. Nucleic Acids Res. 18:6531-6535.

Wang GL, Mackill DJ, Bonman JM, McCouch SR, Nelson RJ (1994) RFLP mapping of genes conferring complete and partial resistance to blast in a durably resistant rice cultivar. Genetics, 136:1421-1434.

Wang GL, Holsten TE, Song WY, Wang HP, Ronald PC (1995) Construction of a rice bacterial artificial chromosome library and identification of clones linked to the *Xa21* disease resistance locus. The Plant Journal 7:525-533.

Wang ZX, Yano M, Yamanouchi U, Iwamoto M, Monna L, Hayasaka H, Katayose Y, Sasaki T (1999) The *Pib* for rice blast resistance belongs to the nucleotide binding and leucine-rich repeat class of plant disease resistance genes. The Plant J. 19:55-64.

Yu ZH Mackill DJ, Bonman, JM, Tanksley SD (1991) Tagging genes for blast resistance in rice via linkage to RFLP markers. Theo. Appl. Genet. 81:471-476.

FINE GENETIC MAPPING AND PHYSICAL DELIMITATION OF THE RICE BLAST RESISTANCE GENE *Pi5(t)* TO A 170-kb DNA SEGMENT OF THE RICE GENOME

*J.-S. Jeon[1], D. Chen[2], G.-H. Yi[3], G. L. Wang[4] S. Kawasaki[5] and P. C. Ronald[6]

[1]*Plant Metabolism Research Center and Graduate School of Biotechnology, Kyung Hee University, Yongin 449-701, Korea;*

[2]*DNA LandMarks, Saint-Jean-sur-Richelieu, Quebec, J3B 6Z1, Canada;* [3]*National Youngnam Agricultural Experiment Station, Rural Development Administration, Milyang 627-130, Korea;*

[4]*Department of Plant Pathology, The Ohio State University, 201 Kottman Hall, 2021 Coffey Road, Columbus, OH 43210, USA;*

[5]*National Institute of Agrobiological Sciences, Kannon-dai, Tsukuba, Ibaraki 305-8602, Japan;*

[6]*Department of Plant Pathology, University of California, One Shields Avenue, Davis, CA 95616, USA*

**E-mail: jjeon@ucdavis.edu*

Abstract: AFLP-derived markers linked to *Pi5(t)*-mediated resistance were isolated using bulked segregant analysis of F_2 populations generated by crossing three recombinant inbred lines (RILs), RIL125, RIL249, and RIL260 with the susceptible line CO39. The most tightly linked AFLP marker, S04G03, was positioned on chromosome 9 of the fingerprint-based physical map of Nipponbare, a well-characterized rice genotype. Flanking BAC-based Nipponbare markers were generated for saturation mapping using four populations, the three initial RILs and an additional one derived from a cross between M202 and RIL260. A BIBAC (binary BAC) library was constructed from RIL260. Using these resources *Pi5(t)* was mapped to a 170-kb interval, and a contiguous set of BIBAC clones spanning this region was constructed. It had previously been suggested that *Pi3(t)* and *Pi5(t)* might be allelic, due to their identical resistance spectrum and tight linkage. We therefore compared genomic

S. Kawasaki (ed.), Rice Blast: Interaction with Rice and Control, 87–106.
© 2004 *Kluwer Academic Publishers. Printed in the Netherlands.*

regions for lines containing *Pi3(t)* using the *Pi5(t)*-linked markers. DNA gel-blot analyses indicated that the region around *Pi3(t)* is identical to that of *Pi5(t)*, suggesting that *Pi3(t)* and *Pi5(t)* are the same resistance gene.

1. INTRODUCTION

Rice blast, caused by the fungus *Magnaporthe grisea*, is one of the most destructive diseases of rice, costing farmers $5 billion a year (Moffat 1994). The management of rice blast relies heavily upon the incorporation of single disease resistance genes. However, blast resistance in many cultivars is short-lived in environments that are conducive to the disease (Lee and Cho 1990), due to the high degree of pathogenic variability of the causative organism *M. grisea* (Ou 1979; Bonman et al. 1986). Therefore breeding for cultivars that display broad-spectrum resistance has become a priority for crop improvement. The genetic basis of broad-spectrum resistance is still not well understood. It may be controlled by single genes or multiple genes with cumulative effects (Johnson 1981).

Relatively broad spectrum or durable resistance has been observed in some rice cultivars. For example, the traditional African cultivar Moroberekan has been cultivated for many years in large areas of West Africa without high losses from blast (Bonman and Mackill 1988). ROK16, LAC23, IRAT13, OS6 and some Brazilian upland rice cultivars show durable resistance to blast in upland conditions (Bidaux 1978; Bonman and Mackill 1988; Lee et al. 1989; Ahn 1994; Fomba and Taylor 1994). Tetep, an *indica* rice cultivar, and Pai-Kan-Tao (PKT), a temperate *japonica* cultivar, both exhibit broad-spectrum resistance to rice blast (Yu et al. 1987; Mackill and Bonman 1992; Ahn 1994, 2000). Many of these rice lines have been used as resistance donors in breeding programs (Mackill and Bonman 1992; Inukai et al. 1994). Phenotypic evaluation of near-isogenic lines (NILs) suggested that as many as four major resistance genes could be identified in Tetep, three in PKT including *Pi3(t)*, two in Moroberekan and two in LAC23 (Mackill and Bonman 1992; Inukai et al. 1994; Wang et al. 1994). Isolation and characterization of the genes conferring resistance in these cultivars should provide insight into the genetic basis of broad-spectrum resistance and may be useful in developing new genotypes.

For this purpose, a recombinant inbred (RI) population consisting of 281 F_7 lines was produced by single-seed descent from a Moroberekan × CO39 cross (Wang et al. 1994). These lines were evaluated for resistance in the greenhouse and field, and analyzed with 127 restriction fragment length polymorphism (RFLP) markers. Two dominant loci, *Pi5(t)* and *Pi7(t)*, that segregated with complete resistance to five blast isolates were identified, while ten quantitative trait loci (QTLs) segregated with partial resistance.

Greenhouse inoculation tests showed that the RI lines (RILs) RIL125, RIL249 and RIL260, which carry *Pi5(t)*, were resistant to at least six races belonging to four lineages in the Philippines (Wang et al. 1994; Chen et al. 2000). These lines were then evaluated for field performance in the Philippines and Indonesia, in locations where the blast fungal populations have been shown to be diverse and broadly virulent (Wang et al. 1994; Chen et al. 1995; Zeigler et al. 1995). The field tests indicated that the *Pi5(t)*-containing RILs displayed resistance to diverse isolates.

When the resistance profiles of the RILs carrying *Pi5(t)* were compared with those of the CO39 NIL lines carrying *Pi3(t)* (C104PKT), the reaction pattern was similar. Genetic and phenotypic analysis of an F_2 population derived from the cross between RIL249 and C104PKT indicated that the *Pi5(t)*resistance gene in RIL249 is allelic, or else closely linked, to *Pi3(t)* (Inukai et al. 1996).

We have initiated a positional cloning approach to clone the *Pi5(t)* locus. Major advances in rice genomics over the last few years have made positional cloning in rice much more efficient. A high-density genetic linkage map and a YAC- and BAC-based contig map have been constructed for the rice cultivar Nipponbare (Harushima et al. 1998; Chen et al. 2002; Wu et al. 2002). Over 110,000 sequence-tagged connectors (STCs) have been generated by sequencing both ends of every BAC clone (Mao et al. 2000). A fingerprint-based contig (FPC) of BAC clones has been anchored with RFLP markers onto the genetic map (Yuan et al. 2000). Using these resources and four mapping populations, we have set up an efficient procedure to construct a genetic and physical map for the *Pi5(t)* locus. We have assembled a 170-kb Binary BAC (BIBAC) contig containing *Pi5(t)*, and shown that the *Pi5(t)* locus is identical to the *Pi3(t)* locus, and that PKT and Moroberekan are not the donors of *Pi3(t)* and *Pi5(t)*, respectively.

2. MATERIALS AND METHODS

2.1. Plant materials and mapping populations

Three *Pi5(t)* RI lines, RIL125, RIL249 and RIL260, derived from a cross between Moroberekan and CO39 (Wang et al. 1994) were selected for this study. These lines were completely resistant to the blast isolate PO6-6. Crosses of each of these lines with CO39, the susceptible parent of the RIL population, were performed to develop segregating F_2 populations for DNA marker and resistance segregation analysis. Each of the three F_2 populations consisted of 50 to 70 individual plants. F_3 families were developed from the

F_2 populations for further DNA marker and resistance analysis. An additional cross of RIL260 to M202, a susceptible parent, was performed to develop an F_3 family consisting of over 1000 individuals for high-resolution mapping.

2.2. Inoculation and disease evaluation

The *M. grisea* isolate PO6-6 was used for all phenotypic analyses. All inoculations and disease evaluations were conducted in greenhouses at the International Rice Research Institute (Los Banos, Philippines) and at Ohio State University, as described in Chen et al. (1996).

2.3. BSA and AFLP analysis

Four pooled DNAs representing homozygous resistance (RR) and homozygous susceptible (rr) lines were made for bulk segregant analysis (BSA) (Michelmore et al. 1991) using amplified fragment length polymorphism (AFLP) screening. Each bulk DNA pool contained equivalent amounts of DNA from ten DNA samples extracted from the bulked leaves of the F_3 homozygous families. AFLP markers were used to screen DNA markers linked with the resistance according to Chen et al.(1999).

2.4. RFLP analysis

For RFLP analysis, samples of over 3 μg of rice genomic DNA were digested with restriction enzymes and fractionated by electrophoresis in a 0.8% agarose gel. DNA gel-blot analyses were carried out according to standard procedures under high-stringency hybridization conditions (Sambrook et al. 1989).

2.5. CAPS analysis

For CAPS (cleaved amplified polymorphic sequence) analysis, rice genomic DNA was isolated from young leaves using a simple miniprep method (Chen and Ronald 1999). CAPS analysis was performed in a volume of 30 μl (100 pmol of each primer, 200 μM each of dNTPs, 10 mM TRIS-HCl pH 9.0, 2 mM $MgCl_2$, 50 mM KCl, 0.1% Triton X-100, and 0.5 U of Taq polymerase) using 50 ng of genomic DNA as template. The digested PCR products were subsequently size-fractionated on 2% agarose gels.

2.6. BIBAC library construction

High-molecular-weight (HMW) DNA was prepared from young leaves of RIL260 by the CTAB method (Murray and Thompson 1980). The isolated HMW DNA was partially digested with *Hin*dIII and then size-fractionated using a pulsed-field gel electrophoresis device (CHEF DR II system, Bio-Rad) as described previously (Wang et al. 1995). The purified DNA from the low-melting point agarose slice containing fragments of 30–50 kb was ligated to the *Hin*dIII-digested and dephosphorylated pBIGRZ vector provided by Dr S. Kawasaki (Tsunoda et al. 2000). The ligation mix was transformed by electroporation using a Cell-Porator (Gibco-BRL) into *Escherichia coli* DH10B. About 500 clones were resuspended in 4 ml of the freezing buffer (Peterson et al. 2000) to form a BIBAC minipool.

2.7. BIBAC library screening

Aliquots (50 μL) of *E. coli* cells from each of 186 BIBAC minipools were used for preparation of total BIBAC plasmids. Total BIBAC DNAs were isolated by standard alkaline lysis procedures (Sambrook et al. 1989). Equivalent amounts of DNAs purified from 11 or 12 minipools were mixed to make a super pool, producing 16 super pools. The first round of PCR was performed using 16 super pools and *Pi5(t)* marker-specific primers (see below). In the second round of PCR, the 11 or 12 minipools making up the super pool showing a positive hit were then screened individually. Over 2000 individual clones of the identified mini-pool were further screened by colony blot hybridization (Sambrook et al. 1989) using the products amplified in the second PCR as probes.

2.8. Sequence analysis of BIBAC ends

The isolated plasmids were sequenced using T3 and T7 primers to obtain both end sequences of the cloned insert. The BIBAC-end sequences of each clone were amplified by PCR using specific primers and were subsequently used for chromosome walking.

2.9. Sublibrary construction of BAC or BIBAC clones

Size-fractionated DNAs (~2 kb) were obtained after partial digestion with *Sau* 3AI, ligated to the *Bam*HI site of pBluescriptII SK+, and transformed into *E. coli* DH10B. The inserts were used for RFLP and CAPS analysis.

2.10. Linkage analysis

Linkage analysis was performed using Mapmarker software (Lander et al. 1987) on a Macintosh computer. The segregation dataset generated from the 204-member F_2 population of Black Gora/Labelle (Redoña and Mackill 1996) and the 186 Nipponbare/Kasalath plants (Harushima et al. 1998) were combined with the established AFLP data set for the analysis. Map distances presented in cM between markers were derived using the Kosambi function (Kosambi 1944).

3. RESULTS

3.1. Genetic and phenotypic analysis of RI lines

The resistance and susceptibility phenotypes of the three F_2 populations developed from crosses of RIL125, RIL249, RIL260 with CO39 were determined by monitoring the effects of inoculating the F_2 plants and the corresponding F_3 families. The distribution of resistance in the F_2 population was compatible with a 3:1 segregation ratio for all three populations (Table 1). The three genotypes in the F_3 families (RR, Rr and rr) fit a 1:2:1 segregation ratio (data not shown). In each of the heterozygous F_3 families, the resistance/susceptibility phenotypes followed a 3:1 segregation pattern. The resistance in the additional F_2 population derived from a cross between RIL260 and M202 also segregated in a 3:1 ratio (Table 1). These results indicate that in each of the selected lines a single dominant locus confers resistance to PO6-6.

Table 1. *Chi-square test for the segregation of resistance and susceptibility in F_2 populations inoculated with the rice blast isolate PO6-6. The goodness of fit to a 3:1 ratio is indicated.*

Cross	No. of F_2 plants observed		No. of total plants	X^2
	Resistance (R)	Susceptible (S)		
RIL125/CO39	39	10	49	0.55
RIL249/CO39	53	15	68	0.31
RIL260/CO39	39	11	50	0.24
RIL260/M202	45	14	59	0.05

3.2. Identification of AFLP markers linked with the resistance

Of the 579 AFLP primer combinations screened, one primer pair, S04G03, identified a polymorphic marker associated with resistance in RIL125. Out of 350 primer pairs surveyed, 11 AFLP markers were associated with resistance in the RIL249-derived F_2 population. Out of 750 primers surveyed, 12 markers were associated with resistance in the RIL260-derived F_2 population. These AFLP markers were converted into RFLP markers by cloning the bands into a TA cloning vector. For S04G03 it was difficult to obtain a hybridization signal in DNA gel-blot analyses with the 48-bp AFLP fragment. To obtain a larger fragment including this marker, we isolated a BIBAC clone, JJ5, by screening a Moroberekan library (J.-S. Jeon, D. Chen and P. C. Ronald, unpublished data) with SO4GO3 as a probe. Sequence analysis of JJ7, a 1.9-kb subclone of JJ5, revealed that the clone contains not only SO4G03 but also S10F07, which was isolated from the RIL260-derived F_2 population, another AFLP marker linked to *Pi5(t)*.

Thereafter, the subclone JJ7 was used as a probe to represent the S04G03 and S10F07 markers in the present study.

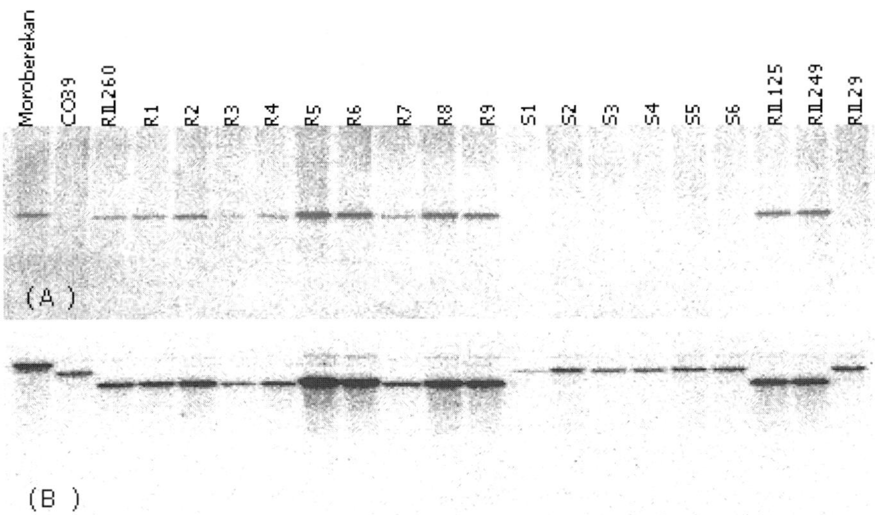

Figure 1. *Co-segregation of two markers, S04G03 (A) and 47J03f (B) with Pi5(t) resistance. Genomic DNA was digested with HindIII and separated in a 0.8% agarose gel. R1 to R9, and S1 to S6 are homozygous resistant and susceptible lines from the RIL260/CO39 F_2 population, respectively. DNA from Moroberekan, CO39, RIL125, RIL249, RIL260 and RIL29 (lacking Pi5(t)) were included as controls.*

To verify that the AFLP markers co-segregate with the resistance phenotype in *Pi5(t)* lines, we carried out DNA gel-blot analysis using F_2 homozygous susceptible and resistant lines. The result showed that S04G03, identified in all three populations, co-segregated with resistance in 15 individuals (Fig. 1A). DNA gel-blot analysis with the co-dominant 47J03f marker (see below) also co-segregated with S04G03 (Fig. 1B). Of all AFLP-derived markers from the three populations, S04G03 was the most tightly linked to *Pi5(t)* in 48 RIL249-derived F_2 lines. The other markers gave one to four recombination events between *Pi5(t)* and each marker (data not shown).

3.3. Pi5(t) is located on chromosome 9

Primer pairs generating the linked markers were used to analyze the parents of the Black Gora/Labell mapping population (Redoña and Mackill 1996). Nine primer pairs (S04G03, S10F07, S08G24, S10M04a, S10M04b, B11G19, R01G23a, R01G23b, P03G07) detected polymorphism between Black Gora and Labell at the loci associated with the resistance to PO6-6. Linkage analysis with Mapmaker showed that all the markers from the

RIL125-, RIL 249- and RIL260-derived populations were located on chromosome 9 (Fig. 2). These results suggested that the loci that conferred resistance to PO6-6 in the three RILs were probably the same as, or tightly linked to, *Pi5(t)* on chromosome 9.

The *Pi5(t)* locus was previously reported to be linked to the DNA marker RG788 on chromosome 4 (Wang et al. 1994). However, DNA gel-blot analysis using RG788 in the F_2 population derived from the RIL249 cross indicated that *Pi5(t)*was not linked to RG788 (data not shown). Our data thus confirmed the results of another study that also showed that the resistance gene in RIL249 was unlinked to RG788 (Inukai et al. 1996).

3.4. High-resolution mapping of Pi5(t): use of Nipponbare BAC clones

Saturation mapping of a small genomic region is an essential tool for positional cloning (Monna et al. 1997). To pinpoint the *Pi5(t)*genomic region, we mapped the co-segregating marker S04G03 onto the linkage map of Nipponbare/Kasalath to identify RFLP markers linked to *Pi5(t)* (Harushima et al. 1998) (Fig. 2). These results indicated that the marker C1454 is tightly linked to S04G03, only 0.8 cM away. We then used C1454 as a probe in colony hybridization experiments with two BAC libraries of Nipponbare provided by the Clemson University Genome Institute (CUGI). These experiments enabled us to identify a large physical region of more than 1000 kb consisting of Nipponbare BAC clones carrying the C1454/SO4G03 genomic region (Chen et al. 2002) (Fig. 2).

Six of these *Pi5(t)*-linked Nipponbare BAC clones were used to develop additional markers that flank *Pi5(t)*. The BAC clones were selected using information based on the primary genetic map of *Pi5(t)*, the location of RFLP markers, and the FPC map consisting of Nipponbare BAC contigs (Fig. 2). They were digested partially with *Sau* 3AI for subcloning. Out of 150 subclones tested, 25 showed clear polymorphisms between the three RILs and CO39 as determined by DNA gel-blot analysis. The remaining clones included those that were monomorphic between the parents, those that gave comparatively weak hybridization signals or those with multiple bands, which were considered inappropriate for further analysis (data not shown). Eight polymorphic markers, 34E14-10, 4D13-2-2, 2P10r, C1454, 76B14f, 40N23r, 47J03f, and 17I18-12, were used because they were evenly distributed or homologous to known resistance genes (Fig. 2). Thus, the amino acid sequences deduced from the DNA sequences of 76B14f and 40N23r showed similarity to a 'nucleotide binding site plus leucine-rich repeat' (NBS-LRR) motif. To use the markers most efficiently in the analysis of a large population, they were converted to CAPS markers using marker-specific primers (Table 2). Because C1454 and S04G03 could not be converted to CAPS markers, we instead amplified the corresponding regions of both markers from RIL260 and M202, and directly compared the sequences of the PCR products. Both markers showed a few nucleotide differences between RIL260 and M202 (Table 2; data not shown).

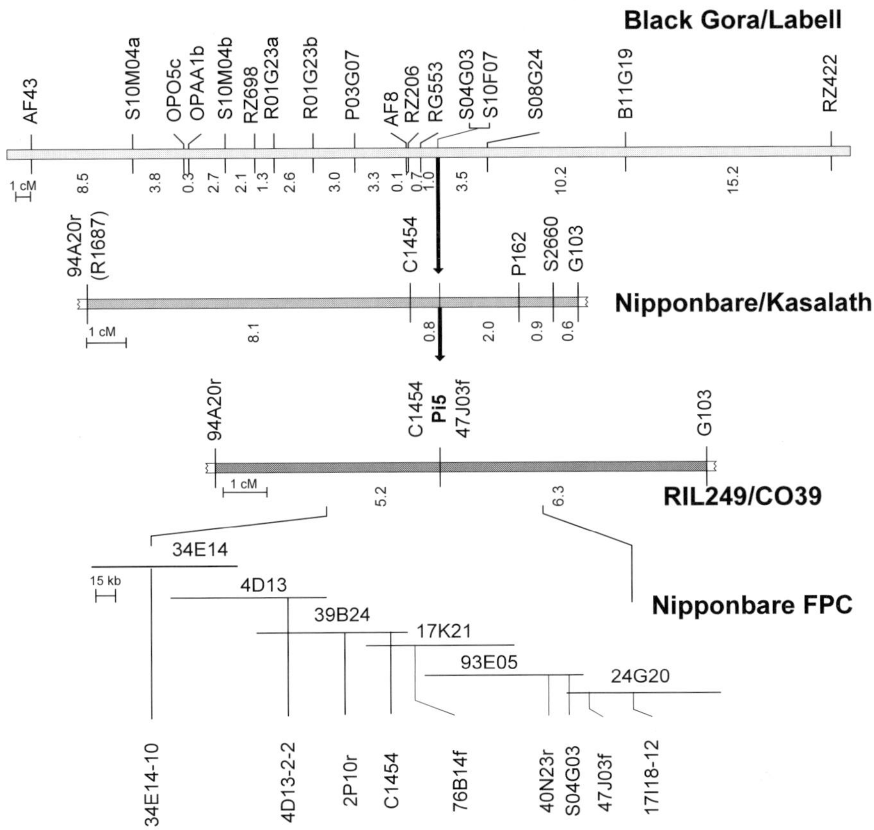

Figure 2. *Genetic and physical maps of the Pi5(t) locus on rice chromosome 9. The identified nine AFLP markers (S10M04a, S10M04b, R01G23a, R01G23b, P03G07, S04G03, S10F07, S08G24, B11G19) were located on the genetic linkage map of Black Gora/Labell (top). The numbers below the map are relative genetic distances in cM. S04G03, which co-segregates with Pi5(t) resistance, was mapped on the high-density genetic map of Nipponbare/Kasalath (middle top). This data was used to identify BAC subclones from the CUGI database that spanned the region (bottom). Markers were developed from these BAC subclones and used to locate Pi5(t) on the RIL249/CO39 map (middle bottom). The key markers used in this study were mapped on the Nipponbare BAC contig as shown at the bottom.*

We used the flanking markers 94A20r and G103 to narrow down the region carrying *Pi5(t)* on the genetic map of RIL249/CO39. This was easily accomplished because the parents of the RIL249/CO39 F_2 population were polymorphic for these markers. The marker 94A20r was developed from a Nipponbare BAC clone that contains the previously mapped marker R1687. R1687 was monomorphic between the parents. The result revealed that *Pi5(t)* mapped to an 11.5-cM interval (11 recombinations in 96 meiotic events) between the markers 94A20r [5.2 cM proximal to *Pi5(t)*] and G103 [6.3 cM distal to *Pi5(t)*] in the RIL249/CO39 F_2 population (Fig. 2).

Table 2. PCR-based markers linked to the *Pi5* locus.

Type	Marker	Forward primers (5'→3')	Reverse primers (5'→3')	Enzyme used	Mapping population
CAPS	94A20r	AATTCCATTCGCCACCGAGTGCTC	TCTCAGTATAGAACACTAACTCTA	*Ava* II	RIL/CO39
	34E14-10	CCTACCACCACAGGACATAACA	GTTTCTTCTCTTATCCCCTCTC	*Ase* I	RIL/CO39
	17I18-12	TACACGAACAACCAAATCGACC	AGGCGTTTGGTTTTGGTGGAGA	*Hin* dIII[a],	RIL/CO39, RIL260/M202
	2P10r	ATTGTCAAGCTCTTCTGCTGTC	TGAACTGATCATCAAATCAATC	*Dpn* II	RIL/CO39, RIL260/M202
	76B14f	GTCTTGGACTTAAAGCACTACC	TGAGAAACTGGTTCAAATTGGC	*Dra* I	RIL260/M202
	40N23r	TGTGAGGCAACAATGCCTATTGCG	CTATGAGTTCACTATGTGGAGGCT	*Eco* RV	RIL260/M202
Sequencing[c]	C1454	CACCTGAAGGCTGAAAATCTGAAT	CCGTTGATAGCGCTTAATGTTCTT	-	RIL260/M202
	S04G03	CTTAACAATCAATGTTTAATGAAA	GTTATATTATACTAATTGTTTATC	-	RIL260/M202

[a]RIL/CO39
[b]RIL260/M202
[c]Sequence analysis of PCR products was used for polymorphism survey.

To identify more recombination events at the *Pi5(t)* locus, two markers flanking *Pi5(t)* (34E14-10 and 17I18-12) within the 11.5-cM region defined by the above analysis, were screened in 731 susceptible individuals and 515 resistant and segregating individuals of BC_2F_4 RIL260/CO39 using CAPS analysis. Since the susceptibility score is more reliable than the resistance estimate (due to escapes from the inoculum), we first analyzed all susceptible plants to make an accurate map, and the remaining plants were analyzed later. In the primary screen, 22 and 8 recombination events were identified between 34E14-10 and *Pi5(t)* and between 17I18-12 and *Pi5(t)*, respectively (Fig. 3).

The recombinant lines were further analyzed using internal PCR markers. The analysis identified a recombination event between 4D13-2-2 and 2P10r. The other six markers, 2P10r, C1454, 76B14f, 40N23r, S04G03 and 47J03, co-segregated with *Pi5(t)*. To identify rare recombination events at the *Pi5(t)* locus, F_2 segregating populations of RIL125/CO39 and

RIL249/CO39 were analyzed further (Fig. 3). This identified a recombination event between 4D13-2-2 and 2P10r and two between 47J03f and 17I18-12 from the RIL125/CO39 population, and one between 34E14-10 and 4D13-2-2 and two between 47J03f and 17I18-12 from the RIL249/CO39 population (Fig. 3). Phenotypic analyses of all recombinant lines identified demonstrated that the resistance locus is most likely to be same in all three RILs, RIL125, RIL249 and RIL260 (Fig. 3). On the basis of the Nipponbare contig (Chen et al. 2002) it is estimated that the genomic region between 17I18-12 and 4D13-2-2 is approximately 300 kb long.

We were not able to localize the gene to a smaller physical region even when the mapping population was enlarged, because skewed recombination events were observed at the *Pi5(t)* locus. Thus, we developed another F_2

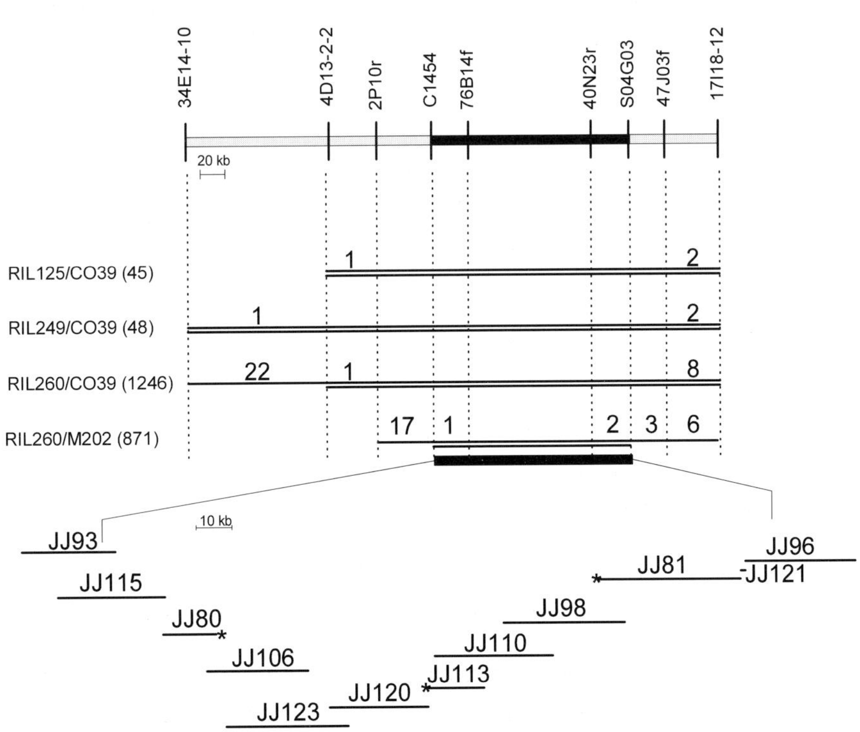

Figure 3. *High-resolution genetic map of the Pi5(t) locus (top). The four populations used for this analysis are shown on the left. The numbers in parenthesis indicate the number of plants analyzed for recombination events at the Pi5(t) locus. The numbers of recombinants obtained are indicated between the relevant markers. The physical region containing Pi5(t) resistance is shown by double lines for each population. The black bar indicates the smallest genomic region carrying Pi5(t) delimited from this analysis from all four mapping populations. This region was physically covered by a contiguous set of RIL260 BIBAC clones (bottom). The three asterisks indicate the position of BIBAC-end markers, JJ80-T3, JJ81-T3, and JJ113-T3, which were used in DNA gel-blot hybridization experiment (see Fig. 5).*

segregating population derived from a cross between RIL260, a resistant cultivar, and M202, a susceptible cultivar, and analyzed 871 F_3 individuals to further delimit *Pi5(t)* to a small physical region. Since we have limited quarantine facilities for blast inoculations, we employed a prescreening strategy to identify plants with rare recombination events around the *Pi5(t)* region using the flanking CAPS markers 2P10r and 17I18-12. These experiments identified 23 recombinants between 2P10r and 17I18-12: 17 recombination events between 2P10r and *Pi5(t)*, and six recombination events between 17I18-12 and *Pi5(t)* (Fig. 3). Phenotypes of all the identified lines displaying these rare recombination events were confirmed in the progeny from each line. Through further analyses of these recombinants with the internal markers, one recombinant between C1454 and *Pi5(t)* and two recombinants between S04G03 and *Pi5(t)* were identified. This high-resolution mapping experiment thus revealed that *Pi5(t)* is located in a ~170-kb interval between the markers S04G03 and C1454. The markers 76B14f and 40N23r co-segregated with *Pi5(t)* (Fig. 3).

3.5. Construction of a BIBAC contig spanning the Pi5(t) locus

Nipponbare does not carry *Pi5(t)*. Therefore we constructed a BIBAC library from RIL260 using the pBIGRZ vector (Tsunoda et al. 2000) to clone the region carrying *Pi5(t)*. To identify positive clones from the library, we adapted a pooling system for a PCR-based procedure (see the Materials and methods for the details). This strategy is efficient for screening a large library with small insert sizes.

To span the physical region containing *Pi5(t)*, four markers, S04G03, C1454, 76B14f, and 40N23r, were initially used for the library screening, yielding the BIBACs JJ96, JJ93, JJ80, and JJ81, respectively (Fig. 3). We sequenced each end of the four isolated BIBAC clones and used this information to generate PCR products to screen the library again in order to extend the region. Using this approach we identified three more clones (JJ98, JJ106, and JJ115) in the region. We repeated this step and identified an additional three clones (JJ110, JJ113, and JJ120) from the region (Fig. 3). Some BIBAC-end sequences were homologous to transposable elements and therefore could not be used as probes for BIBAC library screening due to their repetitive nature. This was the case for the BIBAC-end sequences obtained from JJ106 and JJ120. To close the gap between JJ106 and JJ120, a subclone, JS624-T7, isolated from the *Sau3*AI shotgun library of JJ106, was utilized to find the linking clone JJ123. In another case, a small region (183 bp) between JJ81 and JJ96 was amplified, cloned (JJ121), and its position confirmed by sequence analysis (data not shown). PCR analysis confirmed that both JJ113 and JJ120 are consecutively linked at the *Hin*dIII

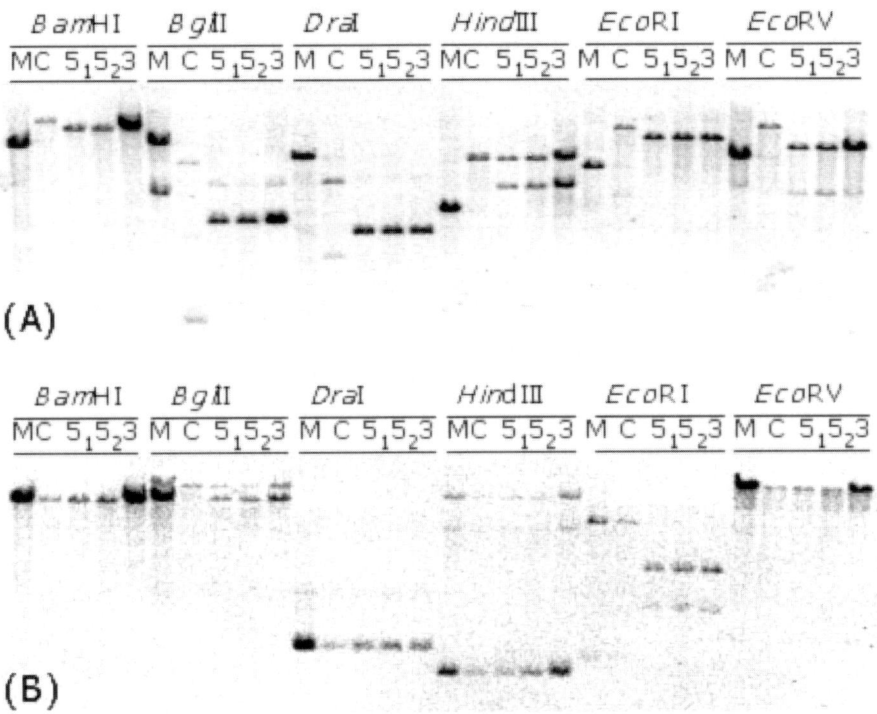

Figure 4. *DNA gel-blot analysis of the Moroberekan (M), CO39 (C), RIL260 (5_1), RIL249 (5_2), and C104PKT (3) genomes. Genomic DNAs were digested with six restriction enzymes, BamHI, BglII, DraI, HindIII, EcoRI, and EcoRV. The Pi5(t)-flanking markers, 17I18-12 (A) and 34E14-10 (B), were used as probes to survey polymorphisms.*

site used for cloning. All 12 clones could be arranged into a single contig spanning the 170-kb region between C1454 and S04G03 (Fig. 3).

3.6. The Pi3(t) and Pi5(t) genomic regions are identical

It has been proposed that *Pi3(t)* is allelic to *Pi5(t)*, as both confer similar resistance spectra to a variety of blast lineages (Inukai et al. 1996). To determine if the genomic regions of *Pi3(t)* and *Pi5(t)* are indeed similar, DNA gel-blot analyses were carried out using the *Pi5(t)*-flanking RFLP markers 34E14-10 and 17I18-12. The hybridization patterns indicated that the region in C104PKT, the *Pi3(t)*-containing line, is identical to that in RIL260 and RIL249 harboring *Pi5(t)* (Fig. 4), but markedly different from those in Moroberekan the putative donor of *Pi5(t)* and CO39, the susceptible parent. Additional DNA gel-blot analyses with 2P10r, 76B14f, C1454, 40N23r, S04G03, and 47J03f indicated that C104PKT, RIL260 and RIL249 are all monomorphic for these markers (data not shown). This suggests that the lines containing the *Pi3(t)* and *Pi5(t)* loci share a common origin. In agreement with this hypothesis, the nucleotide sequences of the *Pi5(t)*-linked marker 40N23r from C104PKT and RIL260 were completely identical (data not shown). The sequence of the 40N23r region differed significantly from that in the presumptive donors of resistance, PKT and Moroberekan. This further confirmed that the *Pi3(t)* and *Pi5(t)* regions are identical, and are not derived from PKT or Moroberekan.

3.7. The Pi5(t) resistance locus in the RI lines does not correspond to the Moroberekan allele

As noted above, all loci in the *Pi5(t)* genomic region differed from their counterparts in Moroberekan, which had been assumed to be the donor of *Pi5(t)*. It has been reported that PKT is susceptible to the rice blast strain PO6-6 (Mackill and Bonman 1992). These data further support our hypothesis that PKT is not the source of *Pi3(t)*. It is not likely that the *Pi5(t)* resistance gene was created by a recombination event in the Moroberekan/CO39 cross because the identical non-parental alleles were observed in RIL125, RIL249, RIL260, and C104PKT, and many dominant

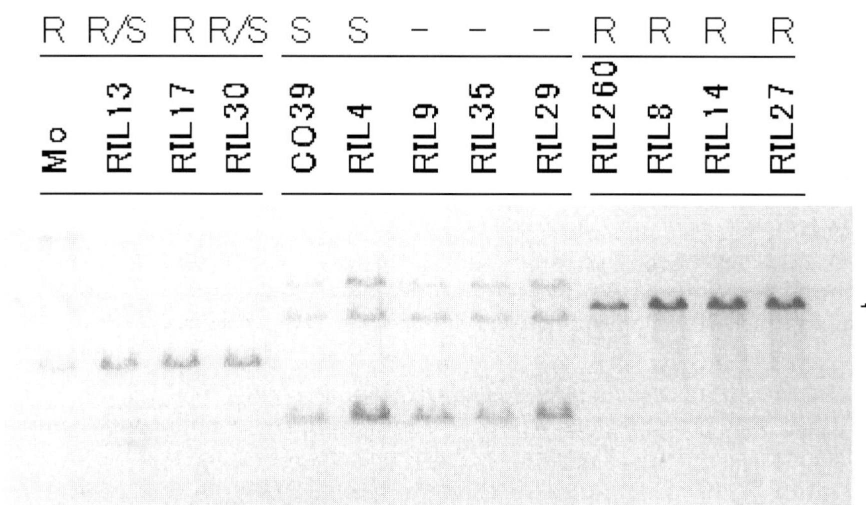

Figure 5. *Genomic DNA gel-blot analysis of RI lines using the Pi5(t) marker 40N23r. Moroberekan (Mo), CO39, RIL260, and ten pre-selected RILs (see text for details) were surveyed for polymorphism at the Pi5(t) locus. Arrowhead indicates hybridizing band not present in CO39 and Moroberekan. R and S signify resistance and susceptibility to M. grisea PO6-6, and R/S signifies a segregating phenotype. RIL9, RIL35, and RIL29 were not used for phenotype analysis.*

sequences were missing in Moroberekan and PKT.

To estimate the frequency of non-parental alleles in the whole population of RILs derived from the Moroberekan/CO39 cross, we analyzed 30 RILs from the 281 F_7 recombinant inbred lines. CAPS analysis using the flanking markers 17I18-12 and 34E14-10 revealed that three (RIL8, RIL14 and RIL27) of 30 lines tested contain the *Pi5(t)*-specific non-parental allele (data not shown). DNA gel-blot analysis with the co-segregating marker 40N23r showed that the non-parental alleles in these three lines are identical to that of *Pi5(t)* (Fig. 5). Furthermore, in an inoculation experiment, RIL13 and RIL30, two of three lines containing Moroberekan alleles at the *Pi5(t)* locus showed a segregating phenotype upon inoculation with PO6-6 (Fig. 5), suggesting that the resistance to PO6-6 in the RI lines containing the *Pi5(t)*-specific non-parental allele is not conferred by the Moroberekan allele.

4. DISCUSSION

We have developed an efficient method for positional cloning in rice using genetic resources that have been developed recently. In the present study, we utilized the genetic resources of Nipponbare, a line lacking *Pi5(t)*, to develop markers required for saturation mapping of the *Pi5(t)* locus. First, an initial marker linked to *Pi5(t)* was mapped on the high-density genetic map of Nipponbare/Kasalath (Harushima et al. 1998). Secondly, Nipponbare BAC clones physically spanning the region were identified using the CUGI database. Third, additional genetic markers for saturation mapping were produced using subclones of the identified BACs. Fourth, a small interval of Nipponbare corresponding to *Pi5(t)* was delimited by determining recombination breakpoints. Finally, a physical map of the *Pi5(t)* genomic region was constructed using flanking markers and a BIBAC library generated from a *Pi5(t)*-containing line.

Our data indicate that the *Pi5(t)* genomic region in RIL260 is identical to the *Pi3(t)* genomic region in C104PKT. This region is markedly different from that in the putative parents, PKT and Moroberekan. Furthermore our data indicate that PKT and Moroberekan do not have a resistance allele at the *Pi3(t)/Pi5(t)* locus. It is not likely that *Pi5(t)* was generated through a genetic event such as a deletion, inversion, duplication, etc., because the same non-parental alleles exist in many of the RIL populations, with a frequency of about 10%, as well as in C104PKT carrying *Pi3(t)*.

Our data suggest that a rice cultivar carrying *Pi3(t)* was outcrossed to CO39 and then backcrossed to CO39 five more times to produce C104PKT. It is possible that the same outcross was performed to generate the RI lines carrying *Pi5(t)*, because the genomic region of interest in these three *Pi5(t)*-

containing RI lines was identical to that in C104PKT for all markers tested. It is also possible that C104PKT contaminated the cross between Moroberekan and CO39, and that C104PKT carrying *Pi3(t)* is the donor for the three *Pi5(t)* RILs. The present study demonstrates the usefulness of genetic markers for discovering the source of a particular resistance gene.

We were not able to identify recombination events in the 170-kb interval encompassing *Pi5(t)* among over 2000 individuals from four different mapping populations. In the RIL260/M202 population, the marker C1454 mapped 0.06 cM away from *Pi5(t)* (1 recombination/1742 meiotic events) and the other flanking marker, S04G03, mapped 0.11 cM away from *Pi5(t)* (2/1742), giving a ratio of over 1000 kb/cM. This is much higher than the average physical/genetic ratio of 260–280 kb/cM estimated for the rice genome as a whole (Wu and Tanksley 1993). This result could be due to lack of pairing and subsequent strand exchange between homologous regions in the RIL260 and M202 parents used for the cross. This is supported by the fact that many of the RIL260 BIBAC-end sequences were not present in M202. Suppression of recombination has been observed in other introgressed regions associated with disease resistance, such as the *Mi* (van Daelen et al. 1993), *Mla* (Wei et al. 1999), and *Pita* [2] (Nakamura et al. 1997) loci. Thus, the *Pi5(t)* locus is highly diverged in disease-resistant and susceptible genotypes.

Over the past decade, a number of dominant *R* genes have been characterized from diverse plant species and their encoded proteins can be grouped into six classes based on structure (Wang et al. 1998; Dangl and Jones 2001). The largest class of *R* genes encodes an NBS-LRR class of proteins that can be further divided into two groups based on their N terminal domains. Sixty percent of the Arabidopsis NBS-LRR proteins carry a domain with homology to the intracellular signaling domains of the Drosophila Toll and mammalian interleukin (IL)-1 receptors (TIR-NBS-LRR), whereas 40% contain putative coiled-coil domains (CC-NBS-LRR) (Dangl and Jones 2001). Over 100 *R* gene sequences of the TIR-NBS-LRR class exist in the Arabidopsis genome, but this subclass has not yet been found in cereals and is not present in available rice sequences (Meyers et al. 1999). To date, five rice *R* genes have been cloned, including the blast resistance genes *Pib* and *Pita* encoding CC-NBS-LRR proteins and the bacterial blight *R* genes *Xa1*, *Xa21*and *Xa21D* which encode a CC-NBS-LRR, a receptor kinase and receptor-like protein, respectively (Song et al. 1995; Wang et al. 1999; Yoshimura et al. 1998; Bryan et al. 2000).

We analyzed ~150 BAC end sequences of Nipponbare corresponding to approximately 70 kb in the *Pi5(t)* locus (Chen et al. 2002). Through similarity searches against sequence data in public databases, we found that the amino acid sequences deduced from two genes in the region show similarity to the conserved NB-and/or LRR motifs. BIBAC end sequence analysis of RIL260 identified a third NBS-LRR sequence. These data indicate that the *Pi3(t)/Pi5(t)* locus contains a cluster of NBS-LRR sequences. These three genes are good candidates for *Pi3(t)/Pi5(t)* and may constitute part of a "natural pyramid" of resistance genes that confer the broad-spectrum resistance. To test this hypothesis, all the isolated BIBAC clones of RIL260 are currently being used in transgenic complementation studies to identify the *Pi3(t)/Pi5(t)* coding region(s). These plants will be inoculated with diverse isolates to determine whether a single gene or multiple genes are required for the *Pi3(t)/Pi5(t)* broad-spectrum resistance.

ACKNOWLEDGEMENTS

Masahiro Yano is specially acknowledged for mapping S04G03 on the Nipponbare/Kasalath genetic map. We thank Dave Mackill for providing DNA filters of the Black Gora/Labell mapping population. J.-S. Jeon was in part supported by the Korean Science and Engineering Foundation (KOSEF) and the Plant Metabolism Research Center (PMRC), Kyung Hee University, Korea. This project was supported by a United States of Department of Agriculture grant to P.C. Ronald

REFERENCES

Ahn, S.W. (1994). International collaboration on breeding for resistance to rice blast. In: Zeigler RS, Leong SA, Teng PS (eds) *Rice blast disease*. CAB International, Wallingford, UK, pp 137–153

Ahn, S.W. (2000). Spatial and temporal stability of genetic resistance to rice blast. In: Tharreau D, Lebrun MH, Talbot NJ, Nottenghem JL (eds) *Advances in rice blast research*. Kluwer, Dordrecht, pp 162–171

Bidaux, J.M. (1978). Screening for horizontal resistance to rice blast (*Pyricularia oryzae*). In: Buddenhagen IW, Persley GJ (eds) *Rice in Africa*. Academic Press, London, pp 159–174

Bonman, J.M., & Mackill, D.J. (1988). Durable resistance to rice blast. *Oryza*, 25, 103–110.

Bonman, J.M., Vergel De Dios, T.I., & Khin, M.M. (1986). Physiological specialization of *Pyricularia oryzae* in the Philippines. Plant Dis., 70, 767–769.

Bryan, G.T., Wu, K.-S., Farrall, L., Jia, Y., Hershey, H.P., McAdams, S.A., Faulk, K.N., Donaldson, G.K., Tarchini, R., & Valent, B. (2000). A single amino acid difference distinguishes resistant and susceptible alleles of the rice blast resistance gene *Pita*. *Plant Cell,* 12, 2033–2045.

Chen, D.-H., & Ronald, P.C. (1999). A rapid DNA minipreparation method suitable for AFLP and other PCR applications. *Plant Mol. Biol. Rep*., 17, 53–57.

Chen, D.-H., Zeigler, R.S., Leung, H., & Nelson, R.J. (1995). Population structure of *Pyricularia grisea* at two screening sites in the Philippines. *Phytopathology*, 85, 1011–1020.

Chen, D.-H., Zeigler, R.S., Ahn, S.W., & Nelson, R.J. (1996). Phenotypic characterization of the rice blast resistance gene *Pi2(t)*. *Plant Dis*., 80, 52–56.

Chen, D.-H., dela Viña, M., Inukai, T., Mackill, D.J., Ronald, P.C., & Nelson, R.J. (1999). Molecular mapping of the rice blast resistance gene, *Pi44(t),* in a line derived from a durably resistant rice cultivar. *Theor. Appl. Genet.,* 98, 1046–1053.

Chen, D.-H., Nelson, R.J., Wang, G.-L., Mackill, D.J., & Ronald, P.C. (2000). Use of DNA markers in introgression and isolation of genes associated with durable resistance to rice blast. In: Tharreau D, Lebrun MH, Talbot NJ, Notteghem JL (eds) *Advances in rice blast research.* Kluwer, Dordrecht, pp 17–27.

Chen, M., et al. (2002). An integrated physical and genetic map of the rice genome. *Plant Cell,* 14, 537–545.

Dangl, J.L., & Jones, J.D. (2001). Plant pathogens and integrated defense responses to infection. *Nature*, 411, 826-833.

Fomba, S.N., & Taylor, D.R. (1994). Rice blast in West Africa: its nature and control. In: Zeigler RS, Leong SA, Teng PS (eds) *Rice blast disease.* CAB International and IRRI, Wallingford, UK, pp 343–355.

Harushima, Y., et al. (1998). A high-density rice genetic linkage map with 2275 markers using a single F2 population. *Genetics,* 148, 479–494.

Inukai, T., Nelson, R.J., Zeigler, R.S., Sarkarung, S., Mackill, D.J., Bonman, J.M., Takamure, I., & Kinoshita, T. (1994). Allelism of blast resistance genes in near-isogenic lines of rice. *Phytopathology*, 84, 1278–1283.

Inukai, T., Zeigler, R.S., Sarkarung, S., Bronson, M., Dung, L.V., Kinoshita, T., & Nelson, R.J. (1996). Development of pre-isogenic lines for rice blast-resistance by marker-aided selection from a recombinant inbred population. *Theor. Appl. Genet.*, 93, 560–567.

Johnson, R. (1981). Durable resistance: definition of, genetic control, and attainment in plant breeding. *Phytopathology*, 71, 567–568.

Kosambi, D.D. (1944). The estimation of map distance from recombination values. *Ann. Eugen.*, 12, 172–175.

Lander, E.S., Green, P., Abrahamson, J., Barlow, A., Daly, M.J., Lincoln, S.E., & Newberg, L. (1987). MAPMAKER: an interactive computer program for constructing primary genetic maps of experimental and natural populations. *Genomics*, 1, 174–181.

Lee, E.J., & Cho, S.Y. (1990). Variation in races of rice blast disease and varietal resistance in Korea. In: Papers presented at the Focus on Irrigated Rice. Seoul, Korea, pp 27–31.

Lee, E.J., Zhang, Q., & Mew, T.W. (1989). Durable resistance to rice disease in irrigated environments. In: Progress in irrigated rice research. International Rice Research Institute, Manila, Philippines, pp 93–110

Mackill, D.J., & Bonman, J.M. (1992). Inheritance of blast resistance in near-isogenic lines of rice. *Phytopathology*, 82, 746–749.

Mao, L., Wood, T.C., Yu, Y.S., Budiman, M.A., Tomkins, M.A., Woo, S.S., Sasinowski, M., Presting, G., Frisch, D., Goff, S., Dean. R.A., & Wing, R.A. (2000). Rice transposable elements: a survey of 73,000 sequence-tagged-connectors. *Genome Res.,* 10, 982–990.

McCouch, S.R., Kochert, G., Yu, Z.H., Wang, Z.Y., Khush, G.S., Coffman, W.R., & Tanksley, S.D. (1988). Molecular mapping of rice chromosomes. *Theor. Appl. Genet.*, 76, 815–829.

Meyers, B.C., Dickerman, A.W., Michelmore, R.W., Sivaramakrishnan, S., Sobral, B.W., & Young, N.D. (1999). Plant disease resistance genes encode members of an ancient and diverse protein family within the nucleotide-binding superfamily. *Plant J.*, 20, 317-332.

Michelmore, R.W., Paran, I., & Kesseli, R.V. (1991). Identification of markers linked to disease resistance genes by bulked segregant analysis: a rapid method to detect markers in specific genomic regions by using segregating populations. *Proc. Natl. Acad. Sci. USA*, 88, 9828–9832.

Moffat, A.S. (1994). Mapping the sequence of disease resistance. *Science*, 265, 1804–1805.

Monna, L., Miyao, A., Zhong, H.S., Yano, M., Iwamoto, M., Umehara, Y., Kurata, N., Hayasaka, H., & Sasaki, T. (1997). Saturation mapping with subclones of YACs: DNA marker production targeting the rice blast disease resistance gene, *Pib*. *Theor. Appl. Genet.*, 94, 170–176.

Murray, M.G., & Thompson, W.F. (1980). Rapid isolation of high molecular weight plant DNA. *Nucleic Acids Res.*, 8, 4321–4325.

Nakamura, S., Asakawa, S., Ohmido, N., Fukui, K., Shimizu, N., & Kawasaki, S. (1997). Construction of an 800-kb contig in the near-centromeric region of the rice blast resistance gene Pi-ta2using a highly representative rice BAC library. *Mol. Gen. Genet.*, 254, 611–620.

Ou, S.H. (1979). Breeding rice for resistance to blast, a critical view. In: Proceedings of the Rice Blast Workshop. International Rice Research Institute, Manila, Philippines, pp 79–137.

Peterson, D.G., Tomkins, J.P., Frisch, D.A., Wing, R.A., & Paterson, A.H. (2000). Construction of plant bacterial artificial chromosome (BAC) libraries: an illustrated guide. *J. Agric. Genomics*, 5, www.ncgr.org/research/jag.

Redoña, E.D., & Mackill, D.J. (1996). Mapping quantitative trait loci for seedling vigor in rice using RFLPs. *Theor. Appl. Genet.*, 92, 395–402.

Sambrook, J., Fritsch, E.F., & Maniatis, T. (1989). Molecular cloning: a laboratory manual (2nd edn). Cold Spring Harbor Laboratory Press, Cold Spring Harbor, N.Y.

Song, W.Y., Wang, G.-L., Chen, L., Zhai, W., Kim, H.K., Holsten, T., Zhu, L., & Ronald, P.C. (1995). The rice disease resistance gene, *Xa21*, encodes a receptor-like protein kinase. *Science*, 270, 1804–1806.

Tsunoda, Y., Jwa, N.-S., Akiyama, K., Nakamura, S., Motomura, T., Kamihara, K., Kodama, O., & Kawasaki, S. (2000). Cloning of the rice blast resistance gene PI-B. In: Tharreau D, Lebrun MH, Talbot NJ, Notteghem JL (eds) *Advances in rice blast research*. Kluwer, Dordrecht, pp 9– 16

Van Daelen, R.A., Gerbens, J.J.F., van Rusissen, F., Aarts, J., Honteleez, J., & Zabel, P. (1993). Long-range physical maps of two loci (Aps-1and GP79) flanking the root-knot nematode resistance gene (*Mi*) near the centromere of tomato chromosome 6. *Plant Mol. Biol.,* 23, 185–192.

Wang, G.-L., Mackill, D.J., Bonman, J.M., McCouch, S.R., Champoux, M.C., & Nelson, R.J. (1994). RFLP mapping of genes conferring complete and partial resistance to blast in a durably resistance rice cultivar. Genetics, 136, 1421–1434.

Wang, G.-L., Holsten, T.E., Song, W.-Y., Wang, H.-P., & Ronald, P.C. (1995). Construction of a rice bacterial artificial chromosome library and identification of clones linked to the Xa-21disease resistance locus. *Plant J.*, 7, 525–533.

Wang, G.-L., Ruan, D.L., Song, W.Y., Sideris, S., Chen, L., Pi, L.Y., Zhang, S., Zhang, Z., Fauquet, C., Gaut, B.S., Whalen, M.C., & Ronald, P.C. (1998). *Xa21D* encodes a receptor-like molecule with a leucine-rich repeat domain that determines race-specific recognition and is subject to adaptive evolution. *Plant Cell*, 10, 765-779.

Wang, Z.-X., Yano, M., Yamanouchi, U., Iwamoto, M., Monna, L., Hayasaka, H., Katayose, Y., & Sasaki, T. (1999). The *Pib* gene for rice blast resistance belongs to the nucleotide binding and leucine-rich repeat class of plant disease resistance genes. *Plant J.*, 19, 55–64.

Wei, F., Gobelman-Werner, K., Morroll, S.M., Kurth, J., Mao, L., Wing, R., Leister, D., Schulze-Lefert, P., & Wise, R.P. (1999). The *Mla* (*powdery mildew*) resistance cluster is associated with three NBS-LRR gene families and suppressed recombination within a 240-kb DNA interval on chromosome 5S (1HS) of barley. *Genetics*, 153, 1929–1948.

Wu. J., et al. (2002). A comprehensive rice transcript map containing 6591 expressed sequence tag sites. *Plant Cell*, 14, 525–535.

Wu, K.S., & Tanksley, S.D. (1993). PFGE analysis of the rice genome: estimation of the fragment sizes, organization of the repetitive sequences and relationship between genetic and physical distances. *Plant Mol. Biol.*, 23, 243–254.

Yoshimura, S., Yamanouchi, U., Katayose, Y., Toki, S., Wang, Z.X., Kono, I., Kurata, N., Yano, M., Iwata, N., & Sasaki, T. (1998). Expression of *Xa1*, a bacterial blight-resistance gene in rice, is induced by bacterial inoculation. *Proc. Natl. Acad. Sci. USA*, 95, 1663-1668.

Yu, Z.H., Mackill, D.J., & Bonman, J.M. (1987). Inheritance of resistance to blast in some traditional and improved rice cultivars. *Phytopathology*, 77, 323–326.

Yuan, Q., Liang, F., Hsiao, J., Zismann, V., Benito, M.I., Quackenbush, J., Wing, R., & Buell, R. (2000). Anchoring of rice BAC clones to the rice genetic map *in silico*. *Nucleic Acids Res*., 28, 3636–3641.

Zeigler, R.S., Cuoc, L.X., Scott, R.P., Bernardo, M.A., Chen, D.-H., Valent, B., & Nelson, R.J. (1995). The relationship between lineage and virulence in *Pyricularia grisea* in the Philippines. *Phytopathology,* 85, 443–451.

TRANSPOSON-INSERTION LINES OF RICE FOR ANALYSIS OF GENE FUNCTION

*Akio Miyao and Hirohiko Hirochika

National Institute of Agrobiological Sciences (NIAS), 2-1-2, Kannondai, Tsukuba, Ibaraki, 305-8602, Japan

**E-mail: miyao@affrc.go.jp*

1. Introduction

Almost the entire sequence of the rice genome is already in the public domain (Sasaki 2002, Feng 2002). There are many resistance-gene-related sequences in the rice genome, but only a few have been characterized as to their function. In addition, almost no genes working in the disease resistance signalling pathways have been identified. Genetic and molecular dissection of these signalling pathways utilizing mutant lines will be the most powerful way to understand the mechanism of disease resistance, as has been nicely demonstrated in *Arabidopsis* and several other plant species.

Mutagenesis of rice genes can be achieved by chemical, radiation, or insertional methods. Chemical and radiation mutagenesis causes base-changes and deletions. In contrast, insertion mutagenesis employs insertion elements, such as T-DNA and transposons, and, consequently, the induced mutations are easily detected by PCR. T-DNA mutagenesis is widely used in *Arabidopsis*. For analysis of rice, both T-DNA and the *Ac* transposon are used for insertion mutagenesis (Enoki 1999, Jeon 2000). T-DNA insertion lines have been made in several laboratories and utilized for analysis of gene function (Rarinov 1999). *Ac* insertion lines are also under construction in several laboratories. However, some difficulties were noted when *Ac* was used to construct disruption lines covering all the rice genes, because the *Ac* transposon transposes only a relatively short distance from the original

S. Kawasaki (ed.), Rice Blast: Interaction with Rice and Control, 107–112.
© 2004 *Kluwer Academic Publishers. Printed in the Netherlands.*

position. To construct disruption lines to cover all the genes, we have chosen another class of transposons, a retrotransposon (Hirochika 2001).

2. FEATURES OF THE RETROTRANSPOSON *TOS17*

Tos17 is an endogenous *copia*-type retrotransposon of rice (Hirochika 1995). Because the reverse-transcribed *Tos17* cDNA is integrated into the rice genome, *Tos17* is transposed by the 'copy and paste mode', whereas DNA-type transposons like *Ac* are transposed by the 'cut and paste mode'. Although most plant retrotransposons are high copy number, *Tos17* exists at a relatively low copy number in the rice genome. Nipponbare, a standard line for rice genome sequence analysis, has only two copies of *Tos17*. This low copy number is one of the advantages of using *Tos17* as a tool for functional analysis of rice genes.

In cultured rice cells, *Tos17* is activated and transposes. When the cultured cells are regenerated into plants, *Tos17* is immediately inactivated and the transposed *Tos17* copies are stably inherited by subsequent generations. Because *Tos17* is an endogenous retrotransposon, regenerated *Tos17*-insertion lines are not considered genetically modified (GM) plants and can be analyzed in rice fields. This is one of the advantages of using *Tos17* for gene disruption as compared with T-DNA or transposons from other plants.

Generally, from 5 to 30 transposed copies of *Tos17* are detected in each regenerated rice plant. The insertion of multiple copies is another advantage of *Tos17*, because the number of insertion lines required for full coverage of the genome is lower than it is for T-DNA mutagenesis, where the average insertion number is one or two. The points of insertion in the rice genome can be determined by sequencing flanking regions of the transposed copies of *Tos17* in each line.

3. PRODUCTION OF LARGE SCALE INSERTION LINES

We produced large scale insertion lines with the *Tos17* retrotransposon. Nipponbare calli were grown in N6 liquid culture for 5 months and then regenerated. After 5 months cultivation, 10 copies of *Tos17* on average are transposed in each regenerated plant. Over the last five years, we have produced 50,000 insertion lines. Phenotypes of a total of 4,300 insertion lines were analysed in the field. The most frequent phenotype was sterile, appearing in nearly 30% of the lines. Dwarf (15% of the lines) and chlorotic mutations (e.g. albino, chlorine; 10% of the lines), were also often observed.

Disease resistance-related phenotypes, e.g. spotted leaf and lesion mimic (5% of the lines), were also observed with relatively high frequency.

DNA from individual lines with observable phenotypes were isolated by the CTAB method and analysed by Southern hybridization using a *Tos17* probe. If the mutant phenotype co-segregated with one of the hybridizing *Tos17*-bands, it was considered likely that the band contains the gene responsible for the mutant phenotype.

Some lines, five percent or fewer, showed the expected co-segregation, strongly suggesting that the observed mutant phenotypes are caused by the insertion of *Tos17* in these lines. For the remaining lines, we have been screening for other transposons that could have caused the observed phenotypes, but we have not yet found such transposons.

4. FLANKING SEQUENCE ANALYSIS

From each insertion line displaying a mutant phenotype in the field, genomic DNA was isolated to determine the flanking sequences of the *Tos17* insertion sites. To isolate DNA flanking the *Tos17* insertion sites, we use thermal asymmetric interlaced (TAIL) PCR and suppression PCR method (Miyao 1998, Liu 1995, Siebert 1995). Because only about 50% of the flanking DNA sequences are recovered using each of these methods, we use them both to increase the efficiency of recovery. In TAIL PCR, flanking sequences are amplified with a *Tos17* 3'-end primer and a randomly chosen short primer. Using a super cycle consisting of one cycle at a low annealing temperature and two cycles at a high annealing temperature, DNA flanking *Tos17* is more efficiently amplified than DNA fragments that have the randomly chosen primer at both ends.

In contrast, suppression PCR is an adaptor ligation-based method. Both ends of digested genomic DNA fragments are ligated with an adaptor. The adaptor consists of 44 bp length DNA and complimentary 8 bp length DNA. The 8 bp length DNA is modified with 5'-P and 3'-NH_3, and lacks the nucleotide sequence which is annealed with adaptor primer. Because the 3'-end of the 8 bp length DNA is blocked with NH_3, the 8bp DNA can not work as a primer for DNA synthesis. In the first PCR cycle, flanking DNA fragments that have sequence complimentary to the adaptor primer are synthesized only from the *Tos17* 3'-end primer. After the first cycle, the flanking DNA is amplified with *Tos17* 3'-end primer and adaptor primer.

Fragments amplified from DNA flanking *Tos17* are separated by agarose gel electrophoresis. Separated DNA fragments are purified with Promega Wizard PCR Preps DNA purification resin. Eluted DNA is directly sequenced with a BigDye cycle sequence kit using *Tos17* 3'-end primer and an ABI 3100 DNA sequencer.

We sequenced over 40,000 fragments from 3,800 lines. Because we use both TAIL PCR and suppression PCR methods on each line, some flanking sequences are redundant. Independent flanking sequences were identified by using an all-against-all BLASTN search. 15,000 independent flanking sequences were identified. This number indicates that we have already disrupted 15,000 loci in the rice genome.

The independent flanking sequences were searched using the BLASTX program (Altschul 1997) against the nr data set provided by NCBI. If significant similarity is detected, the searched fragment may be a piece of the hit gene, and the insertion line from which the flanking sequence was isolated may have a disruption in that gene. We sorted the hit genes into categories according to their putative functions. Genes involved in disease, defence, and signal transduction, especially serine/threonine protein kinases were frequently disrupted. Nearly 25% of the independent flanking sequences that could be assigned a putative function are in the defence and signal transduction categories. Because this value is two fold higher than the proportion of these genes among all annotated genes, we conclude that *Tos17* prefers these loci for insertion.

The independent flanking sequences were searched by BLASTN against the published rice genomic DNA to determine the positions of *Tos17* insertions on the rice genome. 11,000 flanking sequences were mapped on the rice genome sequences. We divided mapped flanking sequences into categories of exon, intron, intergenic, and unknown (un-annotated) regions. The number of insertions in exon and intron regions, i.e. genic regions, is nearly three-fold higher than the number in intergenic regions. This result indicates that *Tos17* prefers genic regions over transcriptionally-inactive intergenic regions.

Two categories of plant retrotransposons have been proposed, one is high copy number, and the other is low copy number (Bennetzen 2000). High copy number retrotransposons tend to make clusters and be located in non-transcribed intergenic regions. Low copy number retrotransposons prefer genic regions, which are transcriptionally active. Given its preference for insertion in genic regions, *Tos17* is best described as a low copy number retrotransposon.

There are many hot spots for *Tos17* insertion in the rice genomic sequences. We calculated the distances between adjacent *Tos17* insertion points. The distribution of these distances is not even, and is highly biased toward short distances. For example, distances of less than 100 bp occurred more than 3000 times on the rice genome sequences. This result indicates that *Tos17* insertion loci in the rice genome are highly aggregated. Although the aggregation of *Tos17* is not ideal for a gene disruption system, over 50% of the flanking sequences have similarity to amino acid sequences from NCBI nr data set. These data indicates that once *Tos17* insertions are detected in the target genes, allelic insertion mutants might be obtained with relatively high frequency.

5. WWW BASED RELATIONAL DATABASE

Phenotype data, along with photographs and flanking sequence related data, have been deposited into a relational database, PostgreSQL. Each seeds stock is kept in plastic bottles and labelled with a barcode label. A portable barcode reader is used for data input in the field and maintenance of seed stocks. Phenotype data in the barcode reader, weight of seed stocks, and digital images are transferred directly into the relational database.

Before import into the database, flanking sequence data are processed to remove the *Tos17* 3'-end sequence and relatively low quality sequences. Imported flanking sequence data are clustered with an all-against-all BLASTN search and their insertion positions identified with a BLASTN search against daily updated rice genomic sequences. On a regular basis, all flanking sequences in the database are also searched with BLASTX against the newest nr dataset. Because BLAST searches require high CPU power and memory, each query is searched on remote PCs. From each search result, the hit protein name, hit score, accession number, etc. are picked up by perl script and then deposited into the relational database.

The computer system is constructed of PCs installed with the FreeBSD operating system and connected to each other with the TCP/IP protocol. Data are served with the Apache WWW server program.

An in silico mutant screening system has been constructed. Users can search a query sequence of interest with BLASTN or tBLASTn against our flanking sequences. When a significant similarity is detected, phenotype data can be obtained via links to the hit sequence name in the BLAST result. Via the link, nucleotide sequence data of the flanking DNA, a primer sequence which is able to amplify the flanking DNA fragment when paired with *Tos17* 3'-end primer, and the history of seed distribution are available.

For scientific study, up to 20 grains each of our mutant lines are available. Because the mutant seeds were from regenerated plants, *Tos17* is segregating in the individual plants from these seeds. To obtain homozygous insertion mutants, DNA from each plant should be analysed by Southern hybridization using a flanking DNA probe. The restriction enzyme *Xba*I is suitable for Southern analysis. Our WWW page for the mutant search is http://tos.nias.affrc.go.jp/. We hope our database will contribute to the study of plant biology.

REFERENCES

Altschul, S.F., Madden, T.L., Schaffer, A.A., Zhang, J., Zhang, Z., Miller, W., and Lipman, D.J. (1997). Gapped BLAST and PSI-BLAST: a new generation of protein database search programs. Nucleic Acids Res. 25, 3389-3402.

Bennetzen, J.L. (2000). Transposable element contributions to plant gene and genome evolution. Plant Mol. Biol. 42, 251-269.

Enoki, H., Izawa, T., Kawahara, M., Komatsu, M., Koh, S., Kyozuka, J., and Shimamoto, K. (1999). *Ac* as a tool for the functional genomics of rice. Plant J. 19, 605-613.

Feng Q, Zhang Y, Hao P, Wang S, Fu G, Huang Y, Li Y, Zhu J, et.al. (2002), Sequence and analysis of rice chromosome 4. Nature, 420, 316-320

Hirochika, H., Sugimoto, K., Otsuki, Y., Tsugawa, H., and Kanda, M. (1996). Retro-transposons of rice involved in mutations induced by tissue culture. Proc. Natl. Acad. Sci. USA 93, 7783-7788.

Hirochika, H. (2001). Contribution of the *Tos17* retrotransposon to rice functional genomics. Curr Opin Plant Biol. 4, 118-122.

Jeon, J., Lee, S., Jung, K., Jun, S., Jeong, D., Lee, J., Kim, C., et.al. (2000). T-DNA insertional mutagenesis for functional genomics in rice. Plant J. 22, 561-570.

Liu, Y., and Whittier, R.F. (1995). Thermal asymmetric interlaced PCR: automatable amplification and sequencing of insert end fragments from P1 and YAC clones for chromosome walking. Genomics 25, 674-681.

Miyao, A., Yamazaki, M., and Hirochika, H. (1998). Systematic screening of mutants of rice by sequencing retrotransposon-insertion sites. Plant Biotech. 15, 253-256.

Parinov, S., Sevugan, M., Ye, D., Yang, W., Kumaran, M., and Sundaresan, V. (1999). Analysis of flanking sequences from Dissociation insertion lines: A database for reverse genetics in Arabidopsis. Plant Cell 11, 2263-2270.

Sasaki, T. ,Matsumoto T., Yamamoto K., Sakata K., Baba T., et.al. (2002), The genome sequence and structure of rice chromosome 1. Nature, 420, 312-316.

Siebert, P.D., Chenchik, A., Kellogg, D.E., Lukyanov, K.A., and Lukyanov, S.A. (1995). An improved PCR method for walking in uncloned genomic DNA. Nucleic Acids Res. 23, 1087-1088.

QUANTITATIVE TRAIT LOCI (QTL) REACTIONS TO RICE BLAST ISOLATES FROM JAPAN AND THE PHILIPPINES

*Y. Fukuta[1], M. J. T. Yanoria[1], D. Mercado- Escueta[1], L. A. Ebron[1], Y. Fujita[2], E. Araki[1] and G. S. Khush[1]

[1]*International Rice Research Institute, DAPO Box 7777, Metro Manila, The Philippines*
[2]*National Agricultural Research Center of Hokkaido Region, Hitsujigaoka, Toyohira, Sapporo, Hokkaido, 062-8555 Japan*

**Email: yfukuta@cgiar.org*

Abstract: fourteen qtl for philippine isolates and 12 qtl for japanese isolates were detected in the 14 regions on nine chromosomes (1, 2, 3, 6, 7, 8, 9, 11, and 12) using recombinant inbred lines derived from a cross between indica-type rice milyang 23 and a japonica-type rice akihikari. Among them, several qtl were detected to be common to philippine and japanese isolates. The qtl in two regions on the short arm of chromosome 1 and in the middle of chromosome 6 showed the resistance to philippine and japanese blast isolates, respectively. These qtl on chromosomes 1 and 6 were estimated to correspond to *pish* and *piz-t*, respectively. Based on the reaction patterns of detected qtl for japanese and philippine blast isolates, we suggested that *pish* and *piz* alleles might have the specified reactions, depending on the origin isolates of temperature or tropical area.

Key words: qyl analysis, japanese and philippine isolate, blast (*pyricularia grisea* sacc), rice (*oryza sativa* l.)

1. INTRODUCTION

Rice blast, caused by *Pyricularia grisea* Sacc., is one of the serious

S. Kawasaki (ed.), Rice Blast: Interaction with Rice and Control, 113–121.
© 2004 *Kluwer Academic Publishers. Printed in the Netherlands.*

diseases that occur in tropical and temperate areas. Genetic analyses have been carried out, and more than 40 resistance genes have been identified. Many resistance genes are expected to come from differentiation and interaction between blast isolates and resistance genes. A differential system based on the gene-for-gene theory (Flor 1956) estimated the resistance genes in rice varieties, but identifying plural resistance genes in one variety was difficult. Quantitative trait loci (QTL) analysis is a useful tool to detect genetic factors and clarify the genetic mechanism of complex agricultural characters related to several genes or minor effect genes. Based on the QTL reaction patterns for two kinds of blast isolates from the Philippines (tropical) and Japan (temperate), we tried to identify the resistance genes and to understand the relationships between blast races and resistance genes.

2. MATERIALS AND METHODS

2.1. Population

QTL analyses were carried out using recombinant inbred (RI) lines derived from a cross between an Indica-ype rice, Milyang 23 (M23), and a Japonica-type rice, Akihikari (AK). This RI population was composed of 191 lines. A DNA marker linkage map was developed using 182 restriction fragment length polymorphism (RFLP) markers (Fukuta et al 1999).

2.2. Evaluation of resistance

Two kinds of differential systems -- a combination between 12 Japanese differential varieties (for blast resistance genes, *Pia, Pib, Pii, Pik, Pik-m, Pik-p, Pik-s, Pit, Pita, Pita-2, Piz (Pi2)*, and *Piz-t*) and 12 isolates from Japan (Kiyosawa 1984, and Yamada et al. 1976); and the other combination between a total of 24 monogenic lines (for *Pia, Pib, Pii, Pik, Pik-m, Pik-h, Pik-s, Pik-p, Pish, Pit, Pita, Pita-2, Piz, Piz-5, Piz-t, Pi1, Pi3, Pi5*(t), *Pi7*(t), *Pi9*(t), *Pi11*(t), *Pi12*(t), *Pi19*(t), and *Pi20*), and 14 isolates from the Philippines (Tsunematsu et al. 2000) -- were used to estimated the genotypes of blast resistance in M23 and AK. The investigations using Japanese differential varieties and monogenic lines were carried out at the Hokuriku National Agricultural Experiment Station, Joetsu, Niigata, Japan, in 1996 and 1997, and at the International Rice Research Institute, Los Baños, Laguna, Philippines in 2000 and 2001, respectively.

The reactions of RI plants to blast isolates from Japan and the Philippines were evaluated using a 0-10 scale for blast nursery (National Agriculture Research Center, 1995) and a 0-9 scale using the *Standard evaluation system for rice* (IRRI 1996).

2.3. QTL analysis

Among the isolates used in investigating the genotypes of parent varieties M23 and AK, five from the Philippines and nine from Japan that showed a different reaction between parents were selected to study the in RI population.

The analysis of QTL associated with the marker for blast resistance was performed by simple interval mapping (1.0 cM) using computer program *QGENE* (Nelson 1997). A LOD score of 2.0 was used as the threshold for detecting QTL locations.

3. RESULTS AND DISCUSSION

3.1. Estimation of resistance genes of parent varieties

In the investigation using the Japanese differential system, M23 showed susceptibility to only three isolates, 007.2, 037.3, and 337.3, and the reaction pattern was the same as that of differential variety BL1 for *Pib*. AK showed resistance to only three isolates, 001, 005, and 045, and the reaction corresponded to a differential variety, Aichiasahi, for *Pia*. M23 and AK indicated simple and clear reactions to 12 Japanese isolates, and were estimated to have resistance genes *Pib* and *Pia*, respectively.

As to 14 isolates from the Philippine, M23 was susceptible to only four isolates, PO6-6, CA89, BN111, and M39. These were moderate or complete resistance to the other 10 isolates. Based on the comparison of reaction patterns of monogenic lines, M23 was estimated to include two resistance genes, *Pib* and at least an unknown one. AK showed moderate or complete resistance to all isolates. This reaction pattern of AK was similar to those of monogenic lines for *Pish, Piz, Piz-t*, or *Pi9*(t). These monogenic lines were including the susceptible reactions to some isolates and different from AK. From these results, AK was estimated to have gene(s) with broad-spectrum resistance (one of *Pish, Piz, Piz-t*, or *Pi9*(t)) and additional unknown one(s). Thus, M23 and AK showed completely different reaction patterns between Japanese and Philippine blast isolates.

3.2. Segregation of RI population

Wide distributions including transgressive segregation were observed in RI population to almost isolates (data not shown). Particularly to Philippines isolates, the transgressive segregation were observed clearly in the both ranges of resistant and susceptible segregation, and several resistance genes were estimated in M23 or AK from the segregation.

3.3. QTL analysis

Fourteen QTL for Philippine isolates and 12 QTL for Japanese isolates were detected in 14 regions on chromosomes, 1, 2, 3, 6, 7, 8, 9, 11, and 12. Several QTL were detected to be common in Philippine and Japanese isolates, and some QTL were detected in the same chromosome regions (Fig. 1, Table 1).

^On chromosome 1, eight QTL were detected by the Philippine and Japanese isolates in three chromosome regions, BR1-1, BR1-2, and BR1-3. Of these, one QTL in BR1-1 located at the distance of 23.8 cM from the end of the short arm was common between Philippine and Japanese isolates. The QTL in BR1-1 was estimated to correspond to *Pit* on the basis of location (Kaji et al 1997). The other seven QTL were specific to the Philippine isolates. In particular, one QTL located between RFLP markers *XNpb147* and *C86* in BR1-3 on the long arm had strong resistance to Philippine isolates with AK allele. No resistance gene has been found in BR1-3 yet.

On chromosome 2, three QTL were detected in BR2-1 of the terminal region of the long arm for two Philippine and eight Japanese isolates, and these QTL were linked tightly. The M23 allele of these QTL localized near the land marker *G1234* in BR2-1, indicated the strongest resistance among the detected QTL. It was assumed that the QTL corresponded to *Pib* on the basis of reaction and chromosomal position (Miyamoto et al. 1996 and Wang et al 1999).

On chromosome 3, two QTL linked tightly for Japanese isolates, (037.1 and 007) and one Philippine isolate (M101), were detected on the long arm of BR3-1. The M23 alleles of these QTL were showed resistance, and no resistance gene has been reported in this chromosome region.

On chromosome 6, one QTL for one Philippine isolate in the terminal region of the long arm and two QTL for seven Japanese isolates located near the centromere were detected. The two QTL located near the centromere were linked closely. These QTL were resistance with the M23 allele and might correspond to one of the *Piz* alleles, *Piz*, *Piz-t*, and *Piz-5* or *Pi9*(t), based on chromosomal position (Zhu et al 1993, Pan et al 1996, and Liu et al 2002).

Two QTL linked tightly in BR8-1 on the middle region of chromosome 8 showed different reactions, depending on the kind of isolates. One Japanese isolate, 437.1, and one Philippine isolate, M64, indicated resistance with AK and M23 allele, respectively. It was not clear if the two different resistance genes are located in BR8-1 or if one resistance gene had multiple effects.

On chromosomes 7, 9, 11, and 12, seven QTL were detected only by one Japanese or Philippine isolate, and these effects were not as big as those of the other QTL. In the estimation using the Japanese differential system, *Pia* was identified in AK. Kiyosawa (1972) and Goto et al (1981) mapped *Pia* in the middle region of chromosome 11. In our analysis, the position of QTL on chromosome 11 was in the terminal region of the short arm, and the QTL did not correspond to *Pia*. Thus, *Pia* was not able to be detected by our QTL analysis. *Pia* is difficult to identify using a differential system because the number of avirulence isolates is small and the other genes mask the reaction easily. *Pib* in M23 can mask the reaction of *Pia.* M23 may also have *Pia* the same allele in AK.

We could detect some new QTL such as those in BR1-3 and BR3-1 on chromosomes 1 and 3 in this study. Among them, the AK allele of QTL in BR1-3 had the strong effect and the reactions were specified to Philippine isolates. In the differential system using monogenic lines and Philippine isolates, AK showed moderate or complete resistance to all Philippine isolates and was estimated to have broad-spectrum resistance gene(s) and unknown one(s). The monogenic lines with *Pish*, *Piz* alleles, *Piz-5* and *Piz* (*Pi2*), and *Pi9*(t) showed similar reactions to almost all isolates. Among broad-spectrum resistance genes for Philippine isolates, *Piz* alleles and *Pi9*(t) had been known to be located in the middle region of chromosome 6 (Zhu et al 1993, Pan et al 1996, and Liu et al. 2002). In this study, two QTL were detected on chromosome 6, and one of them was estimated one of *Piz* alleles or *Pi9*(t) with broad-spectrum resistance from M23, and the QTL found to be not related to the AK reaction and RI segregation. The broad-spectrum QTL detected in BR1-3 on chromosome 1 was conferred by the AK allele to the Philippine isolates. From these results, it was presumed that the QTL with broad-spectrum in BR1-3 was *Pish* from AK.

Additionally, the QTL in BR6-1 showed the resistance to only Japanese isolates and not to Philippine isolates. The QTL in BR6-1 might also be specific to blast isolates. In other words, these results suggest that some resistance genes have specific reactions to temperate and tropical isolates. It is necessary to confirm this reaction specificity using other isolates from Japan, the Philippines, and other areas, and to do fine mapping and to characterize the QTL in detail using substitute or isogenic lines including the chromosome segment of detected QTL.

Although there is still a need to clarify the relationship between the detected QTL and the previously known resistance genes, this information will be useful in understanding the differentiation and interaction among blast races and resistance genes.

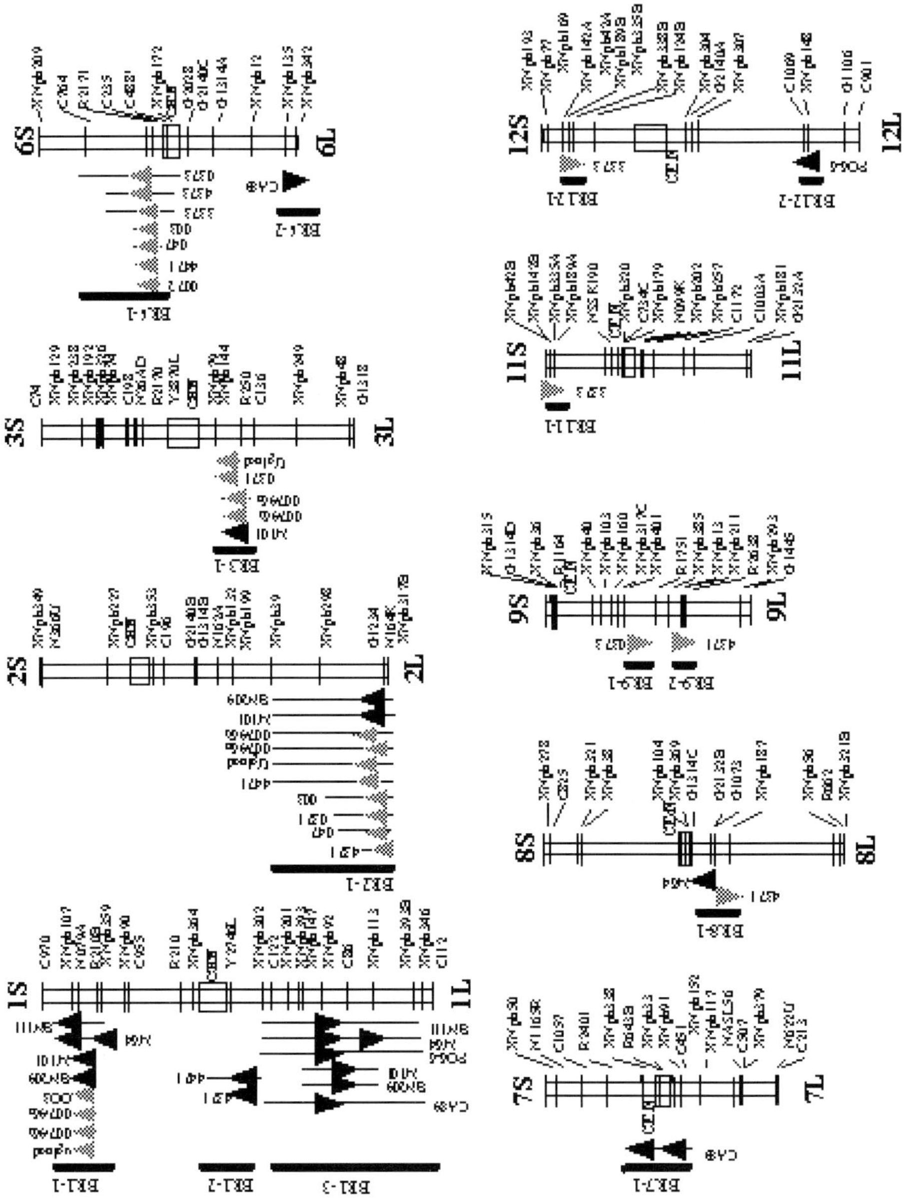

***Figure 1**. QTL mapping for blast resistance using Philippine and Japanese isolates. Triangle indicates QTL position on the chromosome. In each position, isolate name is indicated. Black and gray triangles indicate Philippine and Japanese isolate, respectively. Up and down arrows indicate that Milyang 23 and Akihikari allele was*

resistant to each isolate, respectively. RFLP markers were indicated on the right side of each chromosome.

Table 1. Location and effect of QTL for blast resistance.

Chr. Region[a]	Isolate	Year[b]	Position (cM)[c]	Marker interval[d]	LOD	a^e	R^2 (%)[f]
Chromosome 1							
BR1-1	M64*	2000	17.8	*<u>N079A</u>-R210B*	3.1	-0.53	7.2
	BN111*	2000	18.8	*<u>N079A</u>-R210B*	4.4	-0.66	10.3
	M101*	2000	23.8	*N079A-<u>R210B</u>*	3.3	-0.53	7.6
	BN209*	2000	23.8	*N079A-<u>R210B</u>*	2.8	-0.47	6.5
	3	1997	23.8	*N079A-<u>R210B</u>*	2.8	-0.32	6.6
	7	1996a	23.8	*N079A-<u>R210B</u>*	2.8	-0.9	6.9
	7	1996b	23.8	*N079A-<u>R210B</u>*	3.9	-0.47	9.1
	Upland	1997	23.8	*N079A-<u>R210B</u>*	2.1	-0.34	5.1
	M64*	2000	29.8	*<u>XNpb359</u>-XNpb90*	2.5	-0.48	6
BR1-2	447.1	1997	83.8	*Y2746L-<u>XNpb302</u>*	3.1	-0.38	7.3
	437.1	1997	85.8	*Y2746L-<u>XNpb302</u>*	2.4	-0.17	5.7
BR1-3	M64*	2000	106.8	*XNpb147-<u>XNpb92</u>*	16	1.16	32.2
	BN111*	2000	106.8	*XNpb147-<u>XNpb92</u>*	22.2	1.33	42
	CA89*	2001	106.8	*XNpb147-<u>XNpb92</u>*	15	1.1	32.7
	PO6-6*	2000	107.8	*XNpb147-<u>XNpb92</u>*	15.1	1.1	30.5
	BN209*	2000	107.8	*XNpb147-<u>XNpb92</u>*	3.6	0.56	8.4
	M101*	2000	112.8	*<u>XNpb92</u>-C86*	6.4	0.79	14.3
Chromosome 2							
BR2-1	7	1996b	116	*XNpb298-<u>G1234</u>*	37.2	-1.21	60.4
	Upland	1997	116	*XNpb298-<u>G1234</u>*	36.9	-1.16	59.7
	447.1	1997	116	*XNpb298-<u>G1234</u>*	25.3	-0.88	46.1
	3	1997	117	*<u>G1234</u>-XNpb318B*	24.2	-0.86	44.3
	7	1996a	117	*<u>G1234</u>-XNpb318B*	52.7	-3.01	73.9
	37.1	1997	117	*<u>G1234</u>-XNpb318B*	19.3	-0.65	37.3
	47	1997	117	*<u>G1234</u>-XNpb318B*	15.2	-0.63	30.6
	M101*	2000	117	*<u>G1234</u>-XNpb318B*	46.4	-1.62	67.5
	BN209*	2000	117	*<u>G1234</u>-XNpb318B*	50.1	-1.58	70.5
	437.1	1997	122	*<u>N164K</u>*	3.4	-0.2	8
Chromosome 3							
BR3-1	Upland	1997	60	*XNpb144-<u>R250</u>*	2.4	-0.38	5.7
	37.1	1997	60	*XNpb144-<u>R250</u>*	2.2	-0.25	5.1
	7	1996a	62	*<u>R250</u>-C138*	2.5	-0.85	6.1
	7	1996b	62	*<u>R250</u>-C138*	2.1	-0.35	5
	M101*	2000	63	*<u>R250</u>-C138*	2	-0.44	4.7
Chromosome 6							
BR6-1	37.1	1997	39.9	*C235-<u>C488</u>*	8.5	-0.57	18.6
	437.1	1997	39.9	*C235-<u>C488</u>*	3.4	-0.21	8
	3	1997	41.9	*C488-<u>XNpb172</u>*	3.2	-0.37	7.4
	7.2	1997	41.9	*C488-<u>XNpb172</u>*	2.2	-0.19	5.2
	47	1997	41.9	*C488-<u>XNpb172</u>*	2.9	-0.31	6.7
	337.3	1997	41.9	*C488-<u>XNpb172</u>*	5.1	-0.52	11.6
	447.1	1997	41.9	*C488-<u>XNpb172</u>*	2.6	-0.34	6.2
BR6-2	CA89*	2000	82.9	*XNpb12-<u>XNpb135</u>*	2.1	0.43	5.4

Chr. Region[a]	Isolate	Year[b]	Position (cM)[c]	Marker interval[d]	LOD	a^e	R^2 (%)[f]
Chromosome 7							
BR7-1	CA89*	2000	37.7	*XNpb338-R643*	2.2	-0.45	5.6
	CA89*	2000	51.7	*XNpb152-XNpb117*	2.7	-0.5	6.9
Chromosome 8							
BR8-1	M64*	2000	62.5	*G1073*	2.5	-0.48	5.9
	437.1	2000	64.5	*G1073-XNpb187*	2.4	0.18	5.8
Chromosome 9							
BR9-1	37.1	2000	36	*XNpb401-R1751*	3.5	0.38	8.2
BR9-2	437.1	2000	53	*XNpb13-XNpb211*	2.5	0.17	5.9
Chromosome 11							
BR11-1	337.3	2000	0.9	*XNpb335A-XNpb142*	2.1	0.33	5.1
Chromosome 12							
BR12-1	337.3	2000	9.2	*XNpb189B-XNpb335*	2.4	0.35	5.6
BR12-2	P06-6*	2000	84.2	*XNpb148-G1106*	2.3	-0.52	5.3

Table 1 continued.

* Blast isolate from the Philippines, no marked isolate from Japan.

[a]Degeneration of chromosome regions detected QTL for blast resistance based on chromosome no. and no. of the clustered QTL.

[b]Year of study

[c]Distance from the terminal of the short arm.

[d] Nearest marker to the QTL underlined.

[e]Additive effect of Milyang 23 allele compared with that of Akihikari allele.

[f]Phenotypic variance explained by each QTL

4. REFERENCES

Fukuta Y., K. Tamura, H. Sasahara, and T. Fukuyama (1999) Genetic and breeding analysis using molecular marker. 18, Variation of gene frequency and the RFLP map of the hybrid population derived from the cross between the rice variety Milyang 23 and Akihikari. Breed. Res. (Suppl 2): 176.

Flor H. H. (1956) The complementary genetic systems in flax and flax rust. Adv. Genet. 8: 29-54.

Goto J., Y. L. Jaw and A. A. Baluch (1981) Genetic studies on the resistance of rice plants to the blast fungus. IV. Linkage analysis of four genes, *Pia, Pi-k, Pi-z*, and *Pi-i.* Ann. Phytopathol. Soc. Jpn. 47: 252-254.

IRRI (International Rice Research Institute) (1996) Standard evaluation system for rice 4th edition, INGER, Genetic Resources Center, Manila (Philippines): 17-18.

Kaji R., T. Ogawa, and M. Nishimura (1997) Screening of RAPD markers for blast resistance gene, *Pi-t*, in rice. *In*: Harmonizing agricultural productivity and conservation of biodiversity: breeding and ecology. Proceedings of the 8th SABRAO Congress. The Korean breeding society and The society for advancement of breeding

researches in Asia and Ocieania, Seoul (Korea): 405-406.
Kiyosawa S. (1972) Genetics of blast resistance. In: IRRI Rice breeding. Los Banos, The Philippines, IRRI: 203-225.
Kiyosawa S. (1984) Establishment of differential varieties for pathogenicity test of rice blast fungus. Rice Genet. Newsl. 1: 95-97.
Liu G., G. Lu, L. Zeng and G.-L. Wang (2002). Two broad-spectrum blast resistance genes, *Pi9*(t) and *Pi2*(t), are physically linked on rice chromosome 6. Mol. Genet Genomics 267: 472-480.
Miyamoto M., I. Ando, K. Rybka, O. Kodama and S. Kawasaki (1996). High resolution mapping of the Indica-derived rice blast resitance gene. I. *Pib*. MPMI 9 (1): 6-13.
Nelson J. G. (1997) *QGENE*: software for marker-based genomic analysis and breeding. Mol. Breed 3: 239-245.
National Agricultural Research Center (1995) Rice breeding manual, Miscellaneous publications No.30: 3-19.
Tsunematsu H., M. J. T. Yanoria, L. A. Ebron, N. Hayashi, I. Ando, H. Kato, T. Imbe, and G. S. Khush (2000) Development of monogenic lines of rice for blast resistance. Breed Sci. 50: 229-234.
Pan Q. L. Wang, H. Ikehashi, and T. Tanisaka (1996) Identification of a new blast reistance gene in the indica rice cultivar Kasalath using Japanese differential cultivars and isozyme markers. Genetics 86 (10): 1071-1075.
Yamada M., S. Kiyosawa, T. Yamaguchi, T. Hirano, T. Kobayashi, K. Kushibuchi, and S. Watanabe (1976) Proposal of a new method for differentiating races of *Pyricularia oryza* Cavara in Japan. Ann. Phytopathol. Soc. Jpn 42: 216-219.
Wang Z.-X., M. Yano, U. Yamanouchi, M. Iwamoto, L. Monna, H. Hayasaka, Y. Katayose and T. Sasaki (1999) The *Pib* gene for rice blast resistance belongs to the nucleotide binding and leucine-rich repeat class of plant disease resistance genes. Plant J. 19 (1): 55-64.
Zhu L.-H., Y. Chen, Y.-B. Xu, J.-C. Xu, H.-W. Cai, and Z.-Z. Ling (1993) Construction of a molecular map of rice and gene mapping using a double haploid population of a cross between Indica and Japonica varieties. Rice Genet. Newsl. 10: 132-135.

SELECTION OF CRITERIONAL VARIETIES FOR EVALUATION OF BLAST PARTIAL RESISTANCE OF RICE IN THE TOHOKU REGION OF JAPAN

*Tomomori Kataoka[1], Hiroshi Kato[2], Tadashi Takita[3], Narifumi Yokogami[1] and Masayuki Yamaguchi[1]

[1]*National Agricultural Research Center for Tohoku Region, Yotsuya, Omagari, Akita 014-0102 JAPAN;*
[2]*Miyazaki Agricultural Experiment Station, Minemae 5851, Shimonaka, Sadowara, Miyazaki 880-0212 JAPAN;*
[3]*National Institute of Crop Science, Kannon-dai, Tsukuba, Ibaraki 305-8602 JAPAN*

**E-mail: tkataoka@affrc.go.jp*

1. INTRODUCTION

Blast disease, caused by *Pyricularia grisea* (Cooke) Sacc., is the most devastating disease of rice in Tohoku, a region located in the north-east of Honshu Island. To control blast disease in Japan, breeding programs generally develop varieties with partial (field or quantitative) resistance instead complete (true or qualitative) resistance, since the latter is likely to break down after a few years due to introgression (Watanabe, 1980). Before a resistant line can be recommended as a variety, breeders are normally required to submit information about its complete resistance genes as well as its partial resistance characteristics; in such cases, criterional varieties are indispensable. Because it has been more than 20 years since the establishment of criterional varieties for Tohoku, some of the varieties are no

S. Kawasaki (ed.), Rice Blast: Interaction with Rice and Control, 123–129.
© 2004 *Kluwer Academic Publishers. Printed in the Netherlands.*

longer under commercial cultivation, whilst the partial resistance characteristics of others are not always acceptable to breeders. Therefore, the objectives of the present study were to construct a new series of criterional varieties for the evaluation of partial resistance to leaf and panicle blast in the Tohoku region.

2. SELECTION OF CRITERIONAL VARIETIES FOR PARTIAL RESISTANCE TO LEAF AND PANICLE BLAST

2.1. Evaluation and selection of criterional varieties for resistance to leaf blast

2.1.1. Evaluation of partial resistance to leaf blast of different varieties

Partial resistance to leaf blast was evaluated between 1997 and 2000 for 62 japonica varieties with long or recent histories of cultivation in nine agricultural experiment stations (AESs) (eight in Tohoku and one (Aichi) in Chubu, a mountainous area in mid-Honshu with a blast-conducive climate similar to that of Tohoku). In agreement with the observations of Zenbayashi et al. (2002) that only a limited range of pathotypes exist in Tohoku, the predominant races of blast fungus in experimental field plots were 007 and 037 (data not shown). These races have genotypes compatible with the complete resistance genes Pia, Pii and Pik harboured in the rice varieties tested. In Japan, it is usual to maintain only one or two major resistance genes in cultivars in order to simplify the evaluation of partial resistance. For field tests, varieties could therefore be classified into three groups according to their major genes: group A (+, Pia), group I (Pii, Pia+Pii) and group K (Pik, Pia+Pik). Varieties in upland nurseries were ranked according to diseased leaf area on a scale of 0 (most resistant, lesions absent) to 10 (most susceptible, mortality nearly 100%) after Asaga, 1981. Test materials and methods varied with each AES. Whilst the degree of disease dissemination varied with location and year, the ranking of varieties within them was relatively constant as shown by the strong correlations (data not shown). After removal of outliers, linear regressions were used to normalize rankings at each year/location to standard data from Omagari in 2000 (groups A and I) or Aomori in 1998 (group K). A mean 'disease susceptibility index' was then calculated for each variety. Varieties were further grouped into one of five genotypes according to complete resistance genes '+', 'Pia', 'Pii', 'Pia+Pii' and 'Pik, Pia+Pik', and then ranked into seven levels of resistance (from very high to very low) based on their mean susceptibility indexes. To normalize levels between groups, 'Sasanishiki' and isogenic lines for complete resistance genes, 'Tohoku IL1', 'Sasanishiki BL8' and 'Sasanishiki BL1', were all ranked as slightly low (Table 1).

2.1.2 Selection of criterional varieties for evaluation of partial resistance to leaf blast

Variety	R gene	N	Disease susceptibility			Resistance		
			Mean	SD	CV	I	II	III
Group A								
Tohokumochi 161	+	15	2.92	0.53	18.2			rr
Chubu 32	+	13	3.9	0.55	14.1			rr
Kokoromachi	+	19	5.73	0.51	8.8			r
Sasaminori	+	20	5.81	0.4	6.9	r	r	r
Koganenishiki	+	20	6.01	0.51	8.5	r		(r)
Choukai	+	20	6.29	0.5	7.9	mr		mr
Snow-pearl	+	19	6.81	0.32	4.6			m
Nipponbare	+	20	7.15	0.45	6.3	m		(m)
Tohoku IL1	+	19	7.4	0.43	5.8			ms
Koshihikari	+	19	7.45	0.93	12.4		s	(ms)
Mutsuhikari	+	19	7.85	0.62	7.9		s	s
Ouu 320	*Pia*	27	3.94	0.67	17.1			rr
Reimei	*Pia*	27	5.39	0.58	10.8		r	(rr)
Mutsuhomare	*Pia*	27	5.56	0.51	9.2			r
Toyonishiki	*Pia*	28	6.25	0.54	8.7	r	r	mr
Kiyonishiki	*Pia*	28	6.27	0.53	8.5	mr	mr	mr
Maihime	*Pia*	27	6.41	0.58	9			mr
Yamabiko	*Pia*	28	6.5	0.58	8.9	r		(mr)
Norin 41	*Pia*	28	6.73	0.63	9.4	m	m	m
Sasanishiki	*Pia*	28	7.19	0.42	5.8	ms	ms	ms
Aichiasahi	*Pia*	28	8.13	0.57	7	s		s
Group I								
Chubu 45	*Pii*	24	5.83	0.43	7.3			r
Takaneminori	*Pii*	23	6.29	0.32	5			mr
Todorokiwase	*Pii*	24	6.41	0.48	7.5	r	r	mr
Yoneshiro	*Pii*	24	6.45	0.4	6.1	mr	mr	mr
Manamusume	*Pii*	24	6.64	0.32	4.8			m
Satonouta	*Pii*	17	6.72	0.36	5.4			m
Fujisaka 5	*Pii*	24	6.78	0.33	4.9	m	m	m
Hitomebore	*Pii*	23	7.51	0.49	6.5			ms
Inabawase	*Pii*	24	7.67	0.4	5.2	s	s	s
Hatajirushi	*Pia, Pii*	24	6.15	0.38	6.2			mr
Haenuki	*Pia, Pii*	24	6.82	0.44	6.5			m
Akitakomachi	*Pia, Pii*	24	6.84	0.32	4.7			m
Sasanishiki BL8	*Pia, Pii*	24	7.13	0.33	4.6			ms
Group K								
Himenomochi	*Pik*	13	5.37	0.48	8.9	r	r	r
Sakakimochi	*Pia, Pik*	13	6.12	0.4	6.5	mr		m
Mangetsumochi	*Pik*	13	6.31	0.36	5.6	m	m	m
Sasanishiki BL1	*Pia, Pik*	12	6.5	0.31	4.7			ms
Fukei 69	*Pik*	12	6.82	0.4	5.9			s
Kusabue	*Pik*	13	7.63	0.52	6.9	s	s	(ss)

Table 1. *Disease susceptibility indexes and evaluation of partial resistance to leaf blast.*

(Adapted from Selection of rice standard varieties for leaf blast resistance in Tohoku region, Kataoka et al.,(2001). Tohoku Agric. Res., 54, 15-16.)

Table 1 continued:

Results for some varieties are not included. R gene=Complete resistance gene. N=Number of samples. SD=Standard deviation. CV=Coefficient of variation. I, II and III refer to resistance levels (rr=very high, r=high, mr=slightly high, m=medium, ms=slightly low, s=low, ss=very low) established in 1979, agreed by Minist. Agric. Forest. Fish., Japan (1980) and established in the present study, respectively. Varieties in parentheses were not selected as criterional varieties.

Table 1 sumarizes the suitability of 34 varieties as criterional varieties for evaluation of leaf blast partial resistance based on resistance stability and commercial importance. A new resistance rank, "very high", was determined, in which were placed 'Tohoku 161', 'Chubu 32' and 'Ouu 320'. 'Toyonishiki' and 'Todorokiwase', which were previously ranked as high, were re-evaluated in the present study as only slightly high.

2.2. Evaluation and selection of criterional varieties for resistance to panicle blast

2.2.1 Evaluation of partial resistance to panicle blast of different varieties

Partial resistance to panicle blast was evaluated between 1997 and 2000 for 60 *japonica* varieties with long or recent histories of cultivation in ten AESs (nine in Tohoku and one in Chubu). Varieties were divided into three genotype groups and then subdivided into up to three groups according to time of heading (very early, early and medium-late) because the severity of panicle blast epidemics depends strongly on heading date. Susceptibility of varieties was ranked on a scale of 0 to 10 according to the percentage of diseased spikelets (Asaga, 1981). Disease susceptibility indexes were computed independently for each of the eight groups using linear regressions. The set of data selected as standards for groups A and I were collected from Iwate in 2000, and for group K from south Iwate in 1998. And varieties were then ranked into seven levels of resistance (from very high to very low) based on their mean susceptibility indexes. To normalize levels between groups, 'Sasanishiki' and isogenic lines for complete resistance genes were ranked as low. Results are summarized in Table 2.

2.2.2 Selection of criterional varieties for evaluation of partial resistance to panicle blast

Table 2 summarizes the suitability of 44 varieties as criterional varieties for evaluation of panicle blast partial resistance. Very high resistances to panicle blast were observed for 'Tohoku 161', 'Chubu 32', 'Ouu 357' and 'Iwanan 6'. 'Reimei', 'Sasaminori', 'Toyonishiki' and 'Yoneshiro', varieties previously identified as having high resistances, had lower (slightly high) resistances.

***Table 2**. Disease susceptibility indexes and evaluation of partial resistance to panicle blast*

Variety	R gene	N	Disease susceptibility index			Heading date	Resistance evaluation		
			Mean	SD	CV	Omagari	I	II	III
Group A									
Very early maturity									
Tohokumochi 161	+	19	5.3	0.64	12.1	4-Aug			rr
Aokei 128	*Pia*	22	6	0.49	8.1	3-Aug			r
Reimei	*Pia*	24	6.92	0.52	7.6	3-Aug	r	r	mr
Mutsuhomare	*Pia*	22	7.7	0.42	5.5	1-Aug			m
Maihime	*Pia*	21	7.75	0.49	6.3	1-Aug			m
Mutsuhonami	*Pia*	22	8.08	0.36	4.4	3-Aug		ms	ms
Fukei 94	*Pia*	24	8.77	0.61	7	3-Aug	s	s	s
Early maturity									
Ouu 320	*Pia*	24	5.67	0.5	8.9	5-Aug			r
Kokoromachi	+	24	6.27	0.47	7.5	4-Aug			r
Sasaminori	+	26	6.83	0.49	7.1	5-Aug	r		mr
Hatsunishiki	+	26	7.62	0.6	7.8	4-Aug	m		m
Norin 1	+	26	8.13	0.66	8.1	5-Aug	s		ms
Medium-late maturity									
Chubu 32	+	19	5.13	0.83	16.3	5-Aug			rr
Chiyonishiki	*Pia*	22	6.2	0.51	8.3	10-Aug			r
Ouu 247	+	22	6.27	0.5	8	10-Aug	r		(r)
Toyonishiki	*Pia*	22	6.68	0.52	7.8	8-Aug	r	r	mr
Choukai	+	22	6.86	0.53	7.7	5-Aug	m		mr
Kiyonishiki	*Pia*	22	7.48	0.54	7.2	5-Aug	m	m	m
Snow-pearl	+	22	7.82	0.4	5.1	6-Aug			ms
Sasanishiki	*Pia*	22	8.46	0.72	8.6	9-Aug	s	s	s
Tohoku IL1	+	15	8.49	0.78	9.2	8-Aug			s
Group I									
Very early maturity									
Takaneminori	*Pii*	28	6.8	0.46	6.8	4-Aug			mr
Yoneshiro	*Pii*	27	6.92	0.68	9.8	4-Aug	r	r	mr
Tsugaruroman	*Pia*, *Pii*	18	7.82	0.53	6.8	4-Aug			m
Aokei 131	*Pia*, *Pii*	18	8.04	0.84	10.4	31-Jul			ms
Fujisaka 5	*Pii*	27	8.68	0.78	9	1-Aug	m	m	s
Early maturity									
Hatajirushi	*Pia*, *Pii*	31	6.1	0.61	10.1	5-Aug			r
Satonouta	*Pii*	21	6.91	0.51	7.3	5-Aug			mr
Akitakomachi	*Pia*, *Pii*	31	7.65	0.55	7.2	6-Aug			ms
Inabawase	*Pii*	30	8.83	0.74	8.3	7-Aug	s	s	s
Medium-late maturity									
Ouu 357	*Pia*, *Pii*	32	5.23	0.69	13.1	9-Aug			rr
Iwanan 6	*Pia*, *Pii*	32	5.27	0.85	16.1	10-Aug			rr
Todorokiwase	*Pii*	32	6.06	0.74	12.3	8-Aug	r	r	r
Manamusume	*Pii*	32	6.38	0.62	9.7	10-Aug			r
Haenuki	*Pia*, *Pii*	32	7.57	0.89	11.7	11-Aug			m
Hitomebore	*Pii*	32	7.62	0.69	9	10-Aug			m
Sasanishiki BL8	*Pia*, *Pii*	32	8.98	0.59	6.6	10-Aug			s
Group K									
Early maturity									
Himenomochi	*Pik*	6	4.22	0.26	6.1	4-Aug	r		r
Fukei 69	*Pik*	7	6.87	0.65	9.4	7-Aug	s	s	s
Medium-late maturity									
Chubu 7	*Pia*, *Pik*	4	4.21	0.39	9.3	8-Aug			r
Ouu 321	*Pik*	6	4.25	0.57	13.4	9-Aug			r
Yukigesho	*Pik*	6	5.38	0.52	9.6	7-Aug			mr
Bikei 91	*Pik*	6	6.01	0.57	9.5	15-Aug	r		m
Dewanomochi	*Pik*	6	6.48	0.31	4.8	14-Aug	m		ms
Sasanishiki BL1	*Pia*, *Pik*	6	6.8	0.32	4.7	10-Aug			s

Table 2 continued :
(Adapted from ***Selection of rice standard varieties for leaf blast resistance in Tohoku region,*** *Kato et al., (2001), Tohoku Agric. Res., 54, 13-14).*
Symbols and annotations are same as in Table 1.

3. DISCUSSION

Inoue et al. (1983) suggested that 'Yamabiko' and 'Choukai' had the highest possible levels of partial resistance to leaf blast within traditional Japanese cultivars. Resistance of both varieties was evaluated as 'slightly high' in the present study, but their resistance levels have been dramatically increased by breeding with the upland cultivar 'Sensho' (probably of Chinese origin) and the Taiwanese cultivar 'Miaoli 37'. The high resistance of 'Sensho' is due to an additive effect acquired by pyramiding several quantitative resistance genes (Higashi, 1995), and this is also true in the case of 'Ouu 320'. Yamaguchi et al. (1997) demonstrated that cultivation of the highly resistant 'Ouu 320' in conditions favourable for blast disease reduced yield losses to less than 5%, whilst yield losses due to cultivation of the low-resistant 'Sasanishiki' was more than 50%. Moreover, the effect of variety on yield was greater than that of five applications of fungicide. Yamaguchi et al. (2000) reported similar results for the very highly resistant 'Chubu 32'. The present study indicates that there is still room for improving partial resistance to blast, since several varieties that were previously ranked as having high resistance were re-evaluated here as having only slightly high resistance. By using this new ranking system having higher standards for the evaluation of blast resistance, levels of blast resistance in Tohoku are expected to increase further. And the strict and accurate evaluation of such varieties has already resulted in the development at Omagari 'Ouu 366', a new cultivar that promises a very high level of blast resistance together with good eating quality.

ACKNOWLEDGMENTS

This research was carried out with the cooperation of the following research institutes: Aomori Agric. Exp. Stn.; Fujisaka Branch, Aomori Agric. Exp. Stn.; Iwate Agric. Res. Cent.; Miyagi Prefect. Furukawa Agric. Exp. Stn.; Akita Agric. Exp. Stn.; Shonai Branch, Yamagata Prefect. Agric. Exp. Stn.; Soma Branch, Fukushima Prefect. Agric. Exp. Stn. and Aichi Prefect. Agric. Res. Cent. Mount. Reg. Agric. Inst. We wish to express our sincere appreciation and gratitude to all the researchers in these institutes for supplying data and for their constructive discussions and helpful suggestions.

REFERENCES

Asaga, K. (1981). A procedure for evaluating field resistance to blast in rice varieties. *J. Cent. Agric. Exp. Stn.*, 35, 51-138. (in Japanese, English summary)

Higashi, T. (1995). Genetic studies on field resistance of rice to blast disease. *Bull. Tohoku Natl. Agric. Exp. Stn.*, 90, 19-75. (in Japanese, English summary)

Inoue, M., Morimoto, T., Tanabe, K., Shumiya, A., & Fujii, K. (1983). Breeding of field resistance of rice varieties superior to their parents to blast. *Res. Bull. Aichi Agric. Res. Cent.*, 15, 63-69. (in Japanese, English summary)

Watanabe, S. (1980). Breeding using rice varieties in foreign countries. In: Yamazaki, Y., & Kouzaka, T. (Eds.), *Blast and breeding for resistance* (pp. 34-36). Tokyo: Hakuyu-sha. (in Japanese)

Yamaguchi, M., Kondoh, T., & Higashi, T. (1997). Possibility of reduction in chemical prevention to blast by using field resistant rice varieties. *Tohoku J. Crop Sci.*, 40, 31-33. (in Japanese)

Yamaguchi, M., Saitoh, H., & Higashi, T. (2000). Effect of varietal field resistance for control of rice blast. In: Tharreau, D., Lebrun, M.H., Talbot, N.J., & Notteghem, J.L. (Eds.), *Advances in rice blast research* (pp. 196-202). Netherlands: Kluwer Academic Publishers.

Zenbayashi, K., Dela Peña, F., Ashizawa, T., & Koizumi, S. (2002). Distribution of pathogenic races of rice blast fungus (*Pyricularia grisea*) in Hokkaido and Tohoku districts. *Ann. Rept. Plant Prot. North Japan*, 53, 19-23. (in Japanese)

GENETIC DISSECTION AND MAPPING OF GENES CONFERRING FIELD RESISTANCE TO RICE BLAST IN JAPANESE UPLAND RICE

*Shuichi Fukuoka[1], Takehiko Shimizu[2], Masahiro Yano[1], Kazutoshi Okuno[2] and Tsukasa Nagamine[1]

[1]*National Institute of Agrobiological Sciences (NIAS), JAPAN;*
[2] *Institute of Society for Techno-Innovation of Agriculture, Forestry and Fisheries(STAFF), JAPAN;*
[3]*National Agricultural Research Center for Hokkaido Region (NARCH), JAPAN*

**E-mail: fukusan@affrc.go.jp*

Abstract: Field resistance to rice blast in Japanese upland was analyzed genetically. Chromosomal regions that are involved in field resistance were detected by quantitative trait loci (QTL) analysis using F_4 lines of a cross between resistant upland rice variety *Owarihatamochi* and susceptible irrigated rice variety *Nipponbare*. At two QTLs on chromosome 4 and one on chromosome 12, alleles from *Owarihatamochi* were resistant, while an allele from *Nipponbare* was resistant at one QTL on chromosome 9. Backcrossed progeny lines with each of the resistant alleles from upland rice were developed using the highly susceptible irrigated rice cultivar *Aichiasahi* as the recurrent parent. At all three QTLs, the lines with the *Owarihatamochi* allele showed a significantly higher level of field resistance than did those with the *Aichiasahi* allele. The differences in resistance between two genotypes were largest at the QTL on chromosome 4 and smallest at the QTL on chromosome12, in good accordance with the result of QTL analysis. The QTL of largest effect on chromosome 4 was mapped as a single recessive gene designated as *pi21* by genetic linkage analysis. Using a mapping population consisting of 82 lines, we located *pi21* between marker loci *G271* and *G317* at a distance of 5.0 cM and 8.5 cM, respectively. High-resolution genetic linkage analysis using 1104 individuals identified several DNA markers tightly linked with the *pi21*locus. These markers are being used to identify P1 artificial chromosome (PAC) clones containing this locus for positional cloning of *pi21*.

Key words: upland rice, field resistance, QTL, genetic mapping

S. Kawasaki (ed.), Rice Blast: Interaction with Rice and Control, 131–136.
© 2004 *Kluwer Academic Publishers. Printed in the Netherlands.*

1. INTRODUCTION

Two types of resistance to blast, complete resistance and field (partial) resistance, have been described in rice (Ezuka 1972). Complete resistance induces a hypersensitive reaction and is characterized by a resistant infection type. More than 20 loci for complete blast resistance have been identified in rice (McCouch *et al.* 1994). Several genes for this type of resistance have been isolated (Wang *et al.* 1999, Bryan *et al.* 2000), and their interaction with a fungous avirulence gene has been investigated at the molecular level (Jia *et al.* 2000). In spite of their significant effect against rice blast, complete resistance genes have been overcome by compatible races of the pathogen several years after varieties with this type of resistance were released. Field resistance is usually incomplete and characterized by a susceptible infection type, although it does limit the proportion of diseased leaf area. Field resistance is considered to be durable, but the genetic basis and mechanism of field resistance is not fully understood. Japanese upland rice is a source of the genes that are responsible for field resistance to rice blast. Previous studies have suggested that field resistance in Japanese upland rice is under complex genetic control (Higashi and Saito 1985), and conventional genetic analysis did not allow researchers to identify the genes and their chromosomal locations for field resistance.

2. DETECTION OF CHROMOSOMAL REGIONS INVOLVED IN FIELD RESISTANCE

Quantitative trait loci (QTL) analysis using DNA markers allows for improved understanding of the genetics of the traits controlled by multiple genes (Tanksley 1993). To determine the chromosomal regions involved in field resistance to blast in Japanese upland rice, QTL analysis was carried out using 146 F_4 lines of a cross between Japanese irrigated rice cultivar *Nipponbare* and the upland cultivar *Owarihatamochi* (Fukuoka and Okuno 2001). *Nipponbare* and *Owarihatamochi* have no gene for complete resistance to rice blast, based on screening using Japanese differential blast race. *Nipponbare* has a lower level of field resistance than does *Owarihatamochi*. We assessed field resistance in F_4 lines at the Aichi Prefecture Agricultural Research Center (AARC), Japan, with two replications. The disease severity of 40- to 50-day-old plants was scored using a rating from 0 (highly resistant) to 10 (highly susceptible) based on the diseased leaf area (DLA). We used 111 restriction fragment length polymorphism (RFLP) and 5 single sequence repeat (SSR) marker loci for segregation analysis in the F_4 lines. We used the program MAPMAKER/EXP 3.0 based on the Kosambi function to build an RFLP linkage map. MAPMAKER/QTL ver. 1.0 was used to detect QTLs in a logarithm of the odds (LOD) threshold of 2.0.

The frequency distribution of the field resistance score in the F_4 lines was continuous and ranged from 1 to 10. The scores for *Owarihatamochi* and *Nipponbare* were 2.8 and 8.0, respectively. The QTL analysis revealed four QTLs located on three chromosomes (Fig. 1). The resistant allele on chromosome 9 is derived from *Nipponbare*, while the resistant alleles on chromosomes 4 and 12 come from *Owarihatamochi*. The QTLs on chromosome 9 (*qBR9-1*) and 12 (*qBR12-1*) explained 7.9% and 13.7% of the total phenotypic variation, respectively, while the two QTLs on chromosome 4, close to RFLP marker loci *G271* (*qBR4-1*) and *G177* (*qBR4-2*), explained 45.7% and 29.4% of the phenotypic variation, respectively. All together, the four QTLs explained 66.3% of the total phenotypic variation.

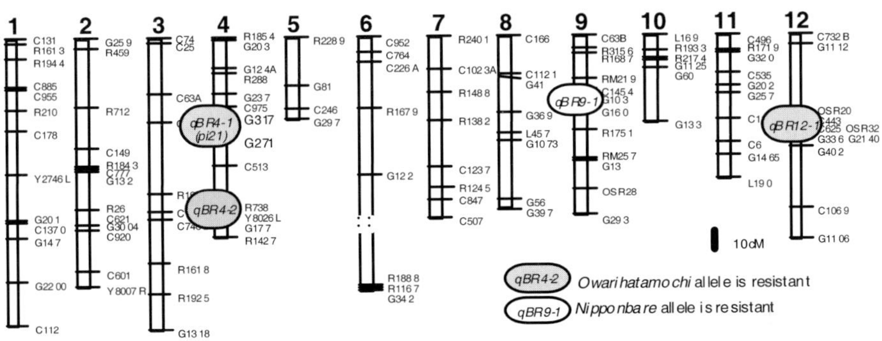

***Figure** 1. Location of the QTLs for field resistance to rice blast in F_4 lines of the cross between irrigated rice Nipponbare and upland rice Owarihatamochi.*

3. MARKER-ASSISTED SELECTION OF FIELD RESISTANCE IN BACKCROSSED LINES

One strategy for analyzing QTL as single Mendelian factors and for characterizing each QTL is the development of backcrossed progeny lines for each QTL by marker-assisted selection (Yano and Sasaki, 1997). To confirm the effect of putative QTLs, backcrossed progeny lines were developed from three F_3 plants identified as having resistant alleles from upland rice based on the result of QTL analysis. *Aichiasahi*, a highly susceptible cultivar, was used as the recurrent parent. The same set of DNA markers in QTL analysis was used to select relevant chromosomal regions. We selected three BC_2F_1 plants that contain just one out of three blast resistance alleles from *Owarihatamochi* (Fig. 2). During backcross and selection the proportion of upland rice chromosomes in these plants decreased to less than 6% based on the genotypes of DNA markers. The field resistance was assessed in 98, 44 and 46 BC_2F_3 lines for *qBR4-1*, *qBR4-2* and *qBR12-1*, respectively, using the same procedures as in QTL analysis. We compared the levels of field resistance among the lines with *Owarihatamochi*-homozygous alleles, heterozygous alleles and *Aichiasahi*-homozygous alleles. Genotypes at each QTL were estimated based on the genotype of DNA markers around QTLs.

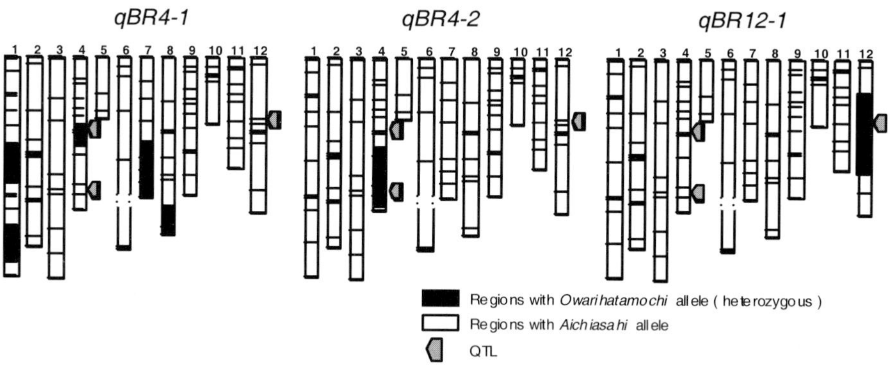

Figure 2. *Graphical genotypes of three BC_2F_1 plants having one of three resistant alleles from upland rice Owarihatamochi.*

The average DLA score of the lines with the *Owarihatamochi* allele was significantly higher than those with the *Aichiasahi* alleles at all three QTLs (Table 1). The difference between two genotypes was largest at the QTL on chromosome 4 and smallest at the QTL on chromosome 12, in good accordance with the result of QTL analysis.

Table 1. *Average score for diseased leaf area of the lines with different genotypes at QTLs*

QTL	O-homozygous	Heterozygous	A-homozygous
qBR4-1	3.1±0.46 (10)	5.7±0.77 (55)	6.9±0.75 (100)
qBR4-2	6.7±0.66 (67)	7.3±0.51 (83)	8.2±0.29 (100)
qBR12-1	6.4±0.29 (80)	6.8±0.64 (93)	7.0±0.57 (100)

O: *Owarihatamochi* (upland rice), A: *Aichiasahi* (irrigated rice).
Percentages of the diseased leaf area compared with A-homozygous lines are indicated in parentheses.

4. FINE GENETIC MAPPING OF A GENE, *PI21* CONFERRING THE GREATEST EFFECT ON FIELD RESISTANCE

Knowing the structure of the field resistance gene is important for studying the differences between field resistance and complete resistance genes. Map-based cloning of target genes is one of the strategies for identifying gene structure. Eighty-two BC_1F_3 lines in which *qBR4-1* was segregated were subjected to genetic linkage analysis to determine the map location of *qBR4-1*. The genotypes at the *qBR4-1* were readily determined based on the difference in diseased leaf area and its segregation among plants within a line. A resistance gene for this QTL, later designated as *pi21*, was mapped as a single recessive gene between RFLP marker loci *G271* and *G317* at a distance of 5.0 cM and 8.5 cM, respectively (Fukuoka and Okuno 2001). Further analysis using expressed sequence tag (EST) clones developed by the RGP (Rice Genome Research Program, Japan) identified RFLP marker locus *RA3591* between RFLP marker locus *G271* and *pi21* at a distance of 0.6 cM to *pi21*. To develop DNA markers around the *pi21* locus, P1 artificial chromosome (PAC) clones were screened from a *Nipponbare* PAC library constructed by the RGP using the sequence-tagged-site (STS) for *RA3591* and *C975*, which is proximal to *G317*. A polymorphism survey of twelve PAC-end fragments or subclones by RFLP and SSCP (single strand conformation polymorphism) analysis identified eleven DNA markers around the *pi21* locus. We developed a mapping population consisting of 1014 lines for fine genetic mapping of *pi21* and selected 182 recombinants between RFLP marker locus *RA3591* and *G317* for a field resistance test to determine genotypes at the *pi21* locus. Linkage analysis revealed that *pi21* is located between the two RFLP marker loci *14T1* and *13S1*. The number of recombinants between these markers and the *pi21* locus were two and one, respectively.

5. CONCLUSION AND FUTURE WORK

Genetic analysis using DNA markers identified three chromosomal regions, two on chromosome 4, one on chromosome 12, that are responsible for field resistance. The analysis using backcrossed lines suggested genes of upland rice included in these regions improve the level of field resistance of an irrigated rice variety. These regions partly overlapped with those for complete or partial resistance previously reported (McCouch et al. 1994). The relationship of alleles for blast resistance in different resistant varieties is relevant for understanding the genetic control of field resistance in rice.

Upland rice in Japan has maintained long-term durability of resistance to blast; however, there is still little information on race-specific reaction to blast. Backcrossed inbred lines or near-isogenic lines are useful to study race specificity of a single QTL (gene) for field resistance underlying resistance to a spectrum of blast races. Fine genetic analysis suggested that a major QTL consisted of a single gene designated as *pi21*. We are close to identifying *pi21* by map-based cloning. Comparing sequences of the region containing the *pi21* locus between the original upland rice variety *Owarihatamochi* and irrigated rice varieties *Nipponbare* or *Aichiasahi* will uncover candidate genes for field resistance. Studies on gene structure and its expression will help us to understand the mechanism of field resistance to blast.

ACKNOWLEDGEMENTS

The authors are gratefully acknowledge to the researchers in Aichi Prefecture Agricultural Research Center (AARC) for kind assistance in the field test.

REFERENCES

1. Ezuka A (1972) Field resistance of rice varieties to rice blast disease. Rev Plant Prot Res 5: 1-21
2. Fukuoka S and K Okuno. 2001. QTL analysis and mapping of *pi21*, a recessive gene for field resistance to rice blast in Japanese upland rice. Theor Appl Genet. 103: 185-190
3. Higashi T, Saito S. 1985. Linkage group of field resistance genes upland rice variety Sensho to leaf blast caused by *Pyricularia oryzae* CAV. Japan J Breed 35 : 438-448
4. Jia Y, McAdams SA, Bryan GT, Hershey HP, Valent B. 2000. Direct interaction of resistance gene and avirulence gene products confers rice blast resistance. EMBO J 19: 4004-4014
5. McCouch SR, Nelson RJ, Tohme J, Zeigler RS (1994) Mapping of blast resistance gene in rice. In Rice Blast Disease (edt by Zeigler RS, Leong SA, Teng PS) pp167-186, IRRI, Manila
6. Tanksley SD 1993 Mapping polygenes. Annual Rev Genet 27 : 205-233
7. Yano M and Sasaki T. 1997. Genetic and molecular dissection of quantitative traits in rice. Plant Mol Biol 35: 145-153
8. Wang ZX, Yano M, Yamanouchi U, Iwamoto M, Monna L, Hayasaka H, Katayose Y and Sasaki T. 1999. The *Pib* gene for rice blast resistance belongs to the nucleotide binding and leucine-rich repeat class of plant disease resistance genes. Plant J 19: 55-64

STUDIES ON PARTIAL RESISTANCE TO RICE BLAST IN THE TROPICS

*H. Kato[1,2], H. Tsunematsu[1,3], L.A.Ebron[1], M.J.T.Yanoria[1], D.M.Mercado[1] and S.Khush[1]

[1]*Plant Breeding, Genetics and Biochemistry Division, International Rice Research Institute, MCPO Box 3127, 1271 Makati City, Philippines;*
[2]*Present address; Miyazaki Agricultural Experiment Station, Shimonaka5851, Sadowara, Miyazaki 880-0212 Japan;*
[3]*Present address; Japan International Research Center, Ohwashi 1-2, Tsukuba, 305-8686 Japan*

**E-mail: hkato@pref.miyazaki.jp*

Abstract: Partial resistance to blast has not been clearly distinguished from major gene resistance in the tropics. Several major genes for resistance to blast were identified recently in IR varieties at the International Rice Research Institute (IRRI) and the reactions of these genes to the Philippine blast isolates were studied. Consequently, suitable blast isolates could be selected to eliminate the effect of these major genes for evaluating partial resistance to blast under field conditions. Seventy-two varieties and breeding lines were inoculated with three blast isolates during two seasons. Serious blast infection was induced by partial shading, continuous irrigation and the application of large amount of fertilizer. Disease damage was evaluated using the Standard Evaluation System (SES) for rice blast. Partial resistance was clearly distinguished from major gene resistance and the differences between moderate levels of resistance conferred by major genes and partial resistance were clarified. Among the materials used, IR64 showed high level of partial resistance. Partial resistance was moderate in IR60 and IR36 and low in IR50 and CO39. These results were consistent with the results of sequential planting done at IRRI earlier. Partial resistance levels of several Japanese varieties at IRRI were also consistent with their evaluations in Japan. Significant positive correlations observed among partial resistance to the three isolates indicate that partial resistance is horizontal in the tropics.

S. Kawasaki (ed.), Rice Blast: Interaction with Rice and Control, 137–143.
© 2004 *Kluwer Academic Publishers. Printed in the Netherlands.*

Distinction between major gene and partial resistance to blast is necessary for identifying donors for developing durably resistant varieties.

1. INTRODUCTION

Rice blast, caused by *Pyricularia oryzae* Cav., is the most important and potentially damaging disease of rice, *Oryza sativa* L. Breeding rice varieties with durable resistance to this disease is the most logical and environment-friendly approach. Major gene and partial resistance to diseases is known in rice and other crops. In major gene resistance there is incompatibility between the resistance gene in the host and the avirulence gene in the pathogen. Partial resistance reduces the extent of pathogen reproduction within the context of a compatible interaction.

Partial resistance to blast in the tropics has not been well studied for two reasons. First, most of the studies concerning partial resistance were carried out in Japan and published in Japanese journals (Yunoki *et al.* 1970, Asaga 1981, Naito and Yaegashi 1997). Second, partial resistance of each variety can be evaluated only when the type and number of major genes for resistance are known. Until recently, information on the presence of major genes in tropical varieties was lacking. For example, the number of major resistance genes in IR36 was not known. Besides interaction of natural or purified blast isolates to major genes for resistance (pathogenicity) was not known in the tropics until recently.

2. IDETIFICATION OF MAJOR GENES IN IR VARIETIES

Imbe *et al.* (2000) and Leodegario *et al.* recently identified major genes for resistance in IR varieties. At the same time, the pathogenicities of Philippine blast isolates were determined against known major resistance genes (Imbe *et al.* 2000, Tsunematsu *et al.* 2000). These important breakthroughs enabled us to carry out our study successfully. By selecting suitable blast isolates, we could eliminate the effect of major resistance genes and evaluate the partial resistance of rice varieties. International differential lines with single resistance genes for rice blast were also developed (Tsunematsu *et al.* 2000). We could monitor the pathogenicity of blast isolates in the field by using these lines.

3. EVALUATION METHODS OF PARTIAL RESISTANCE

Two experiments were conducted at IRRI, Philippines. The first experiment was conducted from January to March 1999, when the average temperatures were relatively low. The second experiment was conducted from October to December 1999. Serious blast infection was induced by partial shading, continuous irrigation, the installation of plastic fence and the application of large amount of fertilizer. Disease intensity was evaluated using the Standard Evaluation System for rice (SES), (IRRI, 1996).

Seventy varieties and breeding lines were used, 34 of which were IRRI or Philippine Seed Board-released varieties. Twenty-four were differential lines with single genes for blast resistance (Tsunematsu *et al.* 2000). By using these lines, we could confirm the pathogenicity of the inoculated blast isolates in the plots. Each plot was inoculated with one of the following blast isolates that were PO6-6, BN111 and M36-1-3-10-1 (M36).

4. EVALUATION OF PARTIAL RESISTANCE

The first lesions were observed 4 to 5 days after inoculation. The clearest differences in blast damage between susceptible check IR50 and the test materials were observed 4 to 5 weeks after inoculation. The cycles of blast sporulation were considered to be 3rd to 6th generations at that time. Resistance level was calculated as the difference between the blast scores of each variety and the adjacent IR50 check. The distribution of resistance to isolate PO6-6 in the first experiment is shown in Figure 1. The lowest score was -1.5 (CO39) and the highest was 6.4 (PSBRc 10). We sorted out partial resistance from major gene resistance. According to our results, all the entries distributed on the right hand side of the arrow had incompatible major resistance genes for PO6-6. There were no known blast-incompatible major resistance genes in all the other varieties and lines on the left hand side of the arrow. Moderate levels of resistance governed by major genes showed a higher resistance level than partial resistance, even though limited level of overlapping was observed among other isolates in our experiments.

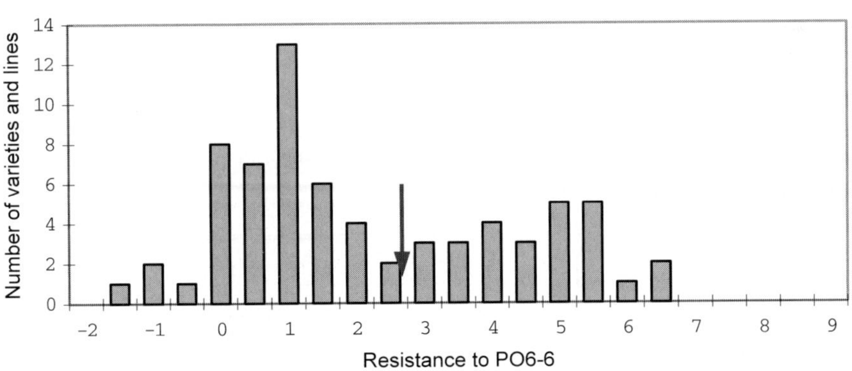

Figure 1. Distribution of resistance to rice blast isolate PO6-6, February 1999.

Bonman and Bandong (1989) studied partial resistance in the tropics. They equated moderate resistant reactions with partial resistance. However, moderate resistant reactions are not always equivalent to partial resistance (Imbe *et al.* 2000). Wang *et al.* (1994) also tried to identify partial resistance in the tropics. They used Moroberekan and CO 39 recombinant inbred lines. Unfortunately, they could not obtain the virulent blast isolates against Moroberakan, a line that is supposed to possess several major resistance genes. Moreover, they used blast isolate PO6-6 to distinguish between plants with major genes for resistance (those with no disease symptoms) and partial resistance (those with moderate resistant reactions). However, the isolate they used showed moderate resistant reactions on varieties with *Pii, Pi3, Piz, Piz-5*, and *Pish* (Imbe *at al.* 2000, Tsunematsu *et al.* 2000). These moderate resistant reactions are due to major genes for resistance and the reactions are specific between PO6-6 and the major genes. As a result, they could not clearly distinguish partial resistance from major gene resistance. Our study is the first report to clearly distinguish partial resistance from major gene resistance in the tropics.

5. HORISONTALNESS OF PARTIAL RESISTANCE

Partial resistance, also known as horizontal resistance, is expressed against all compatible blast isolates (Yunoki *et al.* 1970, Asaga 1981). A positive and highly significant relationship was observed between partial resistance to blast isolates PO6-6 and BN111 in February 1999 experiment (r=0.780**). For this calculation, we selected 32 varieties or lines that are compatible to both isolates. Among the rest of the varieties and lines that have major genes for resistance, there was no significant correlation between the resistance to PO6-6 and BN111 (r=0.277). This is because major genes confer vertical resistance.

Correlation coefficients among partial resistance to different blast isolates over two seasons are shown in Table 1. For this calculation, we

Table 1. *Correlation coefficients among partial resistance to different blast isolates over two seasons*

		Feb, 1999	December, 1999	
		BN111	PO6-6	M36
Feb.	PO6-6	0.79**	0.68**	0.43*
Feb.	BN111		0.69**	0.66**
Dec.	PO6-6			0.78**

selected 24 varieties or lines that are compatible to three isolates. Among the three blast isolates over two seasons, we always observed significant positive correlations. In this study, partial resistance was horizontal as elaborated by Asaga (1981) in Japan. Our results are different from those of Ou (1979) who reported that partial resistance of Japanese varieties was vertical. His results were obtained without the knowledge of the existence of the major gene, *Pish* in Japanese varieties. These results strongly indicate that partial resistance is horizontal also in the tropics.

6. PARTIAL RESISTANCE OF IR VARIETIES

Table 2. *Partial resistance to rice blast in the Philippines*

Order	Cultivar	Scale	Order	Cultivar	Scale
1	IR26	1.82	26	IRBLb-B	0.6
2	IR43	1.69	27	IR29	0.57
3	K59	1.57	28	IR34	0.54
4	IRBLta-K1	1.55	29	IRBLks-F5	0.5
5	IR8	1.45	30	IR62	0.47
6	IR5	1.45	31	IR60	0.46
7	IR64	1.25	32	LTH	0.45
8	IRBLkp-K60	1.25	33	IR72	0.35
9	IRBLkh-K3	1.13	34	IR44	0.3
10	Yashiromochi	1.11	35	IR36	0.29
11	IR24	1.07	36	IR22	0.25
12	Reiho	1.06	37	IR49830-7-1-2-2	0.25
13	IRBL1-CL	0.97	38	IR32	0.16
14	IRBLt-K59	0.94	39	IR38	0.16
15	IR46	0.93	40	IRBL7-M	0.08
16	IR48	0.88	41	IR20	-0.03
17	PSBRc 4	0.86	42	IRBLzt-T	-0.06
18	IRBLa-A	0.76	43	Aichi Asahi	-0.07
19	IRBLk-ka	0.75	44	IR30	-0.2
20	IR58	0.74	45	IRBLa-C	-0.26
21	IR42	0.74	46	IR50	-0.32
22	IRBLks-S	0.71	47	IRBL12-M	-0.43
23	IR52	0.68	48	US-2	-0.53
24	IRBL19-A	0.62	49	IRBLta-CT2	-0.65
25	IR40	0.61	50	CO39	-1.09
				LSD	1.06

There were significant positive correlations between partial resistance for different seasons and isolates, but there were large differences among their averages. By using regression equations, we transformed the partial resistance to M36, BN111 and PO6-6 (December 1999 experiment) to the equivalent of the partial resistance to PO6-6 in the February 1999 experiment (Table 2).

Durable blast resistance was studied at IRRI in sequential planting (IRRI, 1994). Even though information about blast population dynamics was lacking, IR64 was reported to have lower disease severity in all 15 sequential crop cycles. Intermediate levels of infection on IR36 and IR60 were maintained throughout the whole duration of planting. CO 39 and IR50 had higher disease severity. These observations were quite consistent with the results of this study. Among the 50 varieties and lines studied, IR64 ranked 7th in level of partial resistance, IR60 and IR36 ranked 31st and 35th, respectively. IR50 and CO 39 ranked 46th and were the lowest in level. Aichi Asahi is known to have very low partial resistance in Japan and so is US-2. Yashiromochi is also ranked "low" and Reiho is ranked "slightly low" in partial resistance. The ranking of Yashiromochi and Reiho for partial resistance in Japan is not completely consistent with the results of this study. However, the difference is small. We conclude that the results of the present study are quite consistent with the partial resistance evaluations done in Japan.

7. BREEDING STRATEGY FOR DURABLE RICE VARIETIES AGAINST RICE BLAST

The highest level of SES score of partial resistance observed in this study was 4.56 of K59 against M36. Almost all the varieties with blast compatible reactions died in the first experiment, even though they showed the highest level of partial resistance. On the other hand, major gene resistant plants remained with almost no blast damage. However, under field conditions, even low level of partial resistance is expected to be more effective than in this study. In a commercial field when a partially resistant variety is cultivated, the density of blast spores is much less than in this experiment because there is no surrounding susceptible variety such as IR50 as in this experiment. Newly introduced major genes for resistance were easily broken down within several years of intensive cultivation in blast prone areas in Japan (Fujita 1997) and Korea (Chang 1994). Partial resistance is considered to be effective and indispensable in reducing the use of fungicides for blast control in Japan. For evaluating partial resistance of each variety or breeding line, major gene resistance is often considered as an obstacle.

In order to breed varieties with durable resistance to rice blast, it is necessary to identify donors with major genes and partial resistance. First, the pathogenicity of isolates must be determined by using a set of differential lines that have only one gene for resistance (Tsunematsu *et al.* 2000). Second, the major genes for resistance of each variety are identified using these isolates. Finally, by using the information about major genes of each variety, partial resistance is measured. Our study has shown that this can be done. It is possible that unknown blast avirulent genes may exist in other countries. Therefore studies on the pathogenicity of blast isolates and the development of a set of differential lines for each country are necessary.

REFERENCES

Asaga, K. (1981) A procedure for evaluating field resistance to blast in rice varieties. J. Cent. Agric. Exp. Stn. 35,51-138.

Bonman, JM. & Bandong, JM. (1989) Race-specific partial resistance to blast in temperate Japonica rice cultivars. Plant Dis. 73,496-499.

Chang, KK. (1994) Blast management in high input, high yield potential, temperature rice ecosystem. In: R.S. Zeigler *et al.* (eds) Rice blast disease. International Rice Research Institute, P.O. Box 933, 1099 Manila, Philippines. pp 451-463.

Fujita, Y. (1997) Rice blast: research and control. Nihon Bayer Agrochem K.K. Takawa 4-10-8 Minatoku, Tokyo, Japan 108. pp 36-40

Imbe, T., Tsunematsu, H., Kato,H., & Khush GS. (2000) Genetic analysis of blast resistance in IR varieties and resistant breeding strategy. In: D.Tharreau *et al.* (eds.) Advances in rice blast research. Kluwer Academic Publishers. Printed in the Netherlands. pp 1-8.

International Rice Research Institute (1996) Standard evaluation system for rice. 4th ed. IRRI. MCPO Box3127, 1271 Makati City, Philippines

Leodegario, AE., Fukuta, Y., Imbe, T., Kato, H., Yanoria, JMT., Tsunematsu, H., Khush, GS. & Yokoo, M. Identification of blast resistance genes in elite Indica varieties of rice (*Oryza sativa* L.) in preparation

Naito, H., & Yaegashi, H. (1997) Rice blast: Research and control. Nihon Bayer Agrochem K.K. Takawa 4-10-8 Minatoku, Tokyo, Japan 108.

Ou, SH. (1979) Breeding rice for resistance to blast – a critical review. In: Proceedings of the rice blast workshop. International Rice Research Institute (IRRI), Los Baños, Laguna, Philippines. pp 81-137.

Tsunematsu, H., Yanoria, MJT., Ebron, LA., Hayashi, N., Ando, I., Kato, H., Imbe, T., & Khush, GS. (2000) Development of monogenic lines of rice for blast resistance. Breed. Sci. 50,229-234

Yunoki, T., Ezuka, A., Morinaka, T., Sakurai, Y., Shinoda, H. & Toriyama, K. (1970) Studies on the varietal resistance to rice blast. 4. Variation of field resistance due to fungus strains. Bulletin of the Chugoku Agricultural Experimental Station. E. No.6, 21-41

Wang, GL., Mackill, DJ., Bonman, JM., McCouch, SR., Champoux, MC. & Nelson, RJ. (1994) RFLP mapping of genes conferring complete and partial resistance to blast in a durably resistant rice cultivar. Genetics 136,1421-1434.

MICROARRAY ANALYSIS OF GENE EXPRESSION IN RICE TREATED WITH PROBENAZOLE, A RESISTANCE INDUCER, IN SPECIAL REFERENCE TO BLAST DISEASE

*M. Nishiguchi[1,4], M. Shimono[1], Y. Eguchi[1], H. Okuizumi[1], J. Yazaki[1], K. Nakamura[2], F. Fujii[1], K. Shimbo[2], Z. Shimatani[2], Y. Nagata[2], A. Hashimoto[2], T. Ohta[1], Y. Sato[1], S. Honda[1], M. Iwano[3], K. Yamamoto[2], K. Sakata[1], T. Sasaki[1], N. Kishimoto[1] and S. Kikuchi[1]

[1]*National Institute of Agrobiological Sciences, 2-1-2 Kan-nondai, Tsukuba, Ibaraki 305-8602, Japan*

[2] *Society for Techno-Innovation of Agriculture, Forestry and Fisheries (STAFF), 446-1 Ippaizuka, Kamiyokoba, Tsukuba, Ibaraki 305-0854, Japan*

[3]*National Agricultural Research Center, 3-1-1 Kan-nondai, Tsukuba, Ibaraki 305-8666, Japan*

[4]*Present Address: Faculty of Agriculture, Ehime University, Matsuyama, Ehime 790-8566, Japan*

**E-mail: mnishigu@agr.ehime-u.ac.jp*

Abstract. Probenazole (3-allyloxy-1,2-benzisothiazole-1,1-dioxide, PBZ) is a chemical inducer of resistance against plant pathogens and has long been practically used for protecting rice plants against a blast fungus, *Magnaporthe grisea*. We have been investigating the expression profiles of rice genes in response to PBZ using cDNA microarray technique. An array composed of 1265 rice expressed sequenced tag (EST) clones from

S. Kawasaki (ed.), Rice Blast: Interaction with Rice and Control, 145–154.
© 2004 *Kluwer Academic Publishers. Printed in the Netherlands.*

various source including shoots, calli etc. was targeted by Cy5 labeled mRNAs from PBZ treated and untreated rice plants, respectively. As a result, ten cDNA clones exhibited greater than three-fold over-expression in the PBZ treated plants compared to untreated controls. Some of these clones have identities with disease resistance, pathogenesis-related and cell maintenance/development–related genes. The enhanced gene expression was confirmed for nine out of ten clones by using Northern blot and reverse transcription-polymerase chain reaction (RT-PCR) analyses. Several clones were further tested for Northern blot analysis using total RNA from blast fungus infected rice. Under the conditions used, three out of seven clones showed enhanced level of mRNA accumulation in infected plants. This may suggest that the enhanced expression of some genes by PBZ may contribute to the disease resistance in host plants. Furthermore cDNA library was constructed using mRNAs from leaves of PBZ treated rice plants and a microarray of the randomly selected 2880 clones from the library was used for choosing clones whose expression was changed in response to PBZ. Thirty nine clones were found to change their expression levels by a factor of 3 folds after PBZ treatment. Characterization including sequencing is now underway for these clones.

1. INTRODUCTION

Plants defend themselves against pathogens' attack through various resistance means. There are some chemicals known to induce resistance in plants to pathogens. One example is probenazole (3-allyloxy-1,2-benzisothiazole-1,1-dioxide, PBZ)[15) , which has been used as a protectant in Japan as well as in other countries. PBZ induces disease resistance by a mechanism thought to be similar to systemic acquired resistance (SAR) in rice (*Oryza sativa* L.)[17)]. This chemical induced resistance is very effective in rice against pathogens such as the rice blast fungus (*Magnaporthe grisea*)[15)] and *Xanthomonas oryzae* pv. *oryzae*, the bacterial leaf blight pathogen[1)]. PBZ is also reported to induce resistance to tobacco mosaic virus (TMV) in tobacco[3)]. It is reported that PBZ is a chemical activator of SAR in *Arabidopsis* plants with the increased accumulation levels of mRNAs for pathogenesis-related (PR) proteins and the salicylic acid levels[17)]. Several genes from rice were induced after the treatment with PBZ[76,8)].

In this study, using a cDNA microarray technology gene expression profile changes were monitored in PBZ-treated rice. The arrayed cDNAs were hybridized with Cy5-labeled complex probes prepared with mRNAs from PBZ-treated and control plants. Some of them were found to change in their expression after PBZ-treatment. Furthermore these genes were analyzed by Northern blot and RT-PCR analyses to confirm their expression profiles and also investigated the relationship of gene expression profiles between PBZ treatment and rice blast fungal infection.

2. MATERIALS AND METHODS

2.1. Plant materials and PBZ-treatments

Oryza sativa L. cvs. Nipponbare and Kantoh BL6 was grown in a commercial soil mixture (Bonsol® Sumitomo Chemical, Osaka, Japan) with tap water in a growth chamber under a 16: 8 h light: dark regimen at 25°C. Twelve day-old rice seedlings were further submerged in 1.0 gl^1 Oryzemate® (24% granules of PBZ, Meiji Seika Kaisha,Tokyo, Japan). Seven days later the leaves were harvested for total RNA preparation.

2.2. Construction of cDNA library

Total RNA was extracted from rice leaves of PBZ treated rice plants as described by Logemann et al.[5)] and used further for mRNA preparation. mRNA was prepared using Oligo-dT column by the method of Kingston[2)]. Library construction of cDNA clones was carried out using ZAP-cDNA synthesis kit (Stratagene) according to the manufacturer's protocol.

2.3. Production of cDNA microarrays

Expressed sequence tags [ESTs] of 1265 were obtained from Rice Genome Research Program (RGP, Japan). The identity and accession number of the 1265 randomly-selected rice EST clones are listed at the http://microarray.rice.dna.affrc.go.jp/ website. Inserted cDNA clones in pBluescript II vectors were amplified by polymerase chain reaction (PCR) in 50 μl reaction mixture using M13 primer M4 (5'-GTTTTCCCAGTCACGAC-3') and M13 primer RV (5'-CAGGAAACAGCTATGAC-3'). The PCR products were purified by QIAquick 96 PCR Purification Kit (QIAGEN). The 1265 rice cDNAs were spotted in duplicate on aluminum-coated and dimethysulfoxide (DMSO) optimized glass slide using an Array Spotter Generation III (Amersham Pharmacia Biotech)[16).]

2.4. RNA extraction and probe synthesis

Total RNA was extracted from PBZ treated and control rice leaves (approximately 2.5 g) as described by Logemann *et al.*[6)]. Using Oligotex™-dT30 Super (Takara) $poly(A)^+$ -enriched RNA was prepared from the total RNA (250 μg) according to the supplier's protocol and used for labeling with Cy5-dCTP (Amersham Pharmacia Biotech) as described previously (Shimono et al., 2003). Before hybridization, the probes were purified by QIAquick PCR purification kit (QIAGEN).

2.5. Hybridization, scanning and data analysis

Hybridization reactions were carried out as described previously [16]. Microarrays of 1226 and 2880 clones were scanned using Generation III Array Scanner (Amersham Pharmacia Biotech) and FLA-8000 (Fuji Photo Film), respectively. The image data were quantified and analyzed with commercial software, Array Vision (Imaging Research) and Spotfire (Spotfire) for 1265 and ArrayGauge (Fuji Photo Film) for 2880 arrays, respectively.

2.6. Northern blot analysis

Northern blot analyses were carried out using the total RNA (15 μg) as described previuously[13]. The cloned DNA fragment was amplified and digested with *Sal* I and *Not* I then extracted using QIAEX II Gel Extraction Kit (QIAGEN). The DNA fragment was labeled with α-^{32}P-dCTP and used as a probe as described previously[13]. The transferred RNA on nylon membrane (Hybond-N+, Amersham Pharmacia Biotech) was hybridized with α-^{32}P-labeled DNA and washed as described previously[13]. The signals were detected and quantified with a BAS-2000 imaging analyzer (Fuji Photo Film).

2.7. RT-PCR analysis

The RT-PCR was carried out using the total RNA (0.5 μg) from PBZ treated and control plants as described previously[11].

2.8. Pathogen inoculation

Conidia of blast fungus (*Magnaporthe grisea* Cavara, race 001) were prepared as described by Watanabe *et al.* (1977)[15]. This race shows compatible response with Nipponbare but incompatible with an isogenic line of Nipponbare, Kantoh BL6 carrying *Pi-b, a resistance gene to blast disease*. The suspension of spores was sprayed onto the 3 week-old rice seedlings of both varieties. The leaves (approximately 1.0g) were harvested appropriately for total RNA extraction and used for Northern blot analysis.

3. RESULTS

3.1. Analysis using1265 microarray

From rice EST clones from tissues of various types and treatments, 1265 clones were selected for microarray. This microarray was used for hybridization with Cy5 labeled cDNA from PBZ-treated or control leaves.

3.1.1. Hybridization signals

To monitor the gene expression profile changes in the PBZ treatment, cDNA arrays were hybridized independently with Cy5-labeled cDNA probe, which was derived from PBZ-treated, or control plants. Fig. 1 shows the scatter plots for a PBZ/control hybridization signals. From these results, 10 clones were selected which exhibited differential expression profiles between the two treatments (Figs. 1 and 2). They exhibited greater than 3-fold over expression in the PBZ-treated probe compared to the control probe. No clones were exhibited greater than 3-fold over expression in probe generated from untreated plants compared to PBZ-treated probe. The putative identities of some PBZ-responding genes include PR, disease resistance[12] and cell maintenance/development genes as reported by Schenk et al. (2000)[10](Fig. 2).

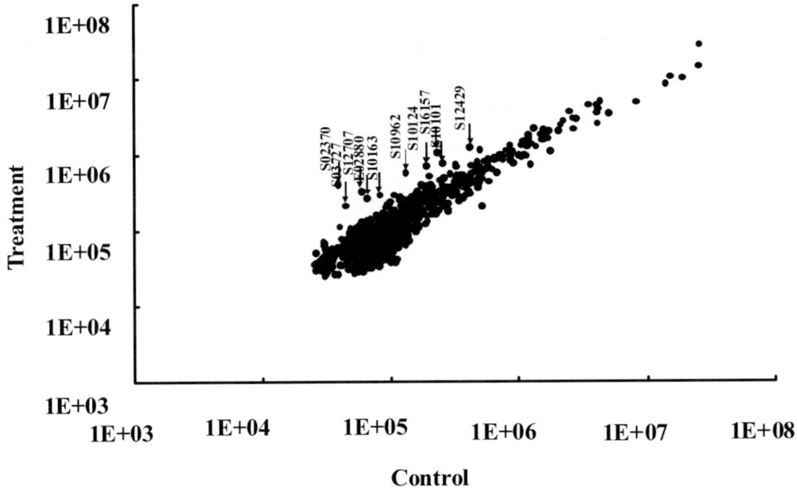

Fig. 1 *Scattered plot analysis of 1265 micoarray hybridization*[11]*. Microarray analysis was performed with the mRNAs (1.0µg) from PBZ treated and control plants. Comparison between the average normalized hybridization signals obtained with probes from PBZ treatment and control plants. Arrows show the clones with a 3-fold or higher variation in hybridization signal for the two probes. Data reprinted with permission.*

3.1.2Northern blot and RT-PCR analyses

To further confirm the results, the expression patterns of the 10 clones were further analyzed by Northern blot analysis. The increased expression patterns were shown for seven clones, S12707, S03727, S16157, E02880, S10124, S10163 and S12429 (Fig. 2). However, it showed the ratio of 1.37 (treatment/control) in S10101 by Northern blot analysis compared to 3.03 in microarray analysis (Fig. 2).

name	C T	Ratio		Putative identity
		Northern blot	DNA array	
S12707 (D47369)*	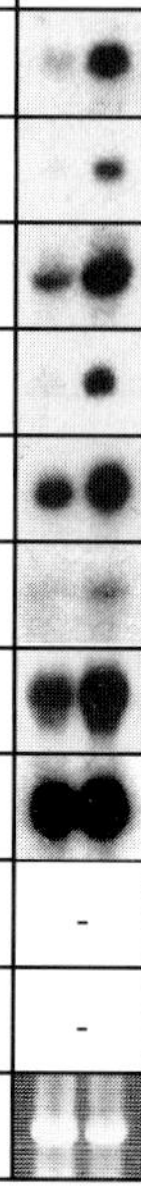	4.88	5.63	Methylenetetrahydrofolate reductase (AF174486)* :9e-67**
S03727 (D41315)*		3.66	4.89	β-1,3-glucanase (AB027429)* :2e-56**
S16157 (AU065955)*		3.27	4.64	*S*-adenosylmethionine synthetase (AJ296743)* :4e-39**
E02880 (AU085932)*		3.53	4.14	Caffic acid 3-*O*-methyltransferase (Q06509)* :4e-50**
S10124 (AU070307)*		2.45	3.74	Ribosomal protein (P45636)* :2e-13**
S10163 (AU083535)*		2.14	3.64	Unknown
S12429 (D47219)*		2.03	3.01	Receptor kinase-like protein (U72724)* :1e-30**
S10101 (AU083517)*		1.37	3.03	ADP-ribosylation factor 1 (AF005238)* :4e-12**
S02370 (AU058100)*	-	-	10.41	Unknown
S10962 (D46347)*	-	-	4.49	Unknown
rRNA		-	-	-

Fig. 2 *List of differentially expressed genes in probenazole treatment by microarray and Northern blot analyses[11]. Highest ten clones in signal ratio of PBZ-treatment to control are listed. The putative identity is indicated from database search as a closely related but different sequence. * and ** represent in DDBJ accession number and E value, respectively. Data reprinted with permission.*

Hybridization signals were not detected for S10962 and S02370 by Northern blot analyses (data not shown). Then RT-PCR analysis was further carried out for these two clones to confirm the microarray data. As a result they showed the increased pattern of gene expression supporting the results using the microarray. Thus the microarray data were verified in 9 out of 10 clones by Northern blot and RT-PCR analyses.

3.2. Analysis using microarray of 2880 cDNAclones from PBZ treated leaves

In order to obtain more cDNA clones whose expression patterns are changed after PBZ treatment, cDNA library was constructed from leaves treated with PBZ. The resultant cDNA library contained $4.8x10^4$ pfu/μl. Total number of 2880 clones was selected for cDNA microarray. These clones contained cDNA insert of approximately 0.4-0.8 kbp (Shimono et al., in preparation).

3.2.1. Hybridization signals

Scatter plots between a PBZ/control samples are shown in Fig.3. Total number of 39 cDNA clones exhibited greater than three-fold overexpression. Only 3 clones showed the decreased level of gene expression by 3 folds. So far the sequences of the inserts were completed for 16 of them. As shown in Table 1, they were grouped into 9 clusters of genes. Homology search showed that some genes have putative identities with probenazole-inducible gene, *PBZ1*, and phenylalanine ammonia-lyase (PAL) gene. *PBZ1* has been reported as one of intercellular PR protein gene family based on the homology[6]. *PAL* was also isolated from rice and its expression was not increased by PBZ treatment[7]. Thus it is suggested that the clone of *PAL* in Table is one of the gene family, whose expression patterns vary in PBZ response. This is also the case of ß-glucanase in Fis.2. All other clones have no significant putative identities with any genes of known functions.

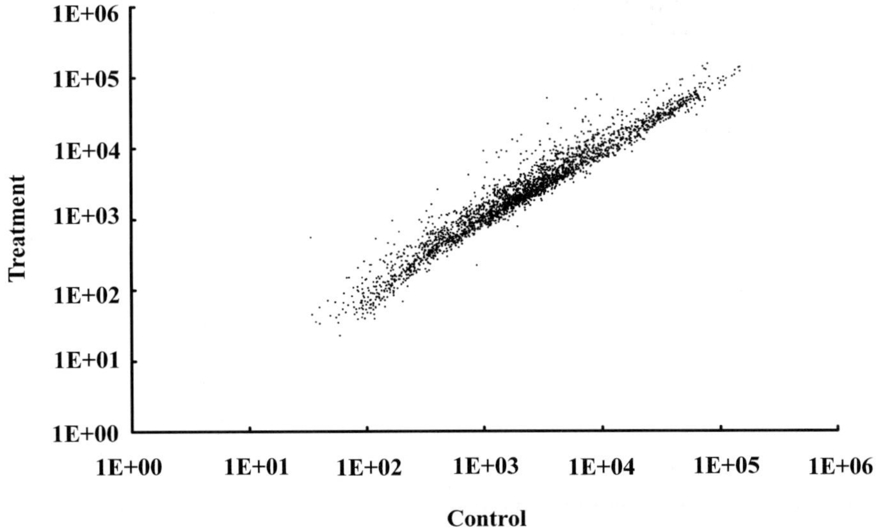

Fig. 3 *Scatter plots for a PBZ-treatment/control by hybridization signals of 2880 clones derived from leaves of PBZ-treated rice (Shimono et al., in preparation).*

3.2.2. Northern blot analysis

Northern blot analysis was carried out for 8 clones out of 39 clones. In all of them the gene expression profiles were more or less similar to those in the microarray analysis (Shimono et al., in preparation). Further analyses were in progress for the other clones.

Table 1. *List of differentially expressed genes in probenazole treatment using 2880 array**

clone no.	Putative identity	Treatment / Control
1	Rice mRNA for probenazole-inducible protein PBZ 1	4.8
26	same as above	5.7
4	Rice gene for phenylalanine ammonia-lyase (PAL)	3.9
6	Unknown	3.6
9	Unknown	9.0
10	same as above	6.5
17	same as above	5.8
18	same as above	5.5
21	same as above	3.9
23	same as above	3.9
12	Unknown	3.1
13	Unknown	3.3
20	same as above	3.2
25	Unknown	6.4
31	Unknown	16.1
33	Unknown	3.7

**Array is composed of 2880 clones derived from leaves of PBZ-treated rice..*
Consecutive "same as above" means a cluster of clones

4. DISCUSSION

4.1. Effect of pathogen infection on expression of PBZ –responding genes

To shed a light on the relationship of PBZ-responding genes between PBZ treatment and pathogen infection, Northern blot analysis was performed using the total RNA from the leaves which had been inoculated with the rice blast fungus in both compatible and incompatible combinations[11). Three (S03727, S10124 and S12429) out of seven clones confirmed by Northern blot analysis to be induced in gene expression in Fig. 2 showed also the similar patterns of gene expression while the rest 4 clones did not (Fig. 4 in Shimono et al. (2003)). Two clones, S03727 and S10124, showed a different expression pattern between compatible and incompatible combinations. In clone S03727, the gene expression level increased only 1 day after inoculation in incompatible combination whereas it increased and reached at a maximum after 1 day and then gradually decreased up to 7 days in compatible combination. In clone S10124, a slight up-regulation was observed 1 to 7 days after inoculation in compatible combination while the clear enhancement of the expression was detected 7 days after inoculation in incompatible combination. In clone S12429, no difference of the mRNA accumulations was observed between compatible and incompatible combinations. It is often reported that the expression of genes induced by pathogen is more rapid in incompatible than in compatible combination[4,6]. However, there is a report that the expression of some gene was induced more rapidly in compatible than in incompatible combinations[9]. Thus the induction pattern of gene expression varies widely depending upon individual genes in both combinations.

ACKNOWLEDGEMENTS

The authors are grateful to Drs. M. Yano and T. Inbe for the seeds of Nipponbare and Kantoh BL6, respectively. This work was supported by a grant from the Ministry of Agriculture, Forestry and Fisheries of Japan (Rice Genome Project MA-2113).

REFERENCES

1. Iwata, M., Suzuki, Y., Watanabe, T., Mase, S. and Sekizawa, Y. (1980). Effect of probenasole on the activity of enzymes related to the resistant reaction in rice plant. Ann. Phytopathol. Soc. Japan 46: 297-306.
2. Kingston, R. E. (1993) . Preparation of poly(A^+) RNA. *In* Current Protocols in Molecular Biology (eds. By F. M. Ausubel et al.), John Wiley & Sons, Inc., New York, 4·5·1-4·5·3.
3. Koganezawa, H., Sato, T. and Sasaya, T. (1998). Effects of probenazole and saccharin on symptom appearance of tobacco mosaic virus in tobacco. Ann. Phytopathol. Soc. Jpn. 64: 80-84.
4. Lee, S.C., Kim, Y.J. and Hwang, B.K. (2001). A pathogen-induced chitin-binding protein gene from pepper: its isolation and differential expression in pepper tissues treated with pathogens, ethephon, methyl jasmonate or wounding. Plant Cell Physiol. 42: 1321-1330.
5. Logemann, J., Schell, J. and Willmitzer, L. (1987). Improved method for the isolation of RNA from plant tissues. Anal. Biochem. 163: 16-20.
6. Midoh, N. and Iwata, M. (1996). Cloning and characterization of a probenazole-inducible gene for an intracellular pathogenesis-related protein in rice. Plant Cell Physiol. 37: 9-18.
7. Midoh, N. and Iwata, M. (1997). Expression of defense-related genes by probenazole or 1,2-benzisothiazole-3(2H)-one 1,1-dioxide. J. Pesticide Sci. 22: 45-47.
8. Sakamoto, K., Tada, Y., Yokozeki, Y., Akagi, H., Hayashi, N., Fujimura T. and Ichikawa, N. (1999). Chemical induction of resistance in rice is correlated with the expression of a gene encoding a nucleotide binding site and leucine-rich repeats. Plant Mol. Biol. 40: 847-855.
9. Sawada, K. and Iwata, M. (2002). Isolation of blast fungal cerebroside elicitor-responsive genes in rice plants. J. Gen. Plant Pathol. 68: 128-133.

10. Schenk, P.M., Kazan, K., Wilson, I., Anderson, J.P., Richmond, T., Somerville, S.C. and Manners, J.M. (2000). Coordinated plant defense responses in *Arabidopsis* revealed by microarray analysis. Proc. Natl. Acad. Sci. USA 97: 11655-11660.
11. Shimono, M., Yazaki, J., Nakamura, K., Kishimoto, N., Kikuchi, S., Iwano, M., Yamamoto, K. Sakata, K., Sasaki, T. and Nishiguchi, M. (2003). cDNA microarray analysis of gene expression in rice plants treated with probenazole, a chemical inducer of disease resistance. J. Gen. Plant Pathol. 69:76-82.
12. Song, W-Y., Wang, G-L., Chen, L-L., Kim, H-S., Pi, L-Y., Holsten, T., Gardner, J., Wang, B., Zhai, W-X., Zhu, L-H., Fauquet, C. and Ronald, P. (1995). A receptor kinase- like protein encoded by the rice disease resistance gene, *Xa21*. Science 270: 1804-1806.
13. Sonoda, S., Koiwa, H., Kanda, K., Kato, H., Shimono, M. and Nishiguchi, M. (2000). The helper component-protenase of sweet potato feathery mottle virus facilitates systemic spread of potato virus X in *Ipomoea nil*. Phytopathology 90: 944-950.
14. Suzuki, Y., Iwata, M., Mase, S., Watanabe, T. and Sekizawa, Y.(1980). Effect of probenazole on bacterial leaf blight of rice plant. Sci. Rep. Meiji Seika Kaisha. 19: 32-40.
15. Watanabe, T., Igarashi, H., Matsumoto, K., Seki, S., Mase, S. and Sekizawa, Y. (1977). The characteristics of probenazole (Oryzemate) for the control of rice blast. J. Pesticide Sci. 2: 291-296.
16. Yazaki, J., Kishimoto, N., Nakamura, K., Fujii, F., Shimbo, K., Otsuka, Y., Wu, J., Yamamoto, K., Sakata, K., Sasaki, T., and Kikuchi, S. (2000). Embarking on rice functional genomics via cDNA microarray: use of 3' UTR probes for specific gene expression analysis. DNA Res. 7: 367-370.
17. Yoshioka, K., Nakashita, H., Klessig, D.F. and Yamaguchi, I. (2001). Probenazole induces systemic acquired resistance in *Arabidopsis* with a novel type of action. Plant J. 25: 149-157

CYTOLOGY OF INFECTION AND HOST RESISTANCE IN RICE BLAST DISEASE

*H. Koga And O. Nakayachi

Ishikawa Agricultural College, Ishikawa, Japan

**E-mail: hkoga@ishikawa-c.ac.jp*

1. INTRODUCTION

Rice blast disease caused by *Pyricularia oryzae* Cav. (telemorph: *Magnaporthe grisea* (Hebert) Barr) is the most widespread and damaging disease of cultivated rice in both tropical and temperate zones. Studies have been made to forecast the occurrence of the disease, breed blast resistant lines, and control the disease by chemicals. The mechanisms underlying pathogenesis and host response must be elucidated for future progress in both applied and basic aspects of rice blast pathology. This chapter reviews the cytology of this disease, including fungal infection structure, change in host ultrastructure induced by infection, and the nature of the fungus-plant interfaces.

2. PRE-PENETRATION EVENTS

Conidia of *M. grisea* germinated and formed appressoria at the tips of germ tubes if the surface was wet. The surface of the epidermal cell wall of the leaf blade is wavy, and wart-like protuberances line up at intervals of 2-7 μm all over the surface of epidermis (Figure 1). It was rather unusual that the apex of conidia attached to the surface of the leaf blade with spore-tip mucilage, which was reported for the attachment on the artificial substrate (Howard, 1994). In most cases, the apex of the slightly developed germ-tube attached to the surface of leaf blade (Figure 1). This was confirmed by the number of conidia fallen from the surface of leaf blade, because conidia

S. Kawasaki (ed.), Rice Blast: Interaction with Rice and Control, 155–161.
© 2004 *Kluwer Academic Publishers. Printed in the Netherlands.*

attached at the tip of a slightly developed germ-tube adhere more strongly than those attached at the apex of spore (Koga and Nakayachi, 2004).

Appressoria and a part of the germ tube were firmly held on the cuticle with mucilage. The wax around the tip of germ tube and appressorium disappeared (Figure 2). The mucilage spread some distance (0.7-1.7 μm) from the appressorium (Figure 3). A melanized layer (Howard, 1994: Koga, 1994 b) was formed around the entire appressorium except on the wall surrounding the penetration peg. It has been demonstrated that the melanin layer acts as a differentially permeable layer that is responsible for the enormous appressorium turgor pressure (Howard, 1994). There were no significant differences in the percentages of spore germination and appressorial formation among six compatible and four incompatible combinations (Koga, 1983). In highly resistant (ZTR) near-isogenic line ($Pi\text{-}z^{t}$), the ultrastructure

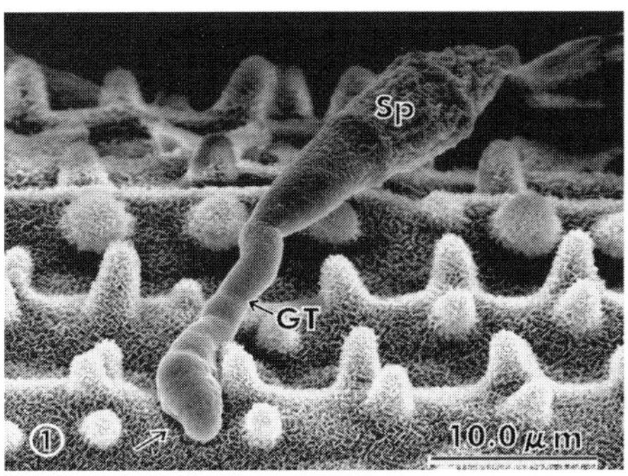

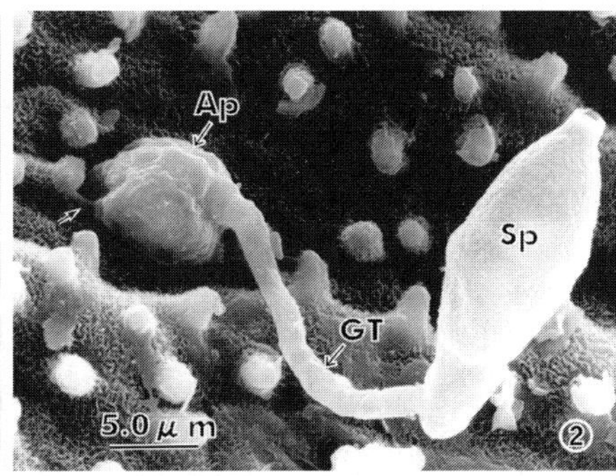

Figure 1. *SEM image of spore and germ tube of M. grisea on the epidermis of rice leaf blade 2.5 h after inoculation. Note that the apex of slightly developed germ-tube attached to the surface of leaf blade, and that the waxes around the germ tube were disarranged (arrow).*

Figure 2. *SEM image of spore, germ tube and appressorium of M. grisea on the epidermis of rice leaf blade. Note that the waxes around the appressorium disappeared (arrow).* Ap, appressorium; GT, germ tube; Sp, spore.

of the host-parasite interaction was the same as that in a susceptible (ZTS) line until the penetration peg pierced the cuticle layer (Koga, 1994 b).

3. INITIAL PENETRATION THROUGH EPIDERMAL CELL

Penetration into the epidermal cell surface was achieved by a penetration peg which emerged from the appressorium. The host cuticle layer was perforated and sometimes folded inward (Figure 3). The inward indentation of the host cuticle suggests that the cuticle is breached mechanically. Penetration pegs were 0.4-1.1 μm in diameter. It is not clear, however, whether enzymatic dissolution is also involved in cuticle penetration. Penetration through cell walls supports previous suggestions (Hashioka et al., 1968) of enzyme action. Micrographs in Figures 3 and 4 showed disorganization of microfibrils around infection pegs and a change in electron density. No ultrastructural changes were observed in the cytoplasm of host epidermal cells during penetration pegs were passing through the cuticle. However, when penetration pegs were passing within host cell wall, collar-like wall appositions were formed between the host cell wall and host plasma membrane (Figure 3). Wall apposition was formed only in living epidermal cells (Figures 3 and 4). Therefore, it is postulated that substances which stimulate the production of apposition material are released at the sites between the penetration peg and cell wall. Wall appositions were pierced by the penetration peg which developed to an invading hypha when it emerged inside the cell lumen (Figure 4). In living host cells invading fungal hyphae were surrounded by host cytoplasm (Figure 4). Most wall appositions were not effective as a barrier to fungal attack as has been reported by Hashioka et al. (1968).

It was clarified that the mode of penetration of blast fungus into the

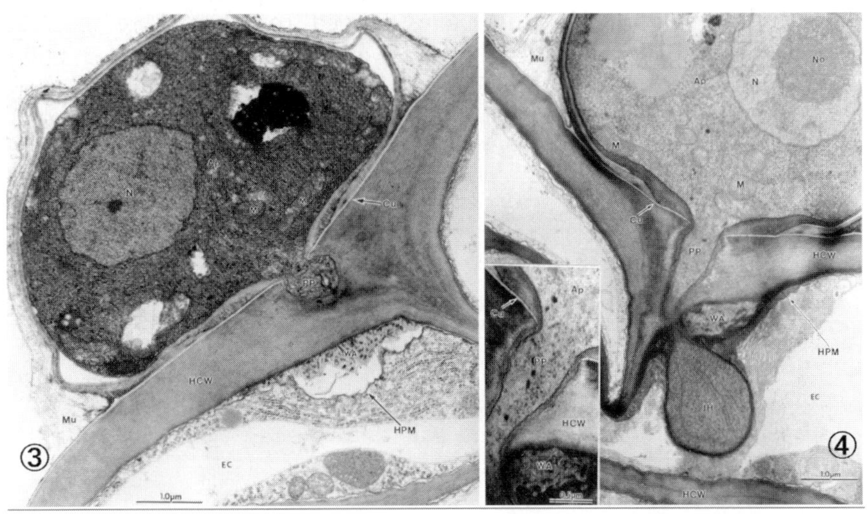

Ap, appressorium; Cu, cuticle; EC, epidermal cell; HCW, host cell wall; HPM, host plasma membrane; IH, invading hypha; M, mitochondrion; Mu, mucilage; N, nucleus; No, nucleolus; PP, penetration peg; WA, wall apposition.

Figure 3. *A medial section through a developing penetration peg. The host cuticle and wall microfibrils appear digested. The penetration peg wall is ill-defined and the host microfibrils surrounding the penetration peg appear disorganized. The wall apposition is formed between the host cell wall and plasma membrane before the host cell wall is completely pierced by the penetration peg. Note that the host cuticle layer is perforated and folded inward. Reprinted from H. Koga (1994). Copyright National Research Council of Canada.*

Figure 4. *Appressorium with penetration peg and invasion hypha formed on an anticlinal host cell wall following piercing of the wall apposition. Inset is a serial section of the penetration peg shown in Figure 4. Reprinted from H. Koga (1994). Copyright National Research Council of Canada.*

epidermal cells of spikelets (Koga, 1995) and panicle necks (Koga, 1994 a) was also similar to that in the epidermal cells of leaf blade (Hashioka et al., 1968) and leaf sheath (Koga, 1994 b, 2001).

4. DEVELOPMENT OF INVADING HYPHAE AND HOST RESPONSES

4.1. Highly Resistant Lines

In the highly resistant ZTR (*Pi-z^t*) line, Toride (*Pi-z^t*), Kanto 51 (*Pi-k*) and Tsuyuake (*Pi-k^m*) cultivars, epidermal cells with the penetration peg entered the cell wall beneath the cuticle or reached into the lumen showed disorganization typical of the hypersensitive reaction (Figures 5 and 6) (Koga and Kobayashi, 1982b; Koga 1994 b, 2001).

The host plasma membrane was fragmented and cell organelles collapsed. As a result debris and vesicles derived from them were observed in the invaded epidermal cells. No wall appositions were found under penetration pegs. Some invading hyphae were moribund, but others were filled with cytoplasm, and cell organelles such as mitochondria (Figure 6).

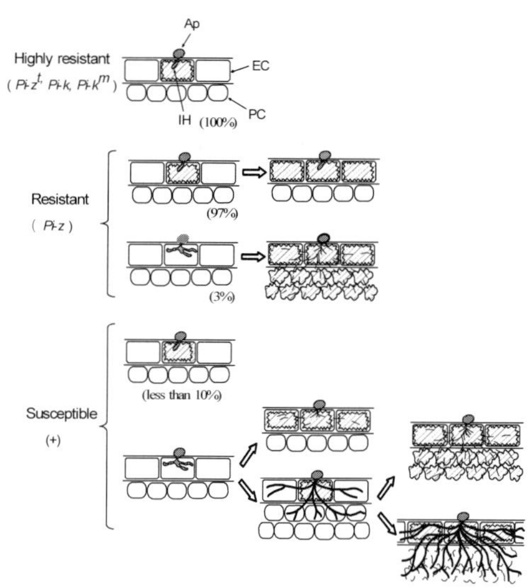

Ap, appressorium;
EC, epidermal cell;
PC, parenchymatous cells.

Figure 5. *Diagrammatic representation comparing the most typical combinations of fungal growth and plant responses seen in intact leaf sheaths of highly resistant (resistant gene: Pi-z^t, Pi-k, Pi-k^m), resistant (Pi-z) and susceptible (+) cultivars and lines of rice inoculated with M. grisea. Oblique line represents browning or autofluorescence, and dead cells are represented by irregular plasma membranes. A broken line of invading hyphae represents moribund or dead. A stippled line of parenchymatous cells represents being digested. The infection sites of unsuccessful penetration without any host reaction have been excluded.*

The invaded host cells died hypersensitively without accompanying wall appositions in the resistant ZTR line (Figure 6), although wall appositions were seen in the cells with the penetration peg incompletely penetrated the cell wall in the susceptible ZTS line (Figure 3). These results suggested that the resistance or susceptibility is determined when the penetration peg incompletely penetrated the cell wall. In the resistant line it is presumed that hypersensitivity was triggered by the release of a 'cell- death- inducing substance (s)'.

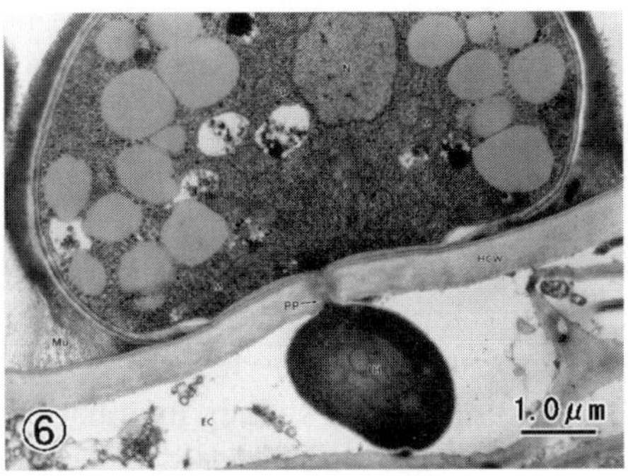

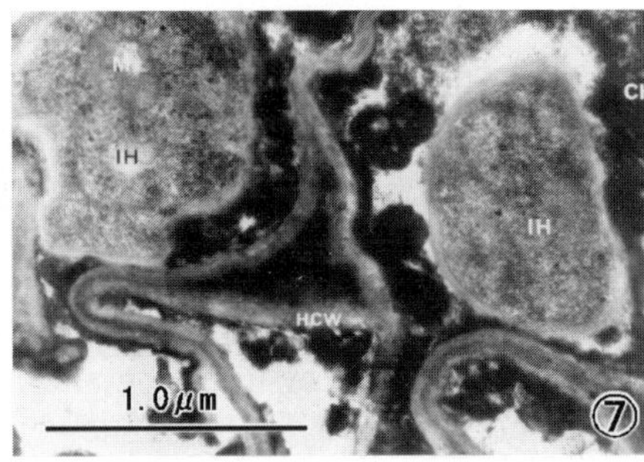

Ap, appressorium; EC, epidermal cell; HCW, host cell wall; IH, invading hypha; M, mitochondrion; Mu, mucilage; N, nucleus; PP, penetration peg

Figure 6. *Penetration of epidermal cell wall of rice leaf sheath of ZTR (resistant line) by M. grisea 27 h after inoculation. Note that the epidermal cell has lost compartmentalization and there is no wall apposition at the site of penetration. Reprinted from H. Koga (1994). Copyright National Research Council of Canada.*

Figure 7. *Invaded hyphae with irregular outlines and swollen mitochondria in degenerated parenchymatous cells of rice leaf blade of Fukunishiki (resistant cultivar) by M. grisea 5 days after inoculation.*
Reprinted from H. Koga (1982). Copyright Phytopathological Society of Japan.

4.2. Resistant Lines

In a resistant cultivar Fukunishiki (*Pi-z*), in which brown lesions appear on the leaf blade, not only a hypersensitive reaction in the epidermal cells but also necrosis of the parenchymatous cells were observed by light microscopy and transmission electron microscopy (Koga et al., 1982 a, c). The frequency of the former was about 97%, and the latter about 3% in the leaf-sheaths inoculated by the intact leaf-sheath inoculation method (Figure 5). In the parenchymatous cells, cytoplasmic membrane and cell organelles appeared to be destroyed by the invasion of blast fungus. Invading hyphae in the necrotic parenchymatous cells were transformed to various degrees.

Mitochondria in the hyphae became swollen, and some of the hyphae seemed to die after granulation of their contents (Figure 7).

In some resistant cultivars such as Pi No. 4 (*Pi-ta*^{2}), most of the invading hyphae developed into the neighbor epidermal cells, but died without invading into parenchymatous cells. On the other hand, in the other resistant cultivars such as Yashiromochi (*Pi-ta*), some of the invading hyphae developed not only in epidermal but also into parenchymatous cells and deep-browned lesions were formed (Koga and Nakayachi, 2000).

4.3. Susceptible Lines

In the susceptible ZTS line, most of the invaded host cells remained alive and frequency of HR observed in the resistant cultivars was less than 10% in the intact leaf sheath (Figure 5). Invading hyphae ceased growing at various stages. Some of them ceased growing at the epidermis and the others at parenchymatous cells which became deeply browned. Finally only a few percentages of infection sites developed to acute-type lesions, which could be observed by using newly devised intact leaf-sheath inoculation method. By this method, the resistance enhanced by aging of plants and treatment with probenazole, which induce systemic acquired resistance, has been cytologically clarified (Koga and Nakayachi, 2002).

5. CONCLUSION

Ultrastructural studies have greatly promoted our understanding of fungal pathogenicity and host response in rice blast disease. It was confirmed electron-microscopically that *M. grisea* is a hemibiotroph, which attacks living tissues in the same way as biotrophs but continues to develop and sporulate after the tissue dies. Timing of death of host cells invaded by the blast fungus seemed to be closely correlated with the resistance expression not only in hypersensitive reactions in cultivars (or lines) controlled with major genes for resistance but also general resistance in susceptible ones.

REFERENCES

Hashioka, Y., Ikegami, H. and Murase, T.(1968). Fine structure of the rice blast. III. The mode of invasion of *Pyricularia oryzae* into rice epidermal cells. Res. Bull. Fac. Agr. Gifu Univ. 26: 23-30.

Howard, R.J. (1994). Cell biology of pathogenesis. In Rice blast disease. Edited by R.S. Zeigen, S. A. Leong and P. S. Teng. CAB International, Cambridge. pp 3-22.

Koga, H. (1983). Comparison of spore germination and appressorial formation of *Pyricularia oryzae* Cav. On rice leaves in compatible and incompatible combinations. Proc. Assoc. Plant Prot. Hokuriku, 31:7-12.

Koga, H. (1994 a). An electron microscopy of early infection processes in the panicle neck of rice inoculated with *Pyricularia oryzae* Cav. Ann. Phytopathol. Soc. Japan. 60:89-98.

Koga, H. (1994 b). Hypersensitive death, autofluorescence and ultrastructural changes in cells of leaf sheaths of susceptible and resistant near-isogenic line of rice (*Pi-z^t*) in relation to penetration and growth of *Pyricularia oryzae*. Can. J. Bot. 72:1463-1477.

Koga, H. (1995). An electron microscopic study of the infection of spikelets of rice by *Pyricularia oryzae*. J. Phytopathology 143:439-445.

Koga, H. (2001) Cytological aspects of infection by the rice blast fungus *Pyricularia oryzae*. Sreenivasaprasad, S. and Johnson, R (eds.) Major Fungal Diseases of Rice Recent Advances, p. 87-110. Kluwer Academic Publishers. Printed in the Netherlands.

Koga, H. and Kobayashi, T. (1982 a). Comparison of the leaf infection process of *Pyricularia oryzae* Cav. in rice leaves of compatible and incompatible combinations. Ann. Phytopathol. Soc. Japan, 48:506-513.

Koga, H. and Kobayashi, T. (1982 b). Cytological reactions in leaves of the rice variety, Toride 1 to the incompatible race of *Pyricularia oryzae* Cav. Proc. Assoc. Plant Prot. Hokuriku, 30:12-18.

Koga, H., Kobayashi, T. and Horino, O. (1982 c). Electron microscopical observation of rice leaves infected with *Pyricularia oryzae* Cav. In compatible and incompatible combinations. I. Fine structure of invaded hyphae in host cells. Ann. Phytopathol. Soc. Jpn.48:281-289.

Koga, H. and Nakayachi, O. (2000). Cytological reactions of rice leaf sheath inoculated with *Magnaporthe grisea* by the intact leaf-sheath inoculation method. Ann. Phytopathol. Soc. Japan. 66:282 (abstract).

Koga, H. and Nakayachi, O. (2002). Cytological analysis of predisposision of rice plant to infection of reactions of *Magnaporthe grisea* by intact leaf-sheath inoculation method. Ann. Phytopathol. Soc. Japan. 68:165 (abstract).

Koga, H. and Nakayachi, O. (2004). Morphological studies on attachment of spores of *Magnaporthe grisea* to the leaf surface of rice. J. Gen. Plant Pathol. 70 (in press).

PROBENAZOLE (ORYZEMATE®) - A PLANT DEFENSE ACTIVATOR

*Michiaki Iwata[1], Kenji Umemura[2] and Naoki, Midoh[3]

[1]*Pharmaceutical Research Center, Meiji Seika Kaisha, Ltd., Yokohama 222-8567,*
[2]*Health & Bioscience Laboratories, Meiji Seika Kaisha, Ltd., Sakado 350-0289,*
[3]*Microbiological Resources & Technology Laboratories, Meiji Seika Kaisha, Ltd., Odawara 250-0852, Japan.*

**E-mail: michiaki_iwata@meiji.co.jp*

1. INTRODUCTION

Most plants have the ability to fight off pathogens by invoking various defense systems, including systemic acquired resistance (SAR). Chemicals that activate SAR are called plant activators. Among the most effective known plant activators is probenazole (3-allyloxy-1,2-benzisothiazole-1,1-oxide ; PBZ, Figure 1), developed by Meiji Seika Kaisha Ltd. Oryzemate® granule, containing PBZ, has been widely used against rice blast by Japanese farmers since 1975. In the past several years, it has been applied in some 25–30% of all paddy fields in Japan to control blast. Since PBZ activates plant defense systems, the study of PBZ's mode of action has helped clarify how rice plant defense systems function. These studies suggest that PBZ is not only an essential protectant for rice cultivation, but also a useful probe in research into plant defense mechanisms.

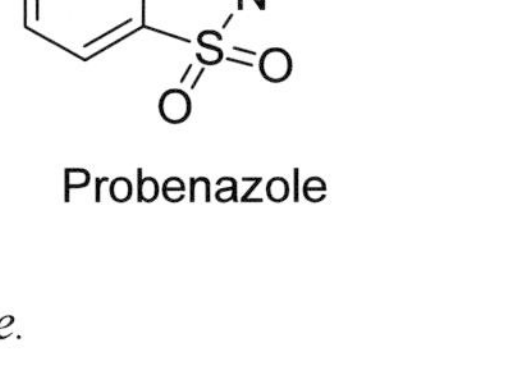

Figure 1. *Structure of probenazole.*

S. Kawasaki (ed.), Rice Blast: Interaction with Rice and Control, 163–171.
© 2004 *Kluwer Academic Publishers. Printed in the Netherlands.*

2. CONTROL EFFECTS OF PROBENAZOLE ON RICE BLAST

Oryzemate® granule provides effective, long-lasting control when applied to paddy fields or in seedling boxes. Following application of granule to rice plants, PBZ is absorbed by the roots, then systemically transferred to the whole plant, almost completely controlling leaf blast for 40–70 days after application. PBZ treatment protects monocotyledonous and dicotyledonous plants from infection by a broad range of pathogens, including bacteria, as well as blast fungus (*Magnaporthe grisea*), e.g., *Cochliobolus miyabeanus* (rice), *Cercospora capsici* (pimiento), *Oidiopsis sicula* (pimiento), *Xanthomonas oryzae* pv. *oryzae* (rice), *X. campestris* pv. *campestris* (cabbage, broccoli), *X. campestris* pv. *vitians* (lettuce), *Burkholderia glumae* (rice), *Pseudomonas syringae* pv. *lachrymans* (cucumber), *P. cichorii* (lettuce), *Erwinia carotovora* subsp. *carotovora* (Chinese cabbage, onion). Despite extensive use over many years, no development of resistance in the target fungus has been observed.

Some of the granule formulations of PBZ contain insecticides to control both blast and rice insects. These formulations reduce both costs and labor in controlling blast and insects.

3. MODE OF ACTION OF PROBENAZOLE

Our experiments on the effects of PBZ on blast fungus have shown that PBZ and its metabolites do not affect the growth or infectivity of the pathogen within the rice tissue or in environments such as paddy water. In addition to 1) formation of resistance-type lesion, it has been observed that treatment of rice plants with PBZ enhances 2) activation of defense-related phenylpropanoid pathway (Iwata *et al.*, 1980), 3) accumulation of fungicidal substances such as unsaturated fatty acids (Shimura, *et al.*, 1983) and phytoalexins (Araki and Kurahashi, 1999), 4) amplification of superoxide production (Sekizawa *et al.*, 1987), 5) induction of defense-related genes (Midoh and Iwata, 1996), and 6) activation of the signal transduction system (Sekizawa *et al.*, 1995), especially in plants inoculated with the blast fungus after treatment, showing incompatible reactions. Based on these observation, we conclude that PBZ activates the natural disease defense system in plants.

3.1. Formation of resistance-type lesion

Small brown spots are formed on rice leaves treated with PBZ after inoculation with a compatible race (Figure 2). This reaction is often observed on leaves following inoculation with an incompatible race. A resistance-type intracellular response was also observed in cells of plants treated with PBZ after inoculation by a compatible race. These results suggest that PBZ induces a disease-resistance reaction in the plant.

3.2. Activation of defense-related phenylpropanoid pathway

We found that treatment of rice plants with PBZ enhances enzyme activity in phenylpropanoid pathway such as phenylalanine ammonia-lyase, tyrosine ammonia-lyase, catechol *O*-methyl transferase, peroxidase, and polyphenoloxidase, especially in plants inoculated with the blast fungus after treatment (Iwata *et al.*, 1980).

The phenylpropanoid pathway plays an important role in the defense system of a plant; when the plant is under attack by a pathogen, lignin is synthesized and acts as a physical barrier against pathogen invasion, and also phytoalexins with antimicrobial activity are produced. These help limit pathogen invasion in plant tissue. Our results show that PBZ activates the phenylpropanoid pathway, thereby enhancing the defense response in the plant.

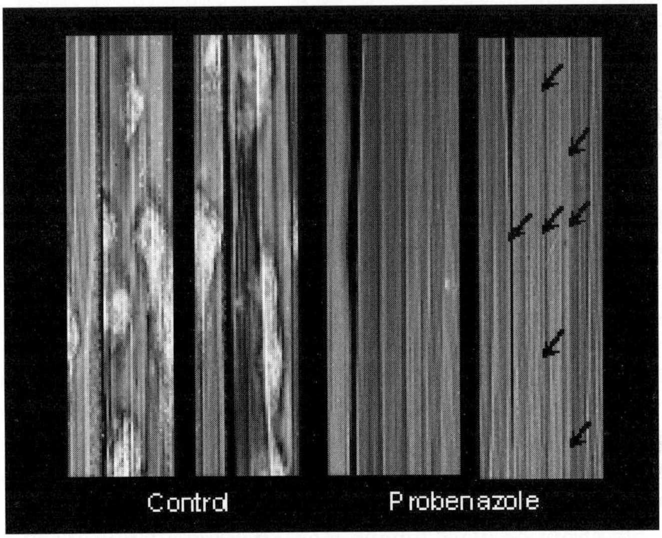

Figure 2. *The blast lesions formed on rice leaves treated with probenazole. The 6th leaf of the plants was inoculated with a compatible Magnaporthe grisea at 7 days after submerging treatment. Arrows indicate small brown lesions.*

3.3. Accumulation of fungicidal substances

We found that fungicidal substances accumulate within the tissues of treated and inoculated rice leaf. We isolated these substances from the leaf. Since PBZ and its metabolites have no fungicidal activity, as described above, we thought that these substances originated from the rice plant. They were identified as hydroxyl unsaturated fatty acids derived from α-linolenic acid (Shimura *et al.*, 1983). We also proposed the following biosynthesis pathway for these hydroxyl unsaturated fatty acids: α-linolenic acid cut off by phospholipase A_2 from phospholipids in the cell plasma membrane is peroxidized into hydroperoxylinolenic acids by lipoxygenase; then the hydroperoxides are rapidly reduced to hydroxides. Activities of both enzymes in the rice leaf were enhanced when the plant was inoculated with a resistant-reaction-inducing, incompatible race of the blast fungus, suggesting the participation of both enzymes in the defense response. The hydroperoxide formed from α-linolenic acid by lipoxygenase is a reactive chemical resembling superoxide and hydrogen peroxide, as described below. Hydroperoxide also shows strong cytotoxicity against the host plant cell, in addition to fungicidal activity.

Hydroperoxide synthesis forms part of the octadecanoid (18-carbon) pathway by which the plant hormone jasmonic acid, an endogenous elicitor of defense gene expression and phytoalexin biosynthesis (Nojiri *et al.*, 1996), is synthesized.

Araki and Kurahashi (1999) have reported that chemicals including PBZ enhance the accumulation of phytoalexins momilactone A and sakuranetin in rice leaves subsequently inoculated with the compatible blast fungus, but not in leaves that are not inoculated. Phytoalexins, which are low molecular antimicrobial compounds, have been investigated in many plant-microbe interaction systems, and phytoalexin accumulation has been posited to be an important process by which plants defend themselves against microbial infection (Darvill and Albersheim, 1984).

3.4. Amplification of superoxide production

In many plants, production of reactive oxygen including superoxide (O_2^-) has been shown to be part of the hypersensitive response, a powerful defense mechanism against pathogen attack (Doke, 1983). In rice plants, superoxide production also appears to be closely related to defense response, since superoxide is more quickly released from rice leaf tissue inoculated with an incompatible strain of blast fungus than after inoculation with a compatible strain. When rice plants treated with 1,2-benzisothiazole-3(2*H*)-one 1,1-dioxide (metabolite of PBZ in rice plants; BIT) were inoculated with a compatible race, superoxide release matched that from untreated plants inoculated with an incompatible race (Sekizawa *et al.*, 1987). Also, superoxide production in a protoplast prepared from rice leaves treated with PBZ was amplified by treatment with a proteoglucomannan elicitor extracted from the blast fungus cell wall, showing that PBZ amplifies superoxide production in leaves attacked by the pathogen. In the protoplast experiment, superoxide was released from the protoplast within several seconds after elicitor treatment, suggesting that superoxide production is one of the earliest defense responses in the rice plant.

After generation from the NADP(H) oxidase system in plant plasma membranes, superoxide is readily dismuted into hydrogen peroxide. Reports have implicated hydrogen peroxide in the killing of invading pathogens, in the cross-linking of cell wall sugar proteins, in the plant cell death process as a cytotoxin, and in the induction of defense gene expression.

3.5. Induction of defense-related genes

As mentioned above, PBZ treatment sensitizes plant disease defense systems. We hypothesized that this sensitization is due to a response involving gene transcription. We screened for rice genes induced by PBZ application to support this hypothesis, thereafter finding a new rice gene *PBZ1* (Midoh and Iwata, 1996). The amino acid sequence estimated from the nucleic acid sequence of the *PBZ1* gene was homologous with the PR (pathogenesis-related)-10 protein. The PR protein is induced after infection by a pathogen, and is thought to be an infection response and defense-participating protein. When rice plants untreated with PBZ were inoculated with the blast fungus, the *PBZ1* gene was also induced in the rice leaf tissue. Expression of the *PBZ1* gene induced by inoculation with the incompatible fungus occurred earlier than with the compatible fungus. These results show that the PBZ1 gene product is a kind of PR protein.

Sakamoto *et al.* (1999) isolated another rice gene, *RPR1* (rice probenazole responsible gene), by a differential display technique. *RPR1* gene transcription was detected 3 days after PBZ treatment, reaching maximum levels in 6-9 days. The mode of *RPR1* expression in PBZ-treated rice plants correlated well with blast protection. The RPR1 protein deduced from the amino acid sequence contains nucleotide binding sites (NBS) and leucine-rich repeats (LRR). Interestingly, NBS and LRR are common characteristics in proteins coded within disease resistance genes isolated from many plants, including rice. These characteristics suggest that the expression of the *RPR1* gene induced by PBZ induces a disease resistance response.

Researchers have reported the induction of numerous defense-related rice genes such as peroxidase, chitinase, and lipoxygenase following PBZ application (Midoh and Iwata, 1997; Schaffrath *et al.*, 2000; Shimono *et al.*, 2000). In an analysis of microarrays composed of 2880 clones from the rice plant, expression levels for 31 clones including defense-related genes were found to increase by a factor of 2.5 after PBZ treatment (Nishiguchi, *et al.*, 2002). That PBZ activates many defense-related genes suggests that individual genes contribute only to part of the defense system of the rice plant. Protection by PBZ treatment may be completed by concurrent activation of a set of genes.

3.6. Activation of the signal transduction system

In a study of the mode of action of PBZ, it was observed that activation of GTPase in the plasma membrane of rice tissue following treatment with the proteoglucomannan elicitor was enhanced by the presence of BIT (Sekizawa *et al.*, 1995). It was suggested that the proteoglucomannan elicitor activates phospholipase C, known as an effector of G-protein in animal cells (Vaughan, 1998), since it temporarily decreases phosphatidylinositol 4,5-bisphosphate (PIP_2) while increasing inositol 1,4,5-trisphosphate (IP_3) in the rice plasma membrane (Kanoh *et al.*, 1993). IP_3 and diacylglycerate are both components of PIP_2; IP_3 releases intracellular Ca^{2+}, while diacylglycerate activates protein kinase C. Moreover, protein kinase C accumulation in rice leaves following inoculation with the blast fungus was enhanced by PBZ treatment (Kiribuchi *et al.*, 1998). These results suggest that the rice disease-defense system activated both by infection of the pathogen and by PBZ treatment may be regulated by the G-protein-mediated signaling pathway.

In higher animals, heterotrimeric G-protein consisting of α, β and γ subunits exists in the plasma membrane (Kaziro *et al.*, 1991). Many hormones, neurotransmitters, chemokines, local mediators, and sensory stimuli exert their effects on cells and organisms by binding to G-protein-coupled receptors (Hamm, 1998). Heterotrimeric G-protein transducing stimuli is dissociated into a GTP-binding α subunit and a βγ complex. By

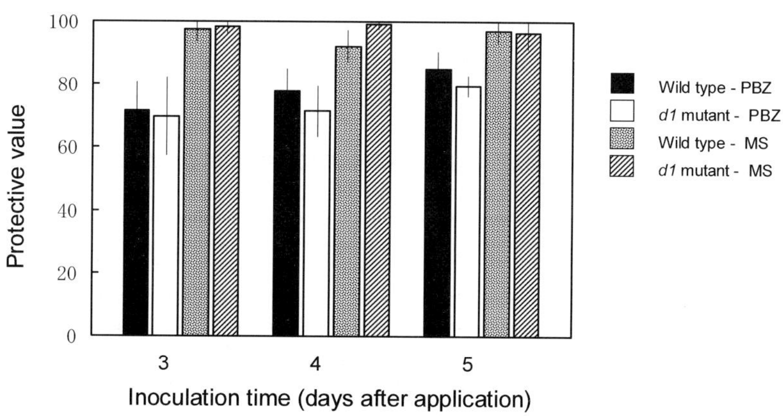

Figure 3. *Control effects of probenazole (PBZ) on rice blast of d1 mutant (defective in the α subunit of the heterotrimeric G protein). The 6th leaf of the plants was inoculated with Magnaporthe grisea at 3, 4, or 5 days after submerging treatment. Protective values were estimated from the number of susceptible-type blast lesions on the leaves. A fungicide metominostrobin (MS) was used as non-plant activating protectant. Bars represent the standard deviation.*

combining a directory with effectors, both the GTP-binding α subunit and the βγ complex regulate functions of the effectors.

Some studies have addressed the function of heterotrimeric G-protein in plants. It was recently reported that a rice plant transformed with antisense cDNA for α subunit protein exhibited dwarf and short-seed phenotypes (Fujisawa *et al.*, 1999). Since the gene of the G-protein α subunit in rice is single-copy (Seo *et al.*, 1995), the dwarf mutants appear to be α subunit knockout rice. Therefore, these mutants are useful experimental tools for exploring the function of the α subunit in rice.

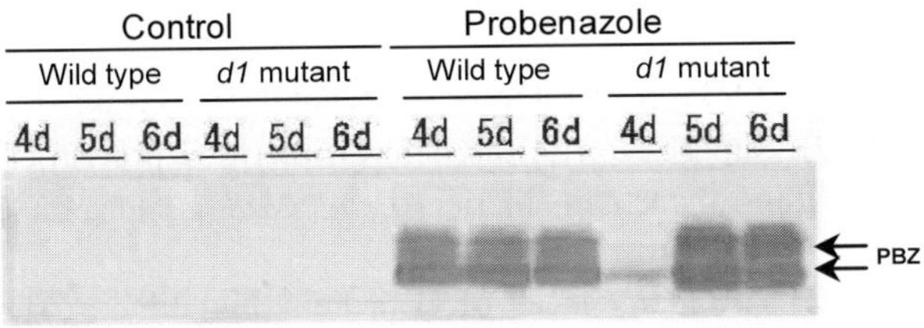

Figure 4. *Delayed PBZ1 protein accumulation in leaves of d1 mutant (defective in the α subunit of the heterotrimeric G protein) treated with probenazole. Protein was sampled at indicated times after treatment and used for western blotting. The upper band is a secreted form PBZ1 protein, while the lower band is a non-secreted form PBZ1 protein (Umemura et al., 2002).*

Most recently, the role of the heterotrimeric G-protein in rice disease resistance was demonstrated by significantly impaired resistance to infection by pathogens on mutants deficient in the G-protein α subunit, and by the delayed activation of PR gene expression in leaves of mutants infected with an incompatible race of rice blast (Suharsono *et al.*, 2002).

To confirm whether heterotirmeric G-protein is involved in the signaling pathway of a defense system induced by PBZ, we investigated protection and *PBZ1* induction using α subunit knockout rice, a dwarf mutant. The rice dwarf mutant was found to be fully protected from blast fungus by PBZ (Figure 3), despite a one-day delay in the induction of *PBZ1* by PBZ (Figure 4). These results suggest that the signaling pathway for protection by treatment of PBZ is not via the G-protein, and that the G-protein is involved in the pathway for the induction of *PBZ1* by PBZ.

We anticipate wider use of plant activators given their many beneficial characteristics, including freedom from biocidal properties, environmental safety, effectiveness against a wide range of plant pathogens, and low associated risk of development of pathogen resistance.

REFERENCES

Araki, Y., and Kurahashi, Y. (1999). Enhancement of phytoalexin synthesis during rice blast infection of leaves by pre-treatment with carpropamid. *J. Pesticide Sci. 24*, 369-374.

Darvill, A. G., and Albersheim, P. (1984). Phytoalexins and their elicitors - A defense against microbial infection in plants. *Annu. Rev. Plant Physiol. 35*, 243-275.

Doke, N. (1983). Involvement of superoxide anion generation in the hypersensitive response of potato tuber tissues to infection with an incompatible race of *Phytophthora infestans* and to the hyphal wall components. *Physiol. Plant Pathol. 23*, 345-357.

Fujisawa, Y., Kato, T., Ohki, S., Ishikawa, A., Kitano, H., Sasaki, T., Asahi, T., and Iwasaki, Y. (1999). Suppression of the heterotrimeric G protein causes abnormal morphology, including dwarfism, in rice. *Proc. Natl. Acad. Sci. USA 96*, 7575-7580.

Hamm, H. E. (1998). The many faces of G protein signaling. *J. Biol. Chem. 273*, 669-672.

Iwata, M., Suzuki, Y., Watanabe, T., Mase, S., and Sekizawa, Y. (1980). Effect of probenazole on the activities of enzymes related to the resistant reaction in rice plant. *Ann. Phytopath. Soc. Japan 46*, 297-306

Kanoh, H., Haga, M., Iwata, M., and Sekizawa, Y. (1993). Transmembrane signaling operated at rice blade cells stimulated by blast fungus elicitor I. Operation of the phospholipase C system. *J. Pesticide Sci. 18*, 299-308.

Kaziro, Y., Itoh, H., Kozasa, T., Nakafuku, M., and Satoh, T. (1991). Structure and function of signal-transducing GTP-binding proteins. *Annu. Rev. Biochem. 60*, 349-400.

Kiribuchi, K., Dunlap, D.Y., Matsumura, F., and Yamaguchi, I. (1998). Protein kinase C as a biomarker for assessing the effect of environmental stress and fungal invasion on plant defense mechanism. *J. Pesticide Sci. 23*, 123-128.

Midoh, N., and Iwata, M. (1996). Cloning and characterization of a probenazole-inducible gene for an intracellular pathogenesis-related protein in rice. *Plant Cell Physiol.* 37, 9-18.

Midoh, N., and Iwata, M. (1997). Expression of defense-related genes by probenazole or 1,2-benzisothiazole-3(2*H*)-one 1,1-dioxide. *J. Pesticide Sci. 22*, 45-47.

Nishiguchi, M., Shimono, M., Eguchi, Y., Okuizumi, H., Yazaki, J., Nakamura, K., Fujii, F., Shinbo, K., Shimatani, Z., Nagata, Y., Hashimoto, A., Ohta, T., Sato, Y., Honda, S., Iwano, M., Yamamoto, K., Sakata, K., Sasaki, T., Kishimoto, N., and Kikuchi, S. (2002). Microarray analysis of gene expression in rice treated with probenazole, a resistance inducer, in special reference to blast disease. *3rd International Rice Blast Conference abstracts*. No.66.

Nojiri, H., Sugimori, M., Yamane, H., Nishimura, Y., Yamada, A., Shibuya, N., Kodama, O., Murofushi, N., and Omori, T. (1996). Involvement of jasmonic acid in elicitor-induced phytoalexin production in suspension-cultured rice cells. *Plant Physiol. 110*, 387–392.

Sakamoto, K., Tada, Y., Yokozeki, Y., Akagi, H., Hayashi, N., Fujimura, T., and Ichikawa, N. (1999). Chemical induction of disease resistance in rice is correlated with the expression of a gene encoding a nucleotide binding site and leucine-rich repeats. *Plant Mol. Biol. 40*, 847-855.

Schaffrath, U., Zabbai, F., and Dudler, R. (2000). Characterization of RCI-1, a chloroplastic rice lipoxygenase whose synthesis is induced by chemical plant resistance activators. *Eur. J. Biochem. 267*, 5935-5942.

Sekizawa, Y., Aoyama, H., Kimura, M., and Yamaguchi, I. (1995). GTPase activity in rice plasma membrane preparation enhanced by a priming effector for plant defence reactions. *J. Pesticide Sci. 20*, 165-168.

Sekizawa, Y., Haga, M., Hirabayashi, E., Takeuchi, N., and Takino, Y. (1987). Dynamic behavior of superoxide germination in rice leaf tissue infected with blast fungus and its regulation by some substances. *Agric. Biol. Chem. 51*, 763-770.

Seo, H.-S., Kim, H.-Y., Jeong, J.-Y., Lee, S.-Y., Cho, M.-J., and Bahk, J.-D. (1995). Molecular cloning and characterization of *RGA1* encoding a G protein α subunit from rice (*Oryza sativa* L. IR-36). *Plant Mol. Biol. 27*, 1119-1131.

Shimono, M., Yazaki, J., Nakamura, K., Kishimoto, N., Kikuchi, S., Kubo, N., Kadowaki, K., Mochizuki, A., Yamamoto, K., Sasaki, T., and Nishiguchi, M. (2000). Analysis of gene expression in rice plants treated with an inducer of disease resistance, probenazole using DNA microarray. *Ann. Phytopath. Soc. Jpn 66*, 115–116 (Japanese abstract).

Shimura, M., Mase, S., Iwata, M., Suzuki, A., Watanabe, T., Sekizawa, Y.,Sasaki, T., Furihata, K., Seto, H., and Otake, N. (1983). Anti-conidial germination factors induced in the presence of probenazole in infected host leaves. III. Structural elucidation of substances A and C. *Agric. Biol. Chem. 47*, 1983-1989.

Suharsono, U., Fujisawa, Y., Kawasaki, T., Iwasaki, Y., Satoh, H., and Shimamoto, K. (2002). The heterotrimeric G protein α subunit acts upstream of the small GTPase Rac in disease resistance of rice . *Proc. Natl. Acad. Sci. USA. 99*, 13307-13312.

Umemura, K., Ogawa, N., Koga, J., Iwata, M., and Usami, H. (2002). Elicitor activity of cerebroside, a sphingolipid elicitor, in cell suspension cultures of rice. *Plant Cell Physiol. 43*, 778-784.

Vaughan, M. (1998). Signaling by heterotrimeric G proteins minireview series. *J. Biol. Chem. 273*, 667-668.

GENETIC ENGINEERING FOR BLAST DISEASE RESISTANCE IN RICE, USING A PLANT DEFENSIN GENE FROM *BRASSICA* SPECIES

Genetic engineering for blast disease resistance in rice

*Motoshige Kawata[1,2], Toshihiko Nakajima[1], Koichi Mori[1], Tetsuo Oikawa[2] and Shigeru Kuroda[1]

[1] *National Agriculture Research Center, Inada, Joetsu, Niigata 943-0193, Japan*

[2] *Graduate School of Natural Science and Technology, Niigata University, Ikarashi, Niigata, 950-2181, Japan*

**E-mail: kawata@affrc.go.jp*

1. DISEASE RESISTANCE IN RICE

Rice (*Oryza sativa* L.) is one of the most important crops in the world, especially in Asia. Rice suffers serious damage from several pathogens, which reduce yield. One of the most devastating fungal diseases is rice blast (*Magnaporthe grisea*). Bacterial diseases are also important, especially bacterial leaf blight (*Xanthomonas oryzae*), which is one of the most serious diseases of rice in South-East Asia. Rice blast attacks the rice leaf and panicle. Because of the variability of the blast fungus's host range and cultivar specificity, disease control through traditional breeding methods produces only partial success (Krishnamurthy *et al.*, 2001). Several approaches based on molecular genetics have been developed to engineer high levels of rice blast resistance, including determination of genetic loci and analysis of the signal transduction pathways for the expression of disease resistance in cells (Michelmore & Meyers, 1998; Ellis *et al.*, 2000; Dangl & Jones, 2001). Also, an increasing number of antifungal peptides are being discovered in animals, plants, insects, and other organisms (Yun *et al.*, 1997; Zasloff, 2002).

S. Kawasaki (ed.), Rice Blast: Interaction with Rice and Control, 173–178.
© 2004 *Kluwer Academic Publishers. Printed in the Netherlands.*

Anti-pathogenic compounds confer the main protection against pathogens that are not specialized to attack a specific cell. Genes for antifungal activity cloned from several species encode proteins that have motifs in common (Zasloff, 2002). We have focused on antifungal peptides (AFPs) found in vegetables. Plant defensin is a pathogenesis-related protein found in *Brassica* species (Fant *et al.*, 1998). It has antifungal properties and is small, basic, and cysteine rich (Broekaert *et al.*, 1995). Although insect and mammalian defensins have 3 disulfide bonds, all known plant defensins have 4, the same as drosomycin of *Drosophila* and hepcidin of human (Fehlbaum *et al.*, 1994; Park *et al.*, 2001). Rice does not have this peptide. The peptide is encoded as a gene cluster, and its activity is likely to change drastically by amino acid substitution (de Samblanx *et al.*, 1997).

2. PLANT DEFENSIN FROM BRASSICA SPECIES

Among the peptides encoded by the plant defensin gene family, AFP1 and AFP2 show the most active antifungal activity (Terras *et al.*, 1995). They differ by the substitution of 2 amino acids (Gln-5 → Glu and Arg-27 → Asn in Rs-AFP1). We have cloned the genes encoding AFP1 from 8 *Brassica* vegetables and determined their structures in order to estimate the whole aspects of the AFP1 gene family in *Brassicaceae*. Sequence analysis showed structural differences in the intron rather than in the coding region. The degree of genetic diversity is consistent with the classification based on taxonomy (Figure 1). The variation in the amino acid sequences matches the variation in the DNA sequences. The amino acid sequences differ in the

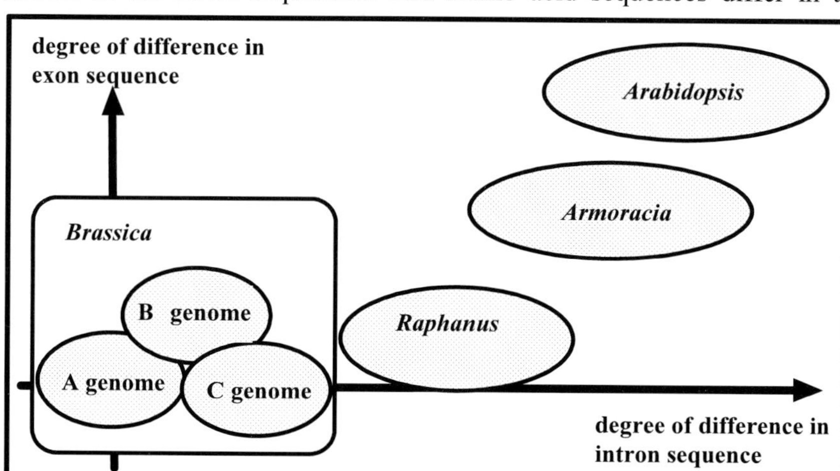

***Figure 1**. Degree of difference in nucleotide sequences among plant defensin genes from Brassicaceae species.*

signal region rather than in the region of the mature peptides. We compared the *in vitro* antifungal activities of the AFP1 peptides of the 8 *Brassica* species in order to determine the most suitable and effective one to use for making rice transformants.

We assessed the antifungal activities of the plant defensins *in vitro* before producing transgenic plants. We successfully recovered purified expressed peptides by making use of the GST-fusion peptide expression system, that is, the pGEX-6P-1 vector system (Amersham Biosciences Corp. Piscataway, NJ, USA) for *E. coli* as a host cell, and the pESP-1 vector system (Stratagene, Inc., La Jolla, CA, USA) for *S. cerevisiae* as a host cell. We performed a variety of experiments to analyze the antimicrobial activity of the peptides against *M. grisea* and *E. coli in vitro*. The antifungal activity of each expressed peptide was estimated, and the concentration that inhibited 50% of the fungus (IE_{50}) was calculated. The defensin peptides expressed by *E. coli* were monomers (5.7 kDa), and had milder antifungal activity than the trimers (17.1 kDa) from *S. cerevisiae* (Figure 2) against *M. grisea*. To increase the defensin activity by modification of the amino acid sequence, we prepared 15 modified DNA clones, and estimated the difference in activity against *M. grisea*. On the basis of *in vitro* antimicrobial activity, we selected the most effective gene for introduction into rice.

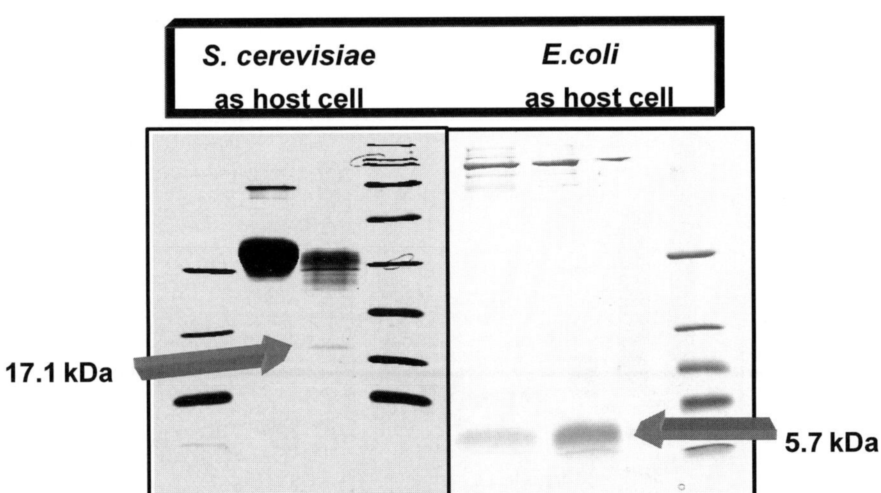

Figure 2. *Fractionation of the expressed defensin peptides from the total proteins of host cells. The defensin peptide from S. cerevisiae formed a trimer (17.1 kDa). The peptide from E. coli formed a monomer (5.7 kDa).*

3. PRODUCTION OF TRANSGENIC RICE BY INTRODUCTION OF DEFENSIN GENE

We established rice transformants with four types of defensin genes that is, from *B. oleracea*, *B. campestris*, and modified genes by making substitution of one amino acid on each original ones, by *Agrobacterium*-mediated transformation. Commercial japonica rice variety Dontokoi was used for establishing transformants. Totally over 1200 transgenic plants were tested for disease resistance and estimated phenotypes. Stability and efficacy of antimicrobial activity are among the most important aims in breeding rice for disease resistance. Estimation of the degree of resistance to rice blast requires skill but is a critical factor in the experimental strategy (Nakajima *et al.,* 2002). Thus, it is important that disease researchers have a critical and accurate system for selecting disease-resistant transformants. Estimation of phenotype is another important factor.

Defensin genes from *B. oleracea* and *B. campestris* and modified genes were individually transformed into rice. Integration of transgene into the genomic DNA was estimated by the PCR method (Kawata *et al.,* in press). Transgenic plants that expressed a gene constitutively throughout their

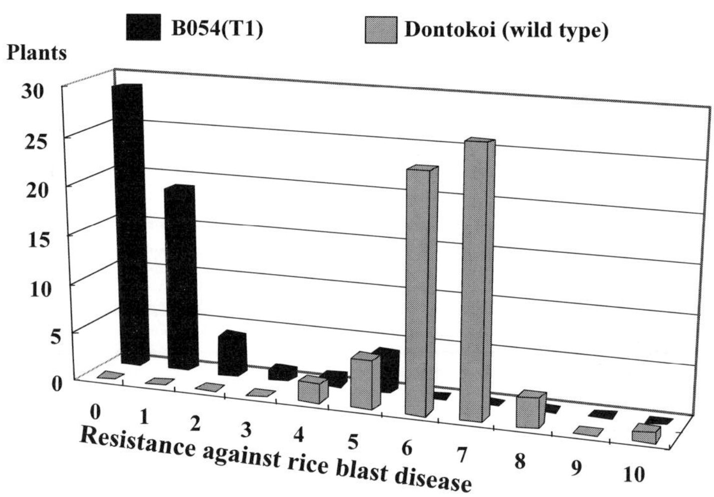

Figure 3. *Resistance to rice blast disease is inherited by progeny of transformants.*

tissues were tested for resistance to rice blast and bacterial leaf blight. Many transformants were highly resistant to rice blast. Most had no phenotypic abnormality. Progeny testing found that the progeny were also highly resistant to rice blast. We also estimated the resistance to bacterial leaf blight. The infected leaves of transgenic rice were highly resistant to leaf blight. One of the modified defensin genes gave much higher resistance than the wild-type genes from *B. oleracea* and *B. campestris*.

Following stringent selection, fewer than 10% of regenerated transformants were selected. Most selected plants contained modified defensin genes. The results indicate that the expressed plant defensins gave effective resistance against both rice blast and bacterial leaf blight. The degree of resistance to rice blast was also tested in the progeny of selected transformants. The progeny inherited the blast resistance (Figure 3). Similar results were obtained in the third generation. We have successfully developed transgenic rice plants with enhanced disease resistance to rice blast and bacterial leaf blight. Furthermore, we have noted that modification of defensin genes conferred the increasing of wide disease resistance spectrum.

Genetic engineering is a powerful tool for the introduction of new traits into rice, and considerable numbers of field trials with transgenic rice have been performed (Kloti & Potrykus, 1999). Our results show that plant defensin is a useful tool for improving disease resistance in rice by transgenic approach. Our goal is to develop this strategy in order to improve rice breeding for disease resistance. Due to its economic importance, a number of studies on rice blast have been developed, and rice blast has been established as a model fungus on analysis of plant pathogen. Our interest is now focused on the degree of resistance against each race of *M. grisea* by introduction of defensin gene into rice. Further analysis is in progress.

REFERENCES

Broekaert, W.F., Terras, F.R., Cammue, B.P., & Osborn, R.W. (1995) Plant defensins: novel antimicrobial peptides as components of the host defense system. *Plant Physiol. 108* (4), 1353–1358.

Dangl, J.L., & Jones, J.D. (2001) Plant pathogens and integrated defense responses to infection. *Nature 415*, 826–833.

de Samblanx, G.W., Goderis, I.J., Thevissen, K., Raemaekers, R., Fant, F., Borremans, F., Acland, D.P., Osborn, R.W., Patel, S., & Broekaert, W.F. (1997) Mutational analysis of a plant defensin from radish (*Raphanus sativus* L.) reveals two adjacent sites important for antifungal activity. *J. Biol. Chem. 272* (2), 1171–1179.

Ellis, J., Dodds, P., & Pryor, T. (2000) Structure, function and evolution of plant disease resistance genes. *Curr. Opin. Plant Biol. 3* (4), 278–284.

Fant, F., Vranken, W., Broekaert, W., & Borremans, F. (1998) Determination of the three- dimensional solution structure of *Raphanus sativus* antifungal protein 1 by 1H NMR. *J. Mol. Biol. 279* (1), 257–270.

Fehlbaum, P., Bulet, P., Michaut, L., Lagueux, M., Broekaert, W.F., Hetru, C., & Hoffmann, J.A. (1994) Insect immunity. Septic injury of *Drosophila* induces the synthesis of a potent antifungal peptide with sequence homology to plant antifungal peptides. *J. Biol. Chem. 269*, 33159–33163.

Kawata, M., Matsumura, Y., Oikawa, T., Kimizu, M., Fukumoto, F., & Kuroda, S. DNA isolation from plant tissue for screening transgenic plants by polymerase chain reaction. *Analytical Biochem., (In press).*

Kloti, A., & Potrykus, I. (1999) Rice improvement by genetic transformation. *In:* Molecular Biology of Rice (Shimamoto, K., ed.), pp. 283–301, Springer-Verlag, Tokyo.

Krishnamurthy, K., Balconi, C., Sherwood, J.E., & Giroux, M.J. (2001) Wheat puroindolines enhance fungal disease resistance in transgenic rice. *Mol. Plant Microb. Interact. 14*, 1255–1260.

Michelmore, R.W., & Meyers, B.C. (1998) Clusters of resistance genes in plants evolve by divergent selection and a birth-and-death process. *Genome Res. 8* (11), 1113–1130.

Nakajima, T., Mori, K., Aoki, H., Kawata, M., Yatou, O., Fukumoto, F., & Kuroda, S. (2002) Alternative method of blast resistance evaluation for transgenic rice plants. *3rd International Rice Blast Conference,* p48.

Park, C.H., Valore, E.V., Waring, A.J., & Ganz, T. (2001) Hepcidin, a urinary antimicrobial peptide synthesized in the liver. *J. Biol. Chem. 276*, 7806–7810.

Terras, F.R., Eggermont, K., Kovaleva, V., Raikhel, N.V., Osborn, R.W., Kester, A., Rees, S.B., Torrekens, S., van Leuven, F., Vanderleyden, J., Cammue, B.P.A., & Broekaert, W.F. (1995) Small cysteine-rich antifungal proteins from radish: their role in host defense. *Plant Cell 7*, 573–588.

Yun, D.J., Bressan, R.A., & Hasegawa, P.M. (1997) Plant antifungal proteins. *Plant Breed. Rev. 14*, 39–88.

Zasloff, M. (2002) Antimicrobial peptides of multicellular organisms. *Nature 415*, 389–395.

TRANSGENIC RICE EXPRESSING WASABI DEFENSIN GENE EXHIBIT ITS ENHANCED RESISTANCE TO BLAST FUNGUS (*MAGNAPORTHE GRISEA*)

*Hiroyuki Kanzaki[1], Shizuko Nirasawa[1], Hiromasa Saitoh[1], Minako Ito[1], Masahiro Nishihara[1], Saburo Yamamura[1], Kazumi Suzuki[1], Ryohei Terauchi[1] and Ikuo Nakamura[2]

[1]*Iwate Biotechnology Research Center.*

[2]*Graduate School of Science and Technology, Chiba University.*

**E-mail: hkanzaki@ibrc.or.jp*

Abstract: A complementary DNA of defensin gene was isolated from wasabi (*Wasabia japonica)* plant. This cDNA was cloned into PVX (*Potato virus X*) vector, and the vector inoculated to *Nicotiana benthamiana* plants to rapidly produce the defensin protein *in planta*. Purified wasabi defensin showed strong antimicrobial activity toward *Magnaporthe grisea*. Thus, wasabi defensin gene fused with a strong maize ubiquitin promoter was used for transformation of rice (*Oryza sativa* cv. Sasanishiki) by *Agrobacterium tumefaciens*-mediated method. Transformants exhibited resistance to rice blast at various levels. Inheritance of blast resistance over the generations was confirmed. Two T3 lines (WT14-5 and WT43-5) stably expressing the defensin protein were challenged with blast fungus by using the press-injured spots inoculation method. The average size of disease lesions of the transgenic line WT43-5 was reduced to about half of that of non-transgenic plants. Transgenic rice plants producing wasabi defensin are expected to possess a durable and wide-spectrum resistance (i.e. field resistance) against various rice blast races.

1. INTRODUCTION

Blast disease caused by *Magnaporthe grisea* results in severe damage of rice plants and a reduced yield worldwide. Much breeding efforts have been made over the decades to generate rice cultivars resistant to the disease. Genetic engineering approach is an effective

S. Kawasaki (ed.), Rice Blast: Interaction with Rice and Control, 179–184.
© 2004 *Kluwer Academic Publishers. Printed in the Netherlands.*

alternative way to create blast-resistant rice. The advantage of the latter is its speed and the ability to introduce disease resistance into elite rice cultivars without changing the original genetic background.

Plant defensins belong to a superfamily of antimicrobial peptides with a cysteine-stabilized $\alpha\beta$ motif that has representatives in vertebrates, invertebrates, and plants (da Silva Conceicao and Broekaert, 1999). Based on SYTOX Green uptake assay, defensins were shown to exert their antifungal activity by increasing the permeability of fungal membranes (Thevissen et al. 1999). We have isolated a cDNA coding for defensin from wasabi (*Wasabia japonica* Matsum.) a crucifer plant species naturally occurring in Japan. To test its antimicrobial activity, we expressed wasabi defensin in *Nicotiana benthamiana* plants by PVX (*Potato virus X*) vector, and purified the protein (Saitoh et al. 2001). Low concentrations of wasabi defensin actually inhibited the growth of rice blast fungus *in vitro* (Figure. 1; half growth inhibition was attained by 5ppm wasabi defensin in the culture medium). This result prompted us to use the gene for transforming rice plants for generating blast resistant cultivars.

2. TRANSGENIC RICE HARBORING WASABI DEFENSIN GENE

2.1 Transformation procedure

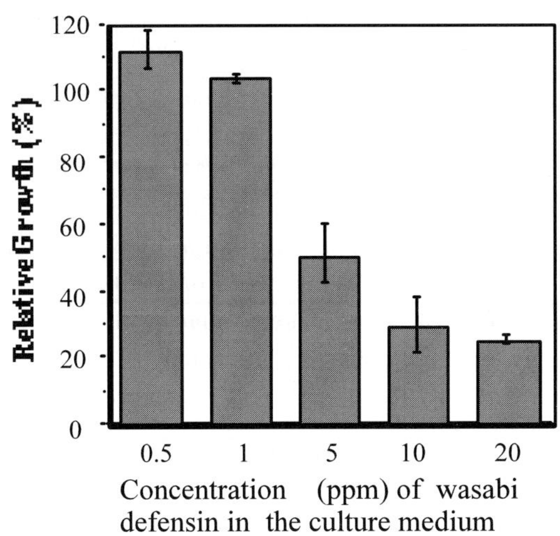

Figure. 1 Inhibitory effect of wasabi defensin protein on the growth of Magnaporthe grisea 36hrs after incubation. Growth was monitored by measuring light absorbance at 595 nm and expressed as a percentage of the growth in the absence of the defensin protein. Standard error ranges from three independent experiments are shown.

The 23 kbp plasmid EKHSubWT harboring the 414bp wasabi defensin cDNA under control of maize ubiquitin-1 promoter (Christensen et al. 1992) was used for rice transformation (Kanzaki et al. 2002;

Figure.2). Maize ubitquitin-1 promoter is extremely strong in monocots and is expected to constitutively produce wasabi defensin in transgenic rice plants. After *Agrobacterium tumefaciens*-mediated transformation, transformants were selected for hygromycin resistance. Eighty-four plants (T0 generation), presumably originating from independent transformation events were grown in a greenhouse. Seeds obtained from the T1 plants were analyzed for segregation in hygromycin resistance, and 22 lines were selected as homozygous lines in the T2 generation.

Figure. 2. T-DNA region of the binary vector pEKHSubWT used for rice transformation. RB and LB are the right and left border sequences of the T-DNA region, respectively. ubiP: maize ubiquitin1 promoter, 35Sp: CaMV 35S promoter, Defensin: wasabi defensin cDNA, nosT: terminator of the nopaline synthase gene, Hpt: hygromycin phosphotransferase gene.

2.2 Blast resistance: T2 generation

Rice cultivar Sasanishiki, a parental line used for transformation, does not possess the resistance (R) gene *Pi-k* against rice blast fungus race 007.0. Therefore, 22 lines of T2 homozygotes harboring the wasabi defensin gene were studied for disease resistance against the race 007.0 by comparing their disease severity with the non-transgenic Sasanishiki control plants. Rice cultivar Tsuyuake, with the R gene *Pi-k* against race 007.0, was used as the resistant control. Transgenic and control plants were inoculated with fungal conidia, and the sizes of disease lesions measured. As shown in Figure.3A, transformants exhibited resistance to rice blast at various levels. Whereas the line WT43-5 was almost as resistant as the resistant cv. Tsuyuake, the line WT21-3 was susceptible to blast at the same level as cv. Sasanishiki. In addition to the rating according to lesion sizes, disease resistance was also inferred from a comparison of the lesion spot numbers between cv. Sasanishiki and transgenic lines WT14-5 and WT43-5 (Figure. 3B). Lesion spot numbers of transgenic lines were about two third (WT14-5) and one third (WT43-5), respectively, of that of non-transgenic cv. Sasanishiki (statistical significance: $P < 0.01$).

Various levels of disease resistance were observed in transgenic plants harboring the wasabi defensin gene (Figure.3A). Different levels of blast resistance in T2 transformants may be caused by two factors. First, transcriptional or post-transcriptional silencing of the transgenes could have affected the level of disease resistance. Second, a position effect of the transgenes in the genome may cause differences in transgene expression (Mayer 1995).

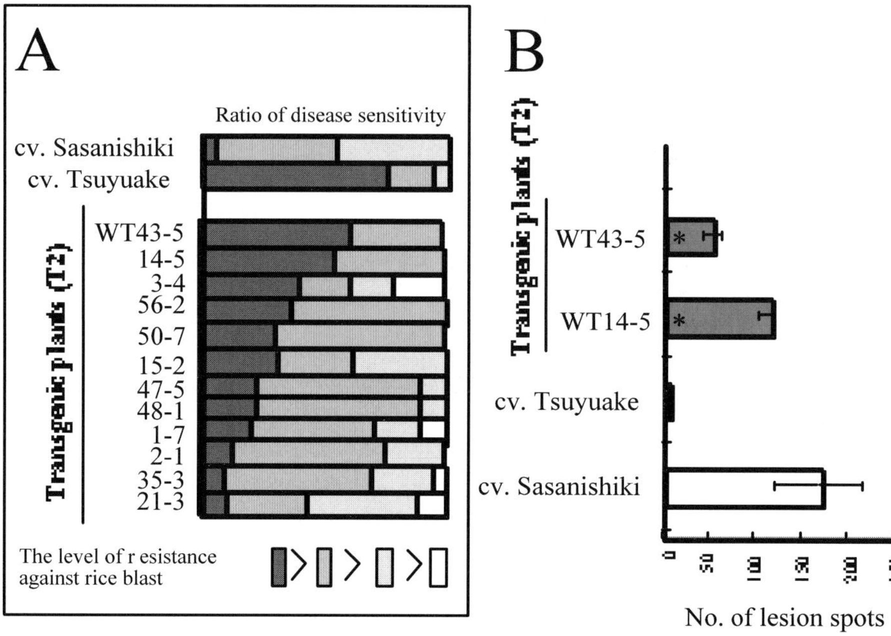

*Figure. 3. Blast disease lesions on leaves of control and T2 transgenic rice plants 10 days after inoculation with rice blast fungus (M. grisea, race 007.0). At the four- to five-leaves stage, rice seedlings were transferred into an inoculation chamber and each pot inoculated with approximately 2.5ml of suspension of blast conidia by spraying. A: Disease severity is expressed as percentage of disease lesions at each level in twenty plants of each transgenic line. B: Numbers of spots on the leaves of control cv. Sasanishiki (without R gene against race 007.0) and cv. Tsuyuake (with R gene against race 007.0), and transgenic rice plants infected by race 007.0, 7 days after inoculation. Bars = ± SE. * : significantly different with P < 0.01*

2.4 Stable inheritance of wasabi defensin expression and blast resistance to T3 generation

To confirm the stable inheritance of wasabi defensin expression to the T3 generation, Western analysis was carried out for the T3 progeny of WT14-5 and WT43-5. The 5kDa peptide, corresponding to the processed form of the wasabi defensin, was detected in them as expected (data not shown). The levels of wasabi defensin expression were variable in different plants. To further investigate whether resistance to rice blast is inherited to the T3 generation, T3 plants of lines WT14-5 and WT43-5 were challenged with blast fungus using press-injured spot inoculation methods. While the average size of lesions of cv. Sasanishiki was 8.1 mm 10 days post inoculation (dpi), those of transgenic plants were 4.1 mm for the line WT14-5 and 4.0 mm for the line WT43-5

(Figure. 4). Average size of the lesions on transgenic line WT43-5 was about half of that of non-transgenic plants, although not as reduced as the lesion size of cv. Tsuyuake. These results demonstrate that wasabi defensin expression was stably inherited to T3 generation, and so for the resistance against blast disease.

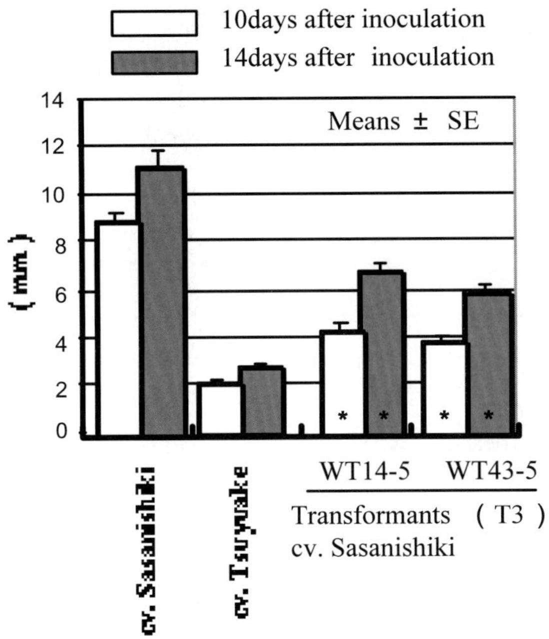

Figure.4 Mean size of lesions in control (non-transformants) and transgenic rice plants inoculated with by the blast fungus (M. grisea; race 007.0) . Buffer suspended spores of blast fungus was applied on press-injured spots made by a specially designed pressing machine.

Rice cultivar Tsuyuake with an R gene against race 007.0 is strongly resistant against this race (Figure. 3, 4). Two resistance genes, *Pi-b* against *M. grisea* (Wang et al. 1999) and *Xa21* against bacterial leaf blight disease (Song et al. 1995), have already been isolated from rice. Since pathogens can overcome R-gene mediated resistance in the field by mutations in the cognate avirulence genes, any transgenic resistance based on R genes may easily break down. On the other hand, this problem probably does not occur in transgenic plants overexpressing antifungal genes. Therefore, overproduction of the wasabi defensin peptide with antifungal activity is expected to confer durable resistance (i.e. field resistance) against a wide variety of rice blast races containing the race 007.0.

Disease resistance of wasabi defensin transgenic rice progeny did not reach the disease resistance level of cv. Tsuyuake (carrying an R gene to rice blast race 007.0). Since R genes are located upstream of the signal transduction cascades of the defense system against pathogen infection, many defense genes are supposed to be activated in response to attacks in cv. Tsuyuake by the rice blast. To obtain a high level of resistance comparable to the R-gene mediated resistance, overexpression of multiple antifungal proteins with different functions may be necessary. The performance of tobacco plants co-expressing the barley transgenes for a class-II chitinase, a class-II β-1,3-glucanase and a type-I ribosome-inactivating protein, in a *Rhizoctonia solani* infection assay was reported to reveal significantly enhanced protection against fungal attack when compared with the protection levels obtained with corresponding isogenic lines expressing a single barley transgenes to a similar level (Jach et al. 1995). Transgenic rice harboring either other antimicrobial gene or multiple ones as well as the wasabi defensin gene, recombined by crossing of different transgenic plants, will show a strong durable protection against *M. grisea.*

REFERENCES

Christensen, A.H., Sharrock, R.A. and Quail, P.H. (1992) Maize polyubiquitin genes: structure, thermal perturbation of expression and transcript splicing, and promoter activity following transfer to protoplasts by electroporation. Plant Mol .Biol. 18 : 675-689.

da Silva Conceicao, A. and Broekaert, W. F (1999) Plant defensins. In: S.K. Datta and S. Muthukrishnan (Eds.) *Pathogenetis-Related Proteins in Plants*. Boca Raton USA CRC Press. (pp 247-260)

Jach, G., Görnhardt, B., Mundy, J., Logemann, J., Pinsdorf, E., Leah, R., Schell, J. and Maas, C. (1995) Enhanced quantitative resistance against fungal disease by combinatorial expression of different barley antifungal proteins in transgenic tobacco. Plant J. 8: 97-109.

Kanzaki, H., Nirasawa, S., Saitoh, H., Ito, M., Nishihara, M., Terauchi, R and Nakamura, I. (2002) Overexpressing of wasabi defensin gene confers enhanced resistance to blast fungus (*Magnaporthe grisea*) in transgenic rice. Theor. Appl. Genet. 105: 809-814.

Meyer, P. (1995) Understanding and controlling transgene expression. Trends Biotechnol. 13: 332-337.

Saitoh, H., Kiba, A., Nishihara, M., Yamamura, S., Suzuki, K. and Terauchi, R. (2001) Production of antimicrobial defensin in *Nicotiana benthamiana* with a potato virus X vector. Mol.Plant-Microbe Interact. 14: 111-115.

Song, W.Y., Wang, G.L., Chen, L.L., Kim, H.S., Piv L.Y., Holsten, T., Gardner, J., Wang, B., Zhai, W.X. and Zhu, L.H. (1995) A receptor kinase-like protein encoded by the rice disease resistance gene, *Xa21*. Science. 270: 1804-6.

Thevissen, K., Terras, F.R.G. and Broekaert, W.F. (1999) Permeabilization of fungal membranes by plant defensins inhibits fungal growth. Appl. Envir. Micro. 65: 5451-5458.

Wang, Z.X., Yano, M., Yamanouchi, U., Iwamoto, M., Monna, L., Hayasaka, H., Katayose, Y. and Sasaki, T. (1999) The *Pib* gene for rice blast resistance belongs to the nucleotide binding and leucine-rich repeat class of plant disease resistance genes. Plant J. 19: 55-64.

RESIDIV: EXPLORING THE GENETIC DIVERSITY OF RICE-BLAST HOST-PATHOGEN INTERACTIONS IN CHINA: A TOOL TO IMPROVE BREEDING FOR RESISTANCE TO BLAST DISEASE

*D. Tharreau [1], C. Kaye [1], I. Fudal [2], H. Böhnert [2], M.H. Lebrun [2], Y. Wang[3], X. Zhu [3], Y. Shen [3], Z. Ling [4], J. Xu [5], L. Zhu [5] and A. Price [3]

[1] *UMR 385, CIRAD, Montpellier, France*

[2] *UMR1932, CNRS-Bayer Crop Science, Lyon, France*

[3] *CNRRI, Hangzhou, China*

[4] *Institute of Crop Breeding and Cultivation, CAAS, Beijing, China*

[5] *Institute of Genetics, Chinese Academy of Sciences, Beijing, China*

[6] *Plant and Soil Science, University of Aberdeen, Aberdeen, UK*

**E-mail: tharreau@cirad.fr*

Abstract. The RESIDIV project is supported by the European Community. The project aims at improving our knowledge on resistance diversity in Chinese rice varieties and on the diversity of the pathogen for its ability to attack resistant varieties (avirulence/virulence). Chinese traditional rice varieties have a large biodiversity and characterization of this material will lead to the identification of new genes conferring resistance to the blast disease This knowledge will be of important value for breeders aiming to introduce the appropriate resistance genes by classical methods or by marker assisted selection to obtain durable resistance. The results of the RESIDIV project will be exploited through conventional breeding methods using natural defense systems.

Resistance genes are usually characterized by inoculation of different isolates of the rice blast fungus *M. grisea*. The choice of isolates is a key point for resistance gene characterization. Because resistance is epistatic,

S. Kawasaki (ed.), Rice Blast: Interaction with Rice and Control, 185–190.
© 2004 *Kluwer Academic Publishers. Printed in the Netherlands.*

two resistance genes in a variety might not be easily distinguished. Our project will use 3 different strategies to improve the identification of resistance genes in rice varieties. First, the most diverse rice varieties and fungal isolates will be identified according to the genetic diversity of the host and of the pathogen in China as determined by molecular markers. The resulting set of varieties will be inoculated with the set of isolates in order to estimate the genetic diversity for resistance. Second, genetic approaches using the pathogen will be used to obtain progeny with various combinations of avirulence genes. These progeny will be inoculated on Chinese rice varieties to better classify rice varieties sharing the same reaction pattern and therefore the same resistance genes. The third method relies on the mapping of resistance genes by inoculation of Chinese isolates of the rice blast fungus. Genes mapping at different loci should be distinct. Resistance genes mapping at the same locus will be assessed for allelism using a wide range of isolates either virulent or avirulent to each of the resistance genes studied. At the same time, avirulence of the Chinese populations of the pathogen will be investigated to study frequency of avirulent/virulent alleles. This information will be very useful to estimate the durability of resistance genes or resistance gene combinations.

1. PROJECT INFORMATION

RESIDIV is the acronym of the project: "Exploring the genetic diversity of rice-blast host-pathogen interactions in China: a tool to improve breeding for resistance to blast disease" (ICA-CT2000-30021). This project is supported by the European Community. RESIDIV is an INCO-DEV project within the 5th Framework Programme. The three year project started in October 2000. RESIDIV is coordinated by CIRAD (Centre de coopération Internationale en Recherche Agronomique pour le Développement) in Montpellier (France) and involves 4 other partners: CNRS (Centre National de la Recherche Scientifique) in Lyon (France), CNRRI (China National Rice Research Institute) in Hangzhou (China), the Chinese Academy of Sciences in Beijing (China) and the University of Aberdeen in Aberdeen (UK).

Detailed information, including annual reports, are available on the RESIDIV website: http://RESIDIV.CIRAD.FR.

2. OBJECTIVES OF THE PROJECT

Rice blast is the most widespread and damaging fungal disease of rice worldwide. Control methods are based on the use of fungicide and of rice resistance, this latter often facing resistance breakdown. Obtaining durable resistance to blast (caused by the fungus *Magnaporthe grisea)* is a main objective in most rice breeding programs. Resistance breakdown is mainly due to incomplete or wrong resistance gene identification and partial knowledge of pathogen population virulence. The research project will take into account the genetic diversity of both the plant and the pathogen.

The objective of the RESIDIV project is to identify and characterize genes conferring resistance to the blast disease (caused by the fungus Magnaporthe grisea) in Chinese rice varieties. Chinese traditional rice varieties have a large biodiversity and characterization of this material will lead to the identification of new resistance genes. The project aims at improving our knowledge on resistance diversity in Chinese rice varieties and on the diversity of the pathogen for its ability to attack resistant varieties (avirulence/virulence). This knowledge will be of important value for breeders to introduce the appropriate resistance genes by classical methods or by marker assisted selection to obtain durable resistance. The results of the RESIDIV project will be exploited through conventional breeding methods using natural defense systems.

The resistance gene diversity in Chinese varieties will be characterized by three different methods: 1) inoculation of a set of isolates with different virulence spectra, 2) inoculation of progeny of crosses between isolates of the fungus, 3) Genetic mapping of resistance genes. The genes identified will be compared to known resistance genes by comparing reaction patterns of these varieties to the reaction patterns of differential varieties with known resistance genes. Fine identification of resistance genes in Chinese rice varieties is expected. Several new resistance genes will probably be identified.

Segregation of blast avirulence genes will be studied and molecular markers tightly linked to these genes will be searched. By exploiting the obtained linked molecular markers, avirulence genes will be cloned and sequenced. These sequences will be used to develop Polymerase Chain Reaction (PCR) markers to estimate the frequency of the different alleles of virulence genes in Chinese populations of the blast fungus.

Characterization of rice resistance and blast avirulence is needed for appropriate resistance gene deployment. The project results will then allow the choice of the most suited resistance genes and the best way to use them. This will ensure resistance durability.

3. ACTIVITIES

The activities of the project are organized in Work Packages (WP), described below.

WP1 Characterization of Chinese rice varieties resistance spectrum to Chinese blast isolates

1. Definition of a working collection of Chinese and international differential rice varieties
2. Characterization of resistance spectrum to blast by inoculation of these rice varieties with isolates of the pathogen
3. Definition of groups of varieties showing the same pattern of reaction to the isolates
4. Comparison of these patterns with patterns of varieties with known resistance genes

WP2 Identification of blast avirulence genes and of corresponding rice resistance genes

1. Achievement of crosses between isolates of the fungus
2. Genetic study of avirulence of several Chinese blast isolates by inoculation of blast progeny from several crosses between these isolates
3. Characterization of resistance gene diversity and distribution
4. Comparison of resistance genes in Chinese varieties with known resistance genes
5. Identification of molecular markers linked to the avirulence genes and construction of genetic maps of crosses to prepare cloning of these genes.

WP3 Molecular characterization of avirulence gene alleles in Chinese populations of the rice blast fungus: Evolution of avirulence genes in relation to the durability of resistance in rice.

1. Positional cloning of avirulence genes mapped in WP2
2. Survey and characterization of alleles at avirulence loci
3. Analysis of the diversity at avirulence loci in the Chinese rice blast fungus populations

WP4 Genetics of resistance of Chinese rice varieties

1. Characterization of Resistance Genes to Blast (RGB) distribution in the parents of rice mapping populations
2. Genetic study of RGBs in different populations
3. Identification of the resistance gene loci responding to different Chinese isolates in different mapping populations
4. Identification of new RGBs by comparison of the RGB loci with the mapping positions of known resistance genes
5. Identification of molecular markers closely linked to RGBs.

4. MILESTONES AND EXPECTED RESULTS

The expected results are:

- a fine characterization of resistance genes in a collection of 100-150 Chinese rice varieties
- comparison of these genes with known resistance genes and identification of new resistance genes
- identification of molecular markers linked to resistance genes
- cloning and sequencing of blast avirulence genes
- characterization of the distribution of avirulence genes in Chinese populations of the blast fungus.

For each Work package, expected results are translated into the milestones described in the following section.

Milestones for WP1

Define a working collection of Chinese rice varieties and a set of *Magnaporthe grisea* isolates with a wide genetic basis.
Characterize the resistance gene diversity in Chinese varieties by use of a set of isolates with different virulence spectra.
Identify resistance genes in Chinese varieties through comparison with resistance spectrum of varieties with known resistance genes.

Milestones for WP2

Use the hybridization potential of the rice blast fungus to generate progeny with different and known avirulence genotypes.
Identify and characterize resistance genes in Chinese varieties with rice blast progeny of crosses.
Genetic mapping and identification of tightly linked markers to avirulence genes of the blast fungus.

Milestones for WP3

Analyze the polymorphism at the Avr1-Irat7 locus in Chinese rice blast populations.
Clone and characterize new avirulence genes from Chinese rice blast populations.
Design primers to survey the polymorphism at these new loci in Chinese rice blast populations.

Milestones for WP4

Map resistance genes to blast in different populations.

Compare newly mapped resistance gene position with positions of known resistance genes.
Provide molecular markers closely linked to resistance genes to facilitate marker-assisted selection.

5. BENEFITS AND BENEFICIARIES

The outcomes of the RESIDIV project are an extensive characterization of resistance gene diversity in Chinese rice varieties, the identification of new resistance genes, and the characterization of avirulence gene diversity in the Chinese populations of the pathogen.

The information provided by the project will be of real importance and easily exploited in resistance breeding. First, results obtained during the project will provide strategic information on the isolates to be used for the characterization of the resistance of Chinese rice varieties. Second, the project will provide a rather complete characterization of resistance genes in Chinese varieties, allowing breeders to improve the choice of resistance donor parents for crosses. Third, the mapping of resistance gene will provide molecular markers closely linked to these resistance genes, preparing the way for Marker Assisted Selection. Finally, the characterization of genetic diversity of avirulence in the fungal population will provide important information on the best way to utilize the resistance gene identified and on the way to combine them to create varieties with durable resistance to blast.

DURABLE CONTROL OF RICE BLAST DISEASE WITH MULTILINES

*S. Koizumi[1], T. Ashizawa[2] and K. S. Zenbayashi[2]

[1]*National Agricultural Research Center, Tsukuba, Ibaraki, 305-8666 Japan*
[2]*National Agricultural Research Center for Tohoku Region, Yotsuya, Omagari, Akita, 014-0102 Japan*

**E-mail: skoizumi@affrc.go.jp*

1. INTRODUCTION

Multilines are mixtures of several near-isogenic lines (NILs) with uniform phenotypic agricultural traits but with different resistance genes to a crop disease. They are known to prevent breakdown of disease resistance and reduce disease development because of diversity of their resistance. Knowledge on the genetic background for both host resistance and pathogen virulence is necessary to develop the multilines. Several multilines have been developed in Japan for rice blast control, since considerable information regarding the disease has already been collected.

This paper describes rice blast control with multilines in Japan and their durability for the control.

2. DEVELOPMENT OF MULTILINES

2.1. Background

Multilines have been developed to control rice blast in Japan for several reasons. The first one is because of the breakdown of complete resistance to blast in newly released rice cultivars. Many instances of the breakdown caused by the multiplication of virulent blast races to the cultivars have been reported in the country (Kiyosawa, 1974; Yamanaka and Yamaguchi, 1987), and the idea of multilines proposed by Jensen

S. Kawasaki (ed.), Rice Blast: Interaction with Rice and Control, 191–199.
© 2004 *Kluwer Academic Publishers. Printed in the Netherlands.*

(1952) and Borlaug (1959) was introduced to inhibit the breakdown (Koizumi, 2001).

The second reason is that we already have enough knowledge on blast resistance in rice cultivars as well as the virulence of the blast fungus to these cultivars. In Japan, 15 complete resistance genes to blast have been identified in rice cultivars (Kiyosawa, 1997; Hayashi *et al.*, 1998). The genotypes of complete resistance and levels of partial resistance are known in almost all rice cultivars, and these must be clarified before their registration. Meanwhile, the distribution of blast races, being identified using 9 or 12 differentials with different single complete resistance genes, is investigated every several years (Naito *et al.*, 1999). Based on this information, NILs have been developed and effective components and their proportions have been decided to control blast with multilines.

Difficulty in combining high levels of partial resistance to blast and good eating quality led to the development of NILs derived from rice cultivars with good eating quality. Furthermore, increased environmental and economic awareness have paved the way to the use of multilines to control rice blast.

2.2. *Near-isogenic lines for multilines*

Among the 15 complete resistance genes to blast, 13 genes (*Pik-s*, *Pik-p*, *Pik-h*, *Pia*, *Pii*, *Pik*, *Pik-m*, *Piz*, *Piz-t*, *Pita*, *Pita-2*, *Pib*, and *Pit*) have been used to develop NILs in Japan. For most of the NILs, the typical back cross method was employed and five and eleven backcrosses have been conducted (Matsunaga, 1996). DNA markers linked to complete resistance genes have also been used for breeding some of them. Currently, NILs with different complete resistance genes to blast have been developed and were bred from 15 recurrent rice cultivars (Table 1). Most of their recurrent parents are leading rice cultivars in Japan. They have good eating quality as requested by consumers, but levels of partial resistance to blast are low.

Multilines from the rice cultivars Sasanishiki and Koshihikari were registered and released for blast control, and are presently cultivated in farmers' fields in Miyagi, Niigata and Toyama Prefectures. The details are described in the following chapters.

2. EFFECTIVENESS OF MULTILINES

3.1 *Mechanisms for blast control in multilines*

Decrease in the population of susceptible plants, the barrier effect by resistant plants and induced resistance are considered to cause rice blast reduction in multilines like other foliar diseases of cereal crops.

Table 1. *Near-isogenic lines for rice blast control with multilines in Japan*

Recurrent parent (RP)	Genotype of RP	Resistance genes in NILs	Breeding place and year developed (released)
Sasanishiki	*Pia*	*Pik-s, Pii, Pik, Pik-m, Piz, Piz-t, Pita, Pita-2, Pib*	Miyagi Pref., 1994, (1995)
Koshihikari	*Pik-s*	*Pia, Pii, Pik, Pik-p, Pik-m, Piz, Piz-t, Pita, Pita-2, Pib*	Niigata, Toyama, and Fukui Pref., 1999, 2000, 2001, (2002)
Hitomebore	*Pii*	*Pik, Pik-m, Piz, Piz-t, Pib, Pita, Pita-2*	Miyagi Pref.
Manamusume	*Pii*	*Pik, Pik-m, Piz-t, Pib, Pita*	Miyagi Pref.
Nipponbare	*Piks/Pia*	*Pii, Pik, Piz, Piz-t, Pita-2, Pib*	Natl. Agric. Res. Cent., 1984
Hokkai 241		*Piz, Pib, Pita-2, Piz-t, Pit*	Hokkaido Natl. Agric. Exp. Stn.,
Toyonishiki	*Pia*	*Pii, Pik, Pita, Pita-2, Piz-t*	Tohoku Natl. Agric. Exp. Stn.,
Hinohikari	*Pia, Pii*	*Pik-m, Pita, Pita-2*	Miyazaki Pref.
Maihime	*Pia*	*Pii, Pik-h, Pik-m, Piz, Pita, Pita-2, Piz-t, Pib*	Aomori Pref.
Hanaechizen	*Piz*	*Pik, Piz-t, Pita, Pita-2, Pib*	Fukui Pref.
Etsunan 157	*Pia*	*Pii, Piz, Piz-t, Pita-2*	Fukui Pref.
Mineasahi	*Pia, Pii*	*Pik, Pik-m, Piz, Piz-t, Pib, Pita, Pita-2*	Aichi Pref.
Chubu 64	*Pii*	*Pik, Pik-m, Piz, Piz-t, Pib, Pita, Pita-2*	Aichi Pref.
Kinuhikari	*Pii*	*Piz-t, Pib, Piz, Pita-2*	Natl. Agric. Res. Cent.
Akitakomachi	*Pia, Pii*	*Pik, Pik-m, Piz, Piz-t, Pita, Pita-2, Pib, Pit*	Akita Pref.

A reduction in the population of susceptible plants dilutes the inoculum and decreases the multiplication foci of the pathogen in cultivar mixtures and multilines (Chin and Wolfe, 1984; Mundt, 2002). The same mechanisms are observed for blast in mixtures of rice cultivars or multilines. Kiyosawa (1995) estimated that the dilution of the inoculum contributed to the reduction of leaf blast in the mixtures of susceptible and resistant rice cultivars.

In mixtures of susceptible rice cultivar Toyonishiki and its resistant NIL, blast severity decreases with an increase in the resistant NIL (Shindo and Horino, 1989), and the equation $r_m = r_s + c\ \log_e m$, proposed by Leonard (1969), where r_m is the apparent infection rate (r) in mixtures of susceptible and resistant rice cultivars, r_s is r in the single planting of susceptible plants, m is the proportion of susceptible plants in the mixtures, and c is a constant, can apply to leaf blast development in upland nursery trials (Koizumi and Kato, 1987; Koizumi 1994). These support the fact that dilution of inoculum and the decrease in the multiplication foci of blast fungus contribute to disease reduction.

The barrier effect by resistant plants is also believed to operate for blast reduction in multilines. In rice canopies of the mixtures of susceptible Sasanishiki and its resistant NIL, the resistant NIL reduces horizontal and vertical development of leaf blast (Koizumi, 2001). The reduction of blast development is assumed to be due to barrier effect by the resistant NILs because the reduction was not observed at initial stages of blast epidemics.

The obstruction to vertical spread of leaf blast with the resistant NIL reduces panicle blast severity, since leaf blast lesions in the upper layer of rice canopies are main inoculum sources for panicle blast (Ashizawa *et al.*, 2000).

Dispersion and deposition of blast spores within the mixtures of Sasanishiki and its resistant NIL were analyzed (Koizumi *et al.*, 1999) using DISBL, a computer model to simulate the dispersion and deposition of the spores in and over rice canopy (Koizumi and Kato, 1991). The analysis suggested that the deposition rate of the spores on the susceptible leaves was higher than that on the resistant leaves and that the barrier effect by resistant plants caused the leaf blast reduction of vertical development.

Induced resistance by pre-inoculation with incompatible strains of the blast fungus reduces blast severity by its compatible strains (Iwano, 1987; Ashizawa *et al.*, 2002). Nakajima *et al.* (1996) consider that the induced resistance depresses blast severity in the Sasanishiki multilines. However, the induced resistance operates only on the leaves pre-inoculated with the incompatible strains for leaf blast (Iwano, 1987), and a high-concentrate of spore suspension of incompatible strains is needed for panicle blast (Ashizawa *et al.*, 2002). The diseased plants by incompatible blast strains are required when the induced resistance operates in multilines, and Japanese farmers expect a high blast reduction in multilines. Hence, the contribution of the induced resistance to blast control in multilines is of less importance compared with that of the reduction in the population of susceptible plants and the barrier effect by resistant plants.

3.2 Effectiveness for blast reduction

The method of Mundt (1994) was employed to summarize blast reduction in rice cultivar mixtures and multilines in Japan and it is presented in Table 2. Blast severity data from the rice cultivar mixtures and multilines of equal proportions of components were used, and the reduction was calculated from the means of blast severity in component pure stands.

Leaf blast reduction was greater than that of panicle blast. Large effects of inter-plot interference; high levels of auto-infection; low levels of barrier effect; and lesion formation by incompatible blast strains are assumed to cause the less effectiveness of panicle blast control in multilines and cultivar mixtures.

The mixture of Sasanishiki and its nine NILs showed stable and high effectiveness in blast control when the mixture was compared with Sasanishiki multilines composed of 2, 3, and 4 components. Distribution of blast races strongly affected the effectiveness of the multilines. The race 007, which can attack three components in the multiline composed of Sasanishiki and its nine NILs, was constantly predominant in it. Race distribution may have contributed to the high effectiveness of the multiline.

Paddy field trials under heavy blast epidemics indicated that NILs with 75% resistance were required in the Sasanishiki multilines to reduce blast severity to the level of fungicide treatments (Koizumi, 2001).

Mundt (1994) showed that plot size, distance between plots, and wind conditions had important effects on stripe rust control measured in wheat cultivar mixture plots. The effectiveness of multilines in blast control is also considered to vary with plot size, weather conditions, blast severity, levels of partial resistance in NILs, amount of initial inoculum, mixing pattern and others. Actually, blast reduction due to mixtures of Sasanishiki NILs was

Table 2 *Blast reduction and yield increase relative to mean of component pure stands for the cultivar mixtures and multilines in Japan*

Number of component in mixture	Number of trials	Reduction (%)						Yield increase (%)		
		Leaf blast			Panicle blast					
		Min.	Max.	Mean	Min.	Max.	Mean	Min.	Max.	Mean
2	27	-51	94	48	- 23	77	24	-11.4	28.1	9.7
3	8	-40	93	60	-15	86	39	2.8	16.5	9.7
4	3	28	33	31	-44	37	0	2.0	7.1	5.1
10	4	59	97	74	16	53	41	0.2	14.1	9.2
Total	42	-51	97	52	-44	86	26	-11.4	28.1	9.2

greater in big plots (150 m^2) than in small plots (5.8m^2), and the reduction was bigger in uniform mixtures of the NILs than in mixtures in alternate rows (Koizumi, 2001).

Most of the trials summarized in Table 2 were conducted using small plots. To know the effectiveness of multilines for blast reduction in farmers' fields, mixture trials should be carried out using farmers' field size plots.

4. FOR DURABLE CONTROL OF RICE BLAST WITH MULTILINES

4.1 Monitoring of blast fungus populations

The effectiveness of multilines for blast control is affected by virulence frequencies of blast populations. Hence, monitoring the virulence frequencies is very important for durable blast control with multilines.

To monitor the virulence frequencies, classical procedure of pathogenic race identification is currently used. The procedure, however, needs much time and labor for mono-conidial isolation of the blast fungus, inoculation, evaluation and others. For easier estimation of virulence frequencies, spore trap using seedlings of NILs and assessing leaf blast severity of NILs in paddies were tested. The tests indicated that there were high positive correlations between the estimated and actual frequencies as shown in Figure 1 (Koizumi and Fuji, 1994; Koizumi and Tani, 1996).

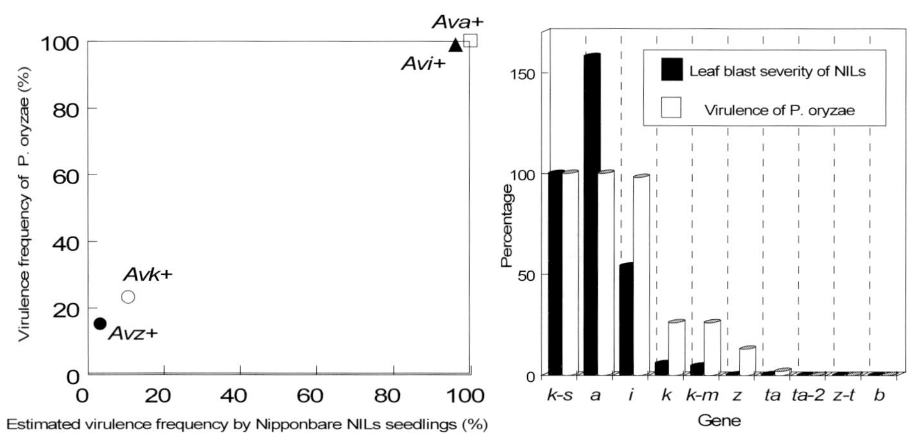

Figure 1. *Estimation of virulence frequencies of rice blast fungus populations by spore trap using NILs seedlings (right) and assessing leaf blast severity of NILs (left).*

Detection of origins of new blast races, which can attack components of multilines, is important in order to reduce them. DNA markers to discriminate blast populations will be very helpful in the detection.

4.2 Decision on components and their proportions

Japanese multilines are currently composed of 3 to 4 NILs, and most of the components are resistant to blast races distributed in their cultivated areas, since high proportions of resistant components cause effective blast control. However, durability of blast resistance in the NILs is not clear because exact prediction of virulence frequencies of blast fungus populations in the multilines is not yet achieved. For durable blast control with the multilines, the prediction system of the virulence frequencies of blast fungus populations in the multilines must be established.

Ashizawa *et al.* (2002) developed a computer model, BLASTMUL, to simulate leaf blast epidemics in multilines. The model can contribute to the exact prediction of virulence frequencies in the multilines.

4.3 Utilization of new resistance genes

Paddy field trials indicate that increase in the number of components in multilines cause effective blast control (Table 2). However, the available genes with complete resistance to blast are limited. New complete resistance genes should be introduced from foreign countries to further develop new NILs.

Partial resistance is also considered to contribute durable blast control with multilines, since the resistance reduces leaf blast development in rice cultivar mixtures (Koizumi, 1994). Genetic analysis of partial resistance as well as complete resistance to blast is progressing (Zenbayashi *et al.*, 2002). The advances in the genetic analysis will contribute to easy development of NILs with partial resistance as well as complete resistance for multilines.

4.4 Production of multiline seeds

It takes two years to change the proportions of components in the Japanese multilines and production of the multiline seeds needs much time and labor compared with common rice cultivars. A simpler and easier method for production and distribution of the multiline seeds should then be developed.

5. CONCLUSION

Multilines increase blast resistance in leading Japanese rice cultivars thus; contribute to the reduction of blasticide applications. However, stability of resistance in the multilines is not clear because their cultivated areas have been relatively small, *e.g.* Sasanishiki multilines are cultivated in less than 5,500 hectares.

In Niigata Prefecture of central Japan, the Koshihikari multilines, which were developed from the most popular rice cultivar and composed of 4 NILs, are planned to be cultivated on 86,000 hectares in 2005. Conceivably, the cultivation in a large scale will make clear the durability of blast control with multilines as experienced in China (Zhu *et al.*, 2000).

REFERENCES

Ashizawa, T., Zenbayashi, K., Fuji, S. and Koizumi, S. (2000) Effects of mixtures of cv. Sasanishiki and its resistant near-isoline on distribution of leaf blast lesions and panicle blast severity in the rice canopy. *Jpn. J. Pytopathol.* 66, 199 (Abstract, in Japanese).

Ashizawa, T., Zenbayashi, K., Fuji, S. and Koizumi, S. (2002) Severity of panicle blast on Sasanishiki near-isogenic lines after inoculation with avirulent isolates of rice blast fungus and disease suppression by pre-inoculation. *Jpn. J. Pytopathol.* 68, 305-308 (in Japanese, English summary).

Ashizawa, T., Zenbayashi, K. and Koizumi S. (2002) Blastmul: A simulation model and forecasting system for leaf blast epidemics in rice multilines. *3rd International Rice Blast Conference*, Abstracts p.75.

Borlaug, N. E. (1959) The use of multilineal or composite varieties to control airborne epidemic diseases of self-pollinated crop plants. In *Proc. 1st Int. Wheat Genet. Symp.*, Winnipeg, pp. 12-26.

Chin, K.M. and Wolfe, M. S. (1984) The spread of *Erysiphe graminis* f. sp. *hordei* in mixtures of barley varieties. Plant Pathol. 33, 161-171.

Hayashi, N., Ando, I. and Imbe, T. (1998) Identification of a new resistance gene to a Chinese blast fungus isolate in the Japanese rice cultivar Aichi Asahi. *Phytopathology* 88, 822-827

Iwano, M. (1987) Suppression of rice blast infection by incompatible strain of *Pyrcularia oryzae* Cav. Bull. Tohoku Natl. Agric. Exp. Stn. 75:27-39 (in Japanese, English summary).

Jensen, N. F. (1952) Inter-varietal diversification in oat breeding. *Agron. J.* 44, 30-34.

Kiyosawa, S. (1974) Studies on genetics and breeding of blast resistance in rice. *Misc. Publ. Bull. Natl. Agric. Sci., Ser.* D 1, 1-58 (in Japanese, English summary).

Kiyosawa, S. (1995) Proposals to the use and research on multiline variety (3). *Nogyo Gijyutsu* 50, 210-212 (in Japanese).

Kiyosawa, S. (1997) Inheritance of resistance to blast disease in rice. In *Plant disease resistance: a new frontier in plant science,* Tokyo, Shujunsha, pp 56-64 (in Japanese).

Koizumi, S. (1994) Effect of field resistance on leaf blast development in mixtures of susceptible and resistant rice cultivars, *Ann. Phytopathol. Soc. Japan* 60, 585-594.

Koizumi, S. (2001) Rice blast control with multilines in Japan. In *Exploiting biodiversity for substainable pest management.* Mew, T. W., Borromeo, E., Hardy, B. eds., Los Banos, Philippines, IRRI, pp. 143-157.

Koizumi, S., Ashizawa, T. and Zenbayashi, K. (1999) Distribution of leaf blast lesions in the rice canopy of a mixture of cv. Sasanishiki and its resistant near-isogenic line. *Ann. Phytopathol. Soc. Jpn.* 65, 349 (Abstract, in Japanese).

Koizumi, S. and Fuji, S. (1994) Effect of mixtures of isogenic lines developed from rice cv. 'Sasanishiki' and 'Nipponbare' on blast development. *Res. Bull. Aichi Agric. Res. Ctr.* 26, 87- 97 (in Japanese, English summary).

Koizumi, S. and Kato, H. (1987) Effect of mixed plantings of susceptible and resistant rice cultivars on leaf blast development. *Ann. Phytopathol. Soc. Jpn.* 53, 28-38.

Koizumi, S. and Kato, H. (1991) Dynamic simulation of blast epidemics using a multiple-canopy spore dispersal model. In *Rice blast modeling and forecasting.* Manila, Philippines, IRRI, pp. 75-88.

Koizumi, S. and Tani, T. (1996) Differences in the effectiveness towards rice blast control among Sasanishiki multilines, cultivars with high-level field resistance and fungicide applications. *Res. Bull. Aichi Agric. Res. Ctr.* 28, 53-68 (in Japanese, English summary).

Leonard, K. J. (1969) Factors affecting rates of stem rust increase in mixed plantings of susceptible and resistant oat varieties. *Phytopathology* 59, 1845-1850.

Matsunaga, K. (1996) Breeding of a multiline rice cultivar Sasanishiki BL and its use for control of blast disease in Miyagi Prefecture. *Nogyo Gijyutsu* 51, 173-176 (in Japanese).

Mundt, C. C. (1994) Use of host genetic diversity to control cereal diseases: Implications for rice blast. In *Rice Blast Disease*, R. S. Zeigler, S. A. Leong and P. S. Teng eds.,Cab International and IRRI, Cambrige, pp. 293-308.

Mundt, C. C. (2002) Use of multiline cultivars and cultivar mixtures for disease management. *Annu. Rev. Phytopathol.* 40, 381-410.

Naito, H., Iwano, M., Fujita, Y. and Ashizawa, T. (1999) Distribution of pathogenic races of rice blast fungus in Japan in 1994. *Misc. Pub. Natl. Agric. Res. Cent.* 39, 1-92 (in Japanese).

Nakajima, T., Sonoda, R., Yaegashi, H., and Saito, H. (1996) Factors related to suppression of leaf blast disease with a multiline of rice cultivar Sasanishiki and its isogenic lines, *Ann. Phytopathol. Soc. Jpn.* 62, 360-364.

Shindo, K., and Horino, O. (1989) Control of rice blast disease by mixed plantings of isogenic lines as multiline cultivars. *Bull. Tohoku Natl. Agric. Exp. Stn.* 39, 55-96 (in Japanese, English summary).

Yamanaka, S. and Yamaguchi, T.(1987) *Rice Blast Disease*, Yokendo, Tokyo (in Japanese).

Zenbayashi, K., Ashizawa, T., Tani, T. and Koizumi, S. (2002) Mapping of the QTL (Quantitative trait locus) confering partial resistance to leaf blast in rice cultivar Chubu 32. *Theor. Appl. Genet.* 104, 547-552.

Zhu, Y. Y., Chen, H.R., Fan, J.H., Wang, Y. Y., Yan, L., Chen, J. X., Fan, J. X., Yang, S. S., Hu, L. P., Leung, H., Mew, T. W. , Teng, P. S., Wang, Z. W. and Mundt, C. C..(2000) Genetic diversity and disease control in rice. *Nature* 406, 718-722.

RICE BLAST CONTROL WITH SASANISHIKI MULTILINES IN MIYAGI PREFECTURE

*Masashi Sasahara[1] and Shinzo Koizumi[2]

[1]*Miyagi Prefectural Furukawa Agricultural Experiment Station, Ohsaki, Furukawa, Miyagi 989-6227 Japan*
[2]*National Agricultural Research Center, Kannondai, Tsukuba, Ibaraki 306-8666, Japan*

**E-mail: smasashi@faes.pref.miyagi.jp*

Abstract. For rice blast control with multilines, nine near-isogenic lines (NILs) with different complete blast-resistance genes were developed from a leading rice cv. Sasanishiki at Miyagi Prefectural Furukawa Agricultural Experiment Station in Japan. Four of these nine NILs, namely, the *Pik, Pik-m, Piz* and *Piz-t* lines were first registered as multilines named Sasanishiki BL (BL) in 1994, and the *Pita-2, Pita* and *Pib* lines were added to the registered NILs in 1997, 1998 and 2001, respectively. BL has been cultivated in farmers' fields in Miyagi Prefecture from 1995, was planted on over 5,000 hectares in 1997. However, its cultivated area is now less than 800 hectares, since Sasanishiki has become unpopular. Resistance of the Sasanishiki multilines to blast has not been broken down for 7 years, and other NILs are also being developed from other leading rice cultivars in our experiment station. New rice multilines are expected to be cultivated for blast control.

Key words: Sasanishiki BL, near-isogenic line, rice blast, pathogenic races, Miyagi Prefecture

1. INTRODUCTION

Sasanishiki is one of the leading rice cultivars in Japan because of its good eating quality. However, since its level of partial resistance to blast is low, the disease has often caused severe damage to the cultivar. To control

S. Kawasaki (ed.), Rice Blast: Interaction with Rice and Control, 201–207.
© 2004 *Kluwer Academic Publishers. Printed in the Netherlands.*

blast in this cultivar, blasticides are commonly applied four to five times per planting season. Thus, farmers request a new labor-saving cultivation method, and consumers demand the reduction of the rate of chemical application for food safety and less environmental pollution. Under these conditions, nine near-isogenic lines (NILs) with different complete blast-resistance genes were developed from Sasanishiki for control of the disease at Miyagi Prefectural Furukawa Agricultural Experiment Station in Japan, and eight of them have been registered as multilines named Sasanishiki BL in 1994. Currently, four lines are mixed and cultivated as multilines in farmers' fields. In this report, rice blast control with Sasanishiki multilines in Miyagi Prefecture, particularly their development, extension process, blast severity level and blast races in fields will be introduced.

2. DEVELOPMENT OF SASANISHIKI BL

In 1986, nine NILs with different complete blast-resistance genes were developed from the recurrent parent Sasanishiki by five to eight backcrosses. Each of these NILs has a different complete blast-resistance gene in addition to the *Pia* gene derived from the recurrent parent, although the *Pia* gene in the *Pita-2*, *Piz-t*, and *Pib* lines has not been confirmed yet. The main characteristics of these lines are almost the same as those of Sasanishiki except for the blast-resistance genes. Among the nine isogenic lines, four lines, namely, the *Pik, Pik-m, Piz* and *Piz-t* lines, were first registered as multilines named Sasanishiki BL (BL) in 1994. BL was the first released multiline cultivar in Japan. Other lines, the *Pita-2, Pita* and *Pib* lines were registered in 1997, 1998 and 2001, respectively (*Table 1*).

Table 1. Components of Sasanishiki BL and Complete Resistance Genotypes

Near-Isogenic Line	Complete Resistance Genotype	Cultivar Sasanishiki BL		
		Nourin-doushitu 327		Release Year
Sasanishiki	*Pia*	-	-	1963
Tohoku IL 1	+	-	-	-
Tohoku IL 2	*Pii, Pia*	-	-	-
Tohoku IL 3	*Pik, Pia*	327-1	Sasanishiki BL 1	1994
Tohoku IL 4	*Pik-m, Pia*	327-2	Sasanishiki BL 2	1994
Tohoku IL 5	*Piz, Pia*	327-3	Sasanishiki BL 3	1994
Tohoku IL 6	*Pita, Pia*	327-6	Sasanishiki BL 6	1998
Tohoku IL 7	*Pita-2, Pia* (?)	327-5	Sasanishiki BL 5	1997
Tohoku IL 8	*Piz-t, Pia* (?)	327-4	Sasanishiki BL 4	1994
Tohoku IL 9	*Pib, Pia* (?)	327-7	Sasanishiki BL 7	2002

3. EXTENSION PROCESS OF SASANISHIKI BL

3.1. Cultivation Guidelines

BL has been cultivated in farmers' fields at Miyagi Prefecture from 1995. Farmers should consider the following guidelines for BL cultivation.

1. BL seeds must be completely disinfected.
2. Blasticides should not be used for leaf blast control.
3. For panicle blast control, a blasticide should be sprayed once at the heading time because incompatible blast races can attack panicles.
4. Farmers must register for BL cultivation.
5. BL seeds must be completely renewable, since the proportion of BL components may be changed, in the case that blast races virulent to BL are predominant in farmers' fields.
6. BL should be cultivated by farmers in groups for effective blast control.

3.2. Seed Production

BL seed is produced as follows. In breeder's stock and original seed farms, the respective components are cultivated separately. The proportion of these components is decided by examination of blast races distributed in the BL-cultivated fields, and the original seeds of these components are first mixed for seed farm cultivation. The BL seeds are produced by mixed cultivation in the seed farms in this manner, and then the seeds produced are supplied to farmers for common cultivation of BL. Thus if we want to change the proportion of BL-components, we have to wait for two years to obtain BL seeds for preparing the new proportion.

3.3. Proportion of the Components and Cultivated Area

Table 2. *Proportions of Components and Cultivated Area*

Year	Proportion of the Components	Cultivated Area
1995	*Pik* : *Pik-m* : *Piz* : *Piz-t* = 4 : 3 : 3 : 0	845
1996	*Pik* : *Pik-m* : *Piz* : *Piz-t* = 3 : 3 : 4 : 0	4,513
1997	*Pik* : *Pik-m* : *Piz* : *Piz-t* = 1 : 1 : 4 : 4	5,454
1998	*Pik* : *Pik-m* : *Piz* : *Piz-t* = 1 : 1 : 4 : 4	3,234
1999	*Pik* : *Pik-m* : *Piz* : *Piz-t* = 1 : 1 : 4 : 4	1,524
2000	*Pik* : *Pik-m* : *Piz* : *Piz-t* = 1 : 1 : 4 : 4	845
2001	*Pik* : *Pik-m* : *Piz* : *Piz-t* = 1 : 1 : 4 : 4	633
2002	*Pik* : *Pik-m* : *Piz* : *Piz-t* = 1 : 1 : 4 : 4	-

3. BLAST SEVERITY OF BL

Figure 1. shows the leaf blast severity of BL and common rice cultivars in farmers' fields. Generally, the common rice cultivars have been treated

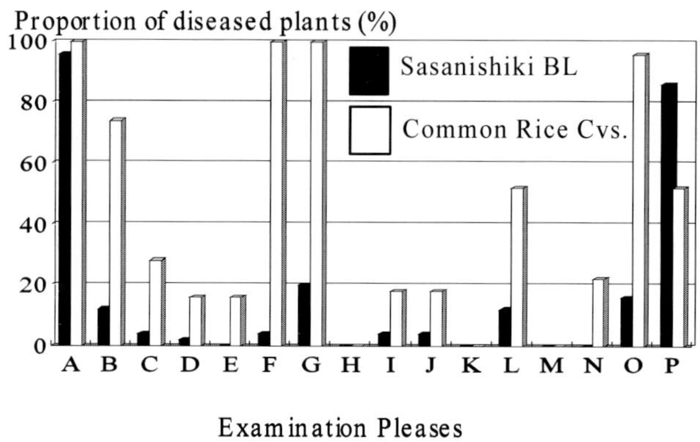

Figure 1. Leaf Blast Severities of the BL and Common Rice Cvs. in Farmers' Fields (1995). Letters indicate examination places in Miyagi Prefecture.

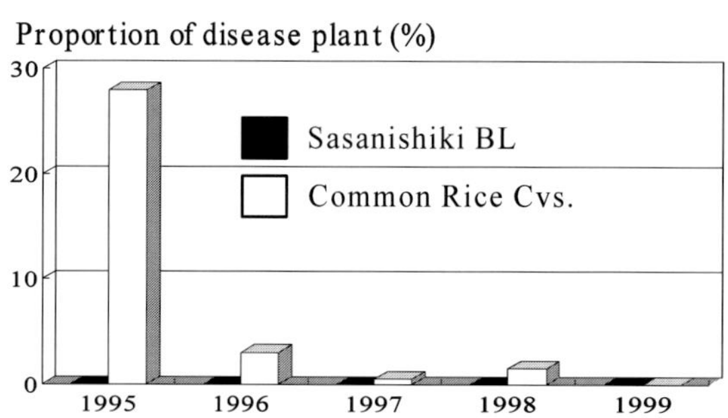

Figure 2. Changes in Panicle Blast Severity of the BL and Common Rice Cvs. in Farmers' Fields from 1995 to 1999.

with blasticides several times for leaf blast control, whereas BL has not been sprayed with any blasticides. The leaf blast severity of BL, however, was lower than that of common rice cultivars. *Figure 2.* shows the changes in the panicle blast severity of BL and common rice cultivars in farmers' fields. BL has been treated with a blasticide only once for leaf and panicle blast control. The panicle blast severity showed similar tendency to leaf blast severity, although we have not observed severe panicle blast epidemics in BL from 1996.

4. BLAST RACES ISOLATED FROM BL-CULTIVATED FIELDS

We examined blast races from BL-cultivated fields and their adjoining fields planted to common rice cultivars. *Table .3* shows the results of our examination. The most predominant blast race isolated from common rice cultivars was 007, which is avirulent to all components of BL. Blast race 037, which is virulent to BL, was isolated from only one place. On the other hand, the most predominant blast race isolated from BL was 037. This race is virulent to the two components of BL (*Pik* and *Pik-m* lines), and blast race 077 or 073 virulent to the three components (*Pik*, *Pik-m* and *Piz* lines) was also isolated at one place. Blast race 007, which is avirulent to BL, was also isolated from BL and its isolation rate was high. The main reasons for the high isolation rate of blast race 007 may be attributed to the facts that incompatible blast races can attack the panicles of BL, and other cultivars have been mixed with BL, such as Sasanishiki or Hitomebore.

Table .3 Races of *Pyricularia grisea* from Panicle Blast Lesions of Common Rice Cvs. (CRCs) and BL in Farmers' Fields (1996)

Place	No. of Isolated Examined	Isolated Japanese Races from CRCs		No. of Isolated Examined	Isolated Japanese Races from BL			
		007	037		007	037	073	077
A	6	6		12		12		
B	6	6		10			1	9
C	6	6		8	6	2		
D	6	6		10	10			
E	6	6		10		10		
F	6	6		10	10			
G	6	6		10	6	4		
H	6	6		10	4	6		
I	6	4	2	10	2	8		
J	6	6		10		10		
K	6	6		10	4	6		
Total	66	64	2	110	42	58	1	9

Table .4 shows the changes in blast races isolated from BL. Races 037and 033, which are virulent to the *Pik* and *Pik-m* lines, were predominant in the BL-cultivated fields in 1995 and 1996. However, from 1997, the number of places where the race 037 was isolated has been reduced. This may be caused by the decreased proportions of the *Pik* and *Pik-m* lines in BL. On the other hand, blast races 047 and 077, virulent to the *Piz* line, appeared in the year after the proportion of the *Piz* line was increased. Blast race 407, which is virulent to the *Piz-t* line, was also isolated from BL two years after the introduction of the *Piz-t* line. Currently, there are no BL-components completely resistant to all of the blast races present in Miyagi Prefecture. However, the race virulent to all components of BL has not yet been detected.

Table .4 Races of *Pyricularia grizea* from Panicle Blast Lesions of the BL in Farmers' Fields

Year	No. of Places Examined	No. of Places Isolated Japanese Races							
		003	007	013	033	037	047	077	407
1995	29	2	11	1	2	13			
1996	39	4	18			15	1	1	
1997	3		2			1			
1998	25	1	20			2	1		1
1999	4		4						
2000	2		2						

CONCLUSION

The area cultivated to Sasanishiki BL has decreased, since Sasanishiki became recently unpopular. However the resistance of BL to blast has not been broken down for seven years, although there were a few cases of slight blast outbreak. Our examination indicated that BL can reduce the rate of blasticide application. Recently new NILs are being developed from other leading rice cultivars with high level of partial resistance to blast in our prefecture, and we expect them to be cultivated as multilines for blast control.

REFERENCES

Ashizawa, T., Zenbayashi, K., Koizumi, S. (2001) Pathogenic Races of *Pyricularia grisea* Isolated from Rice Multiline, 'Sasanishiki BL', in Miyagi Prefecture from 1996 to 2000 and Possible Inoculum Sources. Annual Report of the Society of Plant Protection of North Japan. 52:14-16 [in Japanese].

Ohba, A., Tsuji, H., Sasahara, M. (1999) Occurrence of Field Isolates of *Pyricularia oryzae* Compatible with the Near-Isogenic Lines of Sasanishiki Rice in Miyagi Prefecture in 1998. Annual Report of the Society of Plant Protection of North Japan. 50:12-15 [in Japanese].

Sasaki, T., *et al.* (2002) Multiline Rice Variety of Sasanishiki "Sasanishiki BL". Bulletin of the Miyagi Prefectural Furukawa Agricultural Experiment Station 3: 1-35 [in Japanese].

Sasahara, M., Osada, S., Naito, H. (1998) Distribution of Pathogenic Races *Pyricularia oryzae* in Cultivated Area of Sasanishiki BL of Miyagi Prefecture in 1996. Annual Report of the Society of Plant Protection of North Japan. 49:8-11 [in Japanese].

Tsuji, H., *et al.* (1999) Changes of Pathogenic Races of Rice Blast Fungus on Multiline Cultivars 'Sasanishiki BL' in Recent Years. Annual Report of the Society of Plant Protection of North Japan. 50:16-20 [in Japanese].

DEVELOPMENT AND UTILIZATION OF ISOGENIC LINES *KOSHIHIKARI TOYAMA BL*

*Yoichiro Kojima[1], Takeshi Ebitani[2], Yoshitaka Yamamoto[2] and Tsukasa Nagamine[1]

[1]National Institute of Agrobiological Sciences (NIAS) Kannnon-dai, Tsukuba, Ibaraki 305-8602 JAPAN; [2]Toyama Agricultural Research Center, Yoshioka, Toyama, 939-8153 Japan

**E-mail: ykojima@affrc.go.jp*

Abstract: Six isogenic lines with complete resistance genes (R-genes) on the genetic background of the japonica rice variety *Koshihikari* were developed. BL1 has *Piz-t*, BL2 has *Pii* and Pita-2, BL3 has *Pib*, BL4 has *Pik-p*, BL5 has *Pik-m*, and BL6 has *Piz* and *Pia*, respectively. We adopted these lines for mixed planting, because we expected them to have a barrier effect. Based on the distribution of the blast race, we determined that we should mix BL 1, BL2 and BL4 in a ratio of 4, 4 and 2. After mixing, we investigated the barrier effect under moderate and heavy epidemics of disease. The barrier effect was confirmed under the moderate conditions even when the mix contained 40% of the susceptible variety, but the ratio of the susceptible variety could only be 20% under the heavy conditions. This suggests that the mixing ratio of the susceptible varieties should be restricted to within 20 % under a heavy blast epidemic. Furthermore, BL showed strong resistance to the panicle blast that occurred during the ripening period. In this case, the BL yield was 1.7 times as high as that of *Koshihikari*.

Key words: blast disease, isogenic line, *Koshihikari*, mixed planting

1. INTRODUCTION

Koshihikari is the most popular rice variety in Japan because of its quality and taste. This variety of rice has occupied the largest cultivation area since 1979 in Japan. However, this variety is susceptible to blast disease caused by

S. Kawasaki (ed.), Rice Blast: Interaction with Rice and Control, 209–214.
© 2004 *Kluwer Academic Publishers. Printed in the Netherlands.*

Pyricuralia oryzae Cavara. In 1993 and 1998, there were particularly widespread epidemics of blast disease throughout Toyama prefecture. The economic loss was estimated to be six hundred-twenty million yen in 1993, and three hundred-seventy million yen in 1998.

It has been reported that the mixed planting of isogenic lines is effective in controlling blast disease (J.E.Vanderplank 1984). In addition, the barrier effect has been proposed to be one of the mechanisms for disease control (Nakajima 1996). Some isogenic lines with different R-genes were developed from the rice variety *Sasanishiki* at Miyagi Prefectural Furukawa Agricultural Experiment Station, and have been utilized for decreased fungicide cultivation (Koizumi and Tani 1996). Some isogenic lines for leading varieties, for example, *Hitomebore* and *Akitakomachi* are now ready for use. However, there have been no lines developed for Koshihikari, and we therefore developed isogenic lines with R-genes on the genetic background of *Koshihikari*, and then investigated how to use these lines.

2. DEVELOPMENT OF ISOGENIC LINES

We adopted back-crossings and confirmed the R-genes based on the plant reaction against the inoculation on leaves with the blast spore suspension. Then, six varieties with different R-genes were developed and named *Koshihikari Toyama BL* (BL). The names of these varieties and the R-genes introduced are summarized as follows:

BL1 has *Piz-t*, BL2 has *Pii* and *Pita-2*, BL3 has *Pib*, BL4 has *Pik-p*, BL5 has *Pik-m*, and BL6 has *Piz* and *Pia*, respectively. The agromorphological

Table 1 *Plant characters of BL and variety registration*

Variety	Complete R-genes	Crossing	Heading Date (m.d)	Culm Length (cm)	Taste Quality	Registration (y.m.d)
BL1	*Piz-t*	*Koshihikari**4/Tohoku IL8	7.28	84	0.05	2001.1.16
BL2	*Pita-2*, *Pii*	*Koshihikari**2//Kinuhikari*2 /Tohoku IL7	7.28	84	0.02	Do
BL3	*Pib*	*Koshihikari**5/Tohoku IL9	7.28	84	-0.02	Do
BL4	*Pik-p*	*Koshihikari**5/K60	7.29	82	-0.01	Preparation
BL5	*Pik-m*	*Koshihikari**7/Tohoku IL4	7.28	82	0.10	Do
BL6	*Piz, Pia*	*Koshihikari**7/Tohoku IL5	7.28	83	0.06	Do
Koshihikari	+	-	7.28	83	0.00	-

*Figures in the table are the average of 1998～2001's experiment.
*Values of the taste quality were shown as –2 as "bad", and +2 as "good".
*"Tohoku ILs" are japonica rice variety with a *Sasanishiki* background.

traits of BLs were concluded to be the same as those of *Koshihikari* (Table 1). On January 16, 2001, BL1, BL2, and BL3 were registered.

3. MIXED PLANTING AND MIXING RATIO

Two types of cultivation, mixed planting or alternative planting, were proposed for the use of the isogenic line. We adopted mixed planting and decided the best mixing ratio of the BLs, in order to induce the barrier effect.

We investigated the R-genes of varieties that had been cultivated in Toyama prefecture (Fig.1), and the distribution of the pathogenic races of the blast disease (Table 2).

The cultivation area of the *Pii* variety reached its peak in 1987 and then decreased. The *Piz* variety also reached a peak in 1995 and then decreased. But *Pii* with the *Piz* variety is increasing still now.

The dominant races were 001 and 041 in 1999, and some minor races were detected, which were not observed before. In addition, race 005, which is virulent to *Pii*, was not isolated in 1982, but in 1994 it was observed at a high rate. Race 041, which is virulent to *Piz*, increased after the peak of cultivation of the *Piz* variety. However, race 045, which is virulent to *Pii* and *Piz*, is now rare.

This indicates that the distribution of race depends on the cultivated variety of rice, especially which R-genes this rice has. Then *Piz-t*, *Pita-2* and *Pib* would be available. Because these genotypic varieties have not been cultivated in the past twenty-five years, and a compatible race has not been detected. However, some races virulent to *Pik-p* and *Pik-m* were observed at a low rate. It would be best to reserve *Pi-a* with *Pi-z* for the future, because

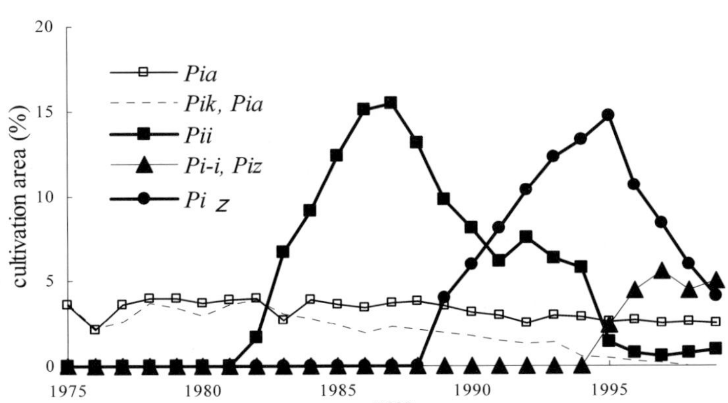

Figure -1. Changes of cultivation area of varieties with different R-genes in Toyama prefecture

the cultivation area of this genotypic variety is increasing.

In summary, we decided to mix BL1, BL2 and BL4 in a ratio of 4, 4 and 2, respectively.

Table 2. *Changes of pathogenic races of rice blast fungi*

Race	Virulent to	1982	1994	1999
		%	%	%
001	+	80.6	55.6	73.6
003	*Pia*	11.9	2.2	0.0
005	*Pii*	0.0	22.2	1.1
007	*Pia, Pii*	1.5	2.2	1.1
033	*Pia, Pik, Pik-m*	1.5	0.0	0.0
037	*Pia, Pii, Pik, Pik-m*	4.5	2.2	0.0
041	*Piz*	0.0	11.1	14.9
043	*Pia, Piz*	0.0	2.2	0.0
045	*Pii, Piz*	0.0	2.2	0.0
101	*Pita*	0.0	0.0	3.4
141	*Piz, Pita*	0.0	0.0	1.1
145	*Pii, Piz, Pita*	0.0	0.0	1.1
013.1	*Pia, Pik, Pik-p*	0.0	0.0	2.3
001.4	*Pit*	0.0	0.0	1.1

4. EFFECTIVENESS OF MIXED PLANTING

We investigated the barrier effect under moderate and heavy blast epidemics. For the tests, *Koshihikari* was added for susceptible variety and its ratio was varied from 0 to 100%. In addition, blasticide was used at one experimental block as a control.

Under moderate blast epidemics, blasticide-free *Koshihikari* showed 8.3-disease severity (Fig.2). '8.3'severity indicated 5 to 10 lesions were observed at 20% of the plants. The barrier effect was estimated based on the comparison of the expected severity when whole *Koshihikari* contained in the mixture was injured by the blast and the observed severity. In the figures, if the circles are over the bar, it shows that the barrier effect really worked. Addition to it, the disease severity in the mix containing 40% *Koshihikari* was equal to that of the blasticide used as a control.

This suggests that it would be possible to add 40% of the susceptible variety to the mixture.

Under the severe blast epidemics, the blasticide-free *Koshihikari* showed a 69-disease severity (Fig.3). A '69'severity indicated that all plants were injured. But at only 20% *Koshihikari*, we were able to observe the barrier effect. This shows that the barrier effect wasn't effective if the mixture contained more than 20% of the susceptible variety of rice.

Then the leaf blast stopped increasing for the sunny days. However, because of a long rainfall, an outbreak of heavy panicle disease suddenly occurred during the ripening period. We will now describe the relationship between panicle disease and yield. Panicle disease occurred heavily at *Koshihikari*, as shown in Table 3. The BL yield was 1.7 times as high as that of *Koshihikari*. That yield was almost the same as the usual yielding of *Koshihikari* in the field in which blasticide was used.

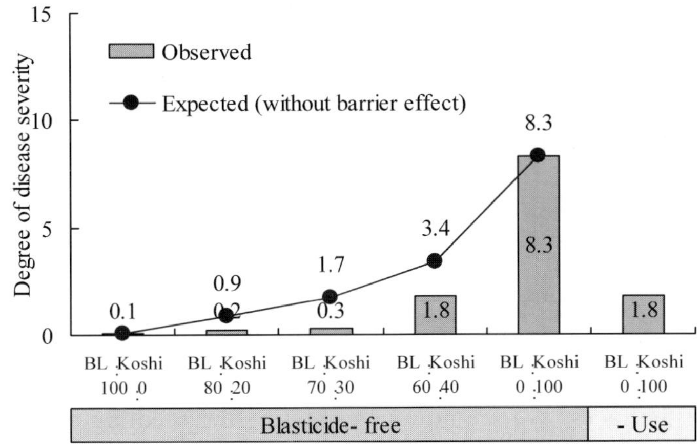

***Figure 2**. Leaf blast occurrence under moderate blast epidemics and the barrier effect*

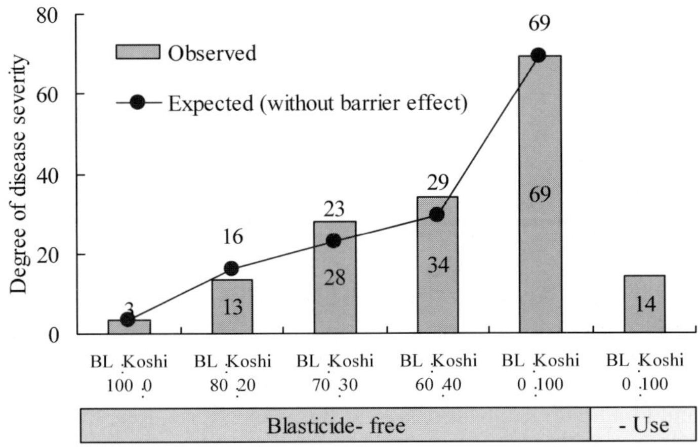

***Figure 3**. Leaf blast occurrence under severe blast epidemics and the barrier effect*

Table 3. *Comparison of panicle disease severity and yields of BL and Koshihikari*

Variety	BL	*Koshihikari*
Degree of disease severity	3.5	9.0
Yields (kg/ 10a)	437	253

* Degree of disease severity is shown from 0 (not observed) to 10 (injured all panicles).

5. CONCLUSION AND FUTURE WORKS

It was suggested that 20% of the susceptible variety was an acceptable ratio for mixed planting for the prevention of blast disease without the use of blasticide. Also, BL showed a strong resistance to panicle disease and maintained its yield. This suggests that when the panicle blasticide is ineffective, BL would be able to prevent the disease and would possibly reduce the appearance of a compatible race. However, the strength of BL in resisting panicle disease must be verified under many more conditions.

The cultivation area of BL increased to 37ha in 2000 and 80ha in 2001, and will be cultivated to a much greater degree in the future. To maintain the effectiveness of it and to avoid breakdown, it is important to trace the distribution of the races. We are now investigating the seedling trap method (Iwano and Asaga 1982) in order to improve the precision with which we can detect the races distributed in the cultivation area.

ACKNOWLEDGEMENTS

The authors are grateful to the farmer T.Sawayasu, who kindly let us use his fields for the tests, and to the researchers in the Plant Protection Center of Toyama for their kind assistance with the field tests.

REFERENCES

1. Iwano M and Asaga K. 1982. Estimation of Racial Composition and Its Proportion of Rice Blast Fungus (*Pyricularia oryzae Cav.*) in the Paddy fields by Seedling Trap Method. Ann. Rept. Plant Prot. North. Japan. 33: 21-24
2. J.E.Vanderplank. 1984. Disease resistance in plants, 2nd ed. Academic Press. Orland. 171-175
3. Koizumi S and Tani T. 1996. Differences in the Effectiveness towards Rice Blast Control among 'Sasanishiki' Multilines, Cultivars with High Level Field Resistance and Fungicide Applications. Res.Bull.Aichi Agric. Res. Ctr. 28: 53-68
4. Nakajima T, Sonoda R, Yaegashi H and Saito H. 1996. Factors Related to Suppression of Leaf Blast Disease with a Multiline of Rice Cultivar Sasanishiki and Its Isogenic Lines. Ann. Phytopathol. Soc. Jpn. 62: 360-364

STUDIES ON THE RICE BLAST PATHOGEN, RESISTANCE GENES, AND IMPLICATION FOR BREEDING FOR DURABLE BLAST RESISTANCE IN COLOMBIA

*F.J. Correa-Victoria[1], D. Tharreau[2], C. Martinez[1], M. Vales[1,2], F. Escobar[1], G. Prado[1] and G. Aricapa[1]

[1]*Centro Internacional de Agricultura Tropical, CIAT, AA 6713 Cali, Colombia,*

[2]*Centre de Coopération Internationale en Recherche Agronomique pour le Développement, CIRAD-CA, 34032 Montpellier, France*

**E-mail: f.correa@cgiar.org*

Abstract: Rice blast disease caused by *Pyricularia grisea,* the anamorph of *Magnaporthe grisea*, is the main rice production constraint in Latin America. Development of resistant cultivars has been the preferred means of controlling this disease; however, blast resistance is defeated by the pathogen shortly after cultivar release, affecting both leaves and panicles of the plant and reducing yields severely. Major efforts are being made at CIAT to understand the high pathogen variation observed, often reported as the main cause of resistance breakdown. We have analyzed extensively the genetic structure of blast pathogen populations using MGR-DNA and rep-PCR fingerprinting techniques and studied the avirulence gene diversity using a set of rice differentials with known resistance genes. The blast pathogen in Colombia has been found to be mainly clonal exhibiting few genetic lineages. At present, there are three lineages predominating in the pathogen population and their frequencies depend upon the susceptibility and planted area of the commercial rice cultivars by farmers. In general, a single haplotype predominates within each lineage. These three genetic lineages exhibit broad spectrum of virulence and together defeat all known blast resistance genes. However, some resistance genes are effective against all members of a lineage, suggesting an association of avirulence genes

S. Kawasaki (ed.), Rice Blast: Interaction with Rice and Control, 215–227.
© 2004 *Kluwer Academic Publishers. Printed in the Netherlands.*

and genetic groups in the pathogen. Avirulence genes vary in frequency in the pathogen population and some are highly frequent in several genetic lineages of the fungus. This suggests that these avirulence genes could play an important role in the pathogen or be associated with pathogenic fitness and then the corresponding resistance genes could be more relevant in breeding for durable resistance. Despite this high virulence diversity, breeders at CIAT have been able to develop durable blast resistant cultivars, indicating that combinations of these resistance genes may confer suitable and durable resistance to the pathogen. Evaluation and selection of breeding lines are routinely carried out under "hot spot" conditions favoring high pathogen pressure and diversity. Our studies are allowing us to identify and predict the durability of resistance gene combinations based on avirulence gene frequencies and the possible association of avirulence genes with pathogenic fitness. We have inferred the possible resistance genes present in our blast susceptible rice cultivars and initiated a backcrossing program to incorporate the desired resistance gene combinations into Latin American rice cultivars through marker assisted selection using mainly microsatellite and scar markers. Rice lines carrying the combination of the resistance genes Pi-1, Pi-2 and Pi-33 are then tested under controlled greenhouse conditions as well as our "hot spot" screening site using a spreader row technique to maintain a high and diverse population of the pathogen in the field. Resistant plants are then selected based on other desirable traits for their distribution to national programs in Latin America.

1. INTRODUCTION

Rice blast caused by *Pyricularia grisea* Sacc., the anamorph of *Magnaporthe grisea* (Hebert) Barr, is the main rice production constraint in the world. The fungus attacks leaves and panicles severely. Rice blast research has concentrated for many years on the identification of races of the pathogen and the incorporation of resistance genes to those races into commercial rice cultivars. However, blast resistance in Colombian commercial rice cultivars is lost in one to three years after cultivar release (Correa-Victoria and Martinez, 1994) with the exception of the commercial cultivars Oryzica Llanos 5 released in 1989 and the cultivar Fedearroz 50 released in 1998 (Table 1). Long-term studies are being conducted at the Centro Internacional de Agricultura Tropical, CIAT, to understand the great variability observed in the blast pathogen, often referred as the main cause of resistance breakdown.

The molecular techniques RFLP using the MGR 586 DNA sequence (Levy *et al.*, 1993) and rep-PCR using the Pot-2 DNA sequence derived primers (George *et al.*, 1998) were used to determine the genetic structure of the same pathogen populations used for virulence studies. The complexity of the pathogen described as many races (Correa-Victoria and Zeigler, 1993a) has been simplified in just six genetic families, named SRL-1 to SRL-6 (Levy *et al.*, 1995) which are mainly compatible with indica type of rice, and a genetic lineage named A-7 (Correa-Victoria *et al.*, 1994) compatible mainly with japonica type. Studies on the relationship between spectrum of virulence and a genetic lineage of the pathogen indicate that the most important aspect of this interaction is the existence of resistance genes, which are effective in controlling all isolates of the same lineage. The rice project at CIAT as objective has been developing a breeding strategy for the development of durable blast resistance. This strategy is based on studies on the composition and frequency of avirulence genes of the pathogen, characterization of the genetic structure, identification and incorporation of resistance gene combinations into commercial rice cultivars effective against populations of each genetic family, and the continuous evaluation and selection of breeding lines under a high disease pressure and pathogen diversity (Correa-Victoria and Zeigler, 1993a, 1993b, 1995).

Table 1 *Rice cultivars released in Colombia, sources of resistance and year of resistance breakdown to Pyricularia grisea*

Cultivar	Source of Resistance	Year of release	Resistance breakdown	Years of Resistance
Cica 4	Peta	1971	1972	1
Cica 6	IR-822-432	1974	1975	1
Cica 7	Colombia 1	1976	1978	2
Cica 9	C 46-15	1976	1977	1
Cica 8	Tetep	1978	1980	2
Metica 1	Colombia 1	1981	1982	1
Oryzica 1	C 46-15, Colombia 1,	1982	1985	3
Oryzica 3	Colombia 1, Tetep	1984	1985	1
Línea 2	C 46-15, Colombia 1,	1988	1989	1
Oryzica Llanos 5	IR 36, 5685, Colombia 1, Cica 9	1989	Not yet	> 13
Oryzica Caribe 8	Tetep, IR 665, Colombia 1, Cica 9	1993	1995	2
Fedearroz 50	IR 665, 5685, Colombia 1, Cica 9	1998	Not yet	> 4

2. CHARACTERIZATION OF VIRULENCE AND GENETIC STRUCTURE OF *PYRICULARIA GRISEA* IN COLOMBIA

The CIAT's Rice Project develops its breeding activities on resistance to *P. grisea* under favored upland conditions in the experimental station "Santa Rosa" from FEDEARROZ in the Meta department, Colombia. This site is characterized for having high blast disease pressure and pathogen diversity. A high disease incidence and severity is maintained in the breeding plots during the entire crop cycle by using spreader rows, which are conformed of a mixture of commercial rice cultivars susceptible to the different genetic lineages of the pathogen. These rows are planted perpendicularly to the rice breeding lines being evaluated for resistance. Large amounts of fungal spores are produced on the spreader rows and then disseminated by the wind and rain onto the breeding lines, increasing the chances of a line to be exposed to the pathogen and reducing the possibility of escape to infection (Figure 1).

Under this condition of evaluation and selection, we have found that the resistance selected is more stable and durable than under conditions of less blast pressure. However, the durability of the resistance depends not only on the plant, but also on the pathogen dynamics and changes in virulence. Establishing a system to understand the pathogenic variability and dynamics is then essential for the development of a strategy to control the blast disease through genetic resistance. Our studies indicate that most isolates exhibit the lost of a large number of avirulence genes, but none is virulent on all the resistance genes (Table 2) suggesting the need to pyramid suitable gene combinations in order to develop a more effective resistance against the pathogen (Correa-Victoria *et al.*, 1994, 2000). Although different genetic families of the fungus share several avirulence genes, we have detected a high specific interaction between some avirulence genes in the pathogen and some resistance genes in the plant. This interaction is characterized by the conservation of some avirulence genes in each genetic family despite all the evolution process occurring in the fungus. This interaction is of much value from a breeding point of view for the development of blast resistant rice cultivars (Table 2). This specific interaction is the base for selecting progenitors to be included in a breeding program for durable blast resistance.

The characterization of the genetic lineages, together with the virulence spectrum and frequency of the avirulence genes in the pathogen population should give us a better estimate of the potential durability of a resistance gene than the individual study of each one of these components. A resistance gene can be defeated by isolates from different genetic lineages, as it is the case of the genes Pi-t and Pi-k^{s} (Table 2). On the other hand, some resistance genes are effective controlling all the pathotypes of a single or several genetic families, even though they may be susceptible to the pathotypes of other different families. In this way, rice cultivars may have complementary resistance genes, which in combination in a cultivar confer resistance to the whole targeted pathogen population. The isogenic lines CT 13432-68, CT 13432-54, and CT 13432-55 which carry the resistance genes Pi-1, Pi-2, and Pi-33, respectively, have complementary resistance genes, which in combination or after pyramided in the isogenic line CT 13432-34 confer resistance to all the pathogen isolates belonging to the different genetic families of the fungus found in Colombia (Table 2).

Figure 1 *Experiment station of Santa Rosa for the evaluation of the resistance to P. grisea using spreader rows for the multiplication and dissemination of the blast pathogen maintaining a high disease*

We have also observed a gain in virulence (loss of avirulence genes) for the most predominant lineages as can be observed in the subdivision of lineage SRL-6 in SRL-6B and lineage SRL-4 in Table 2. The early detection, before having an increase in frequency, of pathotypes with a broad spectrum of virulence is of practical use for the identification of resistance genes that can be used in case that the resistance of a commercial cultivar is lost.

Table 2 *Avirulence genes present in isolates of Pyricularia grisea from Colombia*

		Isolate/ Genetic lineage								
		1	2	3	4	5	6	7	8	9
Source	Resistance Gene	L6B	L6B	L6	L4	L4	L4	L4	L2	L5
CT 13432-68	*Pi-1*			R	R	R	R	R	R	
CT 13432-54	*Pi-2*									R
CT 13432-55	*Pi-33*	R[1]	R	R		R			R	
C 104 PKT	*Pi-3*								R	
C 101 PKT	*Pi-4a*								R	
C 105 TTP4 (L23)	*Pi-4b*								R	
F 124-1	*Pi-ta*									
F 128-1	*Pi-ta*^{2}			R			R	R		
F 80-1	*Pi-k*			R				R		
F 98-7	*Pi-k*^{m}									
F 129-1	*Pi-k*^{p}			R						
F 145-2	*Pi-b*			R		R				
Aichi Asahi	*Pi-a*			R						
K 3	*Pi-k*^{h}			R	R	R	R	R	R	
K 59	*Pi-t*									
Rico 1	*Pi-k*^{s}									
Norin 2	*Pi-sh*			R						R
Nato	*Pi- i*									
OU 244	*Pi-z*	R	R	R					R	R
Toride	*Pi-z*^{t}		R	R						R
CT 13432-34	*Pi-1, Pi-2, Pi-33*	R	R	R	R	R	R	R	R	R

[1]R = Resistant or incompatible reaction as a result of the interaction between an avirulence gene in the pathogen and a resistance gene in the plant

Breeders at CIAT have developed rice cultivars with durable blast resistance lasting more than ten years after their release (Table 1). This is the case of cultivar Oryzica Llanos 5 whose parents exhibit susceptibility to the different genetic lineages of the pathogen (Figure 2). Our studies on blast populations collected in the field from all the five immediate parents show that the lineages recovered from each parent has a specific interaction with the cultivar of origin as shown in Figure 2. Greenhouse inoculations also show that these lineages infect the cultivar of origin, and that each parent carries resistance genes effective against the rest of the lineages of the pathogen. These results suggest that these parents carry complementary resistance genes, which in combination confer the durable resistance observed for more than 13 years in the cultivar Oryzica Llanos 5. According to Johnson, durable resistance is "the resistance that remains effective after being used in a large growing area for a long period of time under favorable environmental conditions to disease development". This cultivar has followed Johnson's definition of durable resistance and it is in our interest to determine the genetic composition of this cultivar to identify the gene combination conferring its durability.

3. IDENTIFICATION AND USE OF RESISTANCE GENE COMBINATIONS TO *PYRICULARIA GRISEA* FOR THE DEVELOPMENT OF RICE CULTIVARS WITH DURABLE BLAST RESISTANCE

A total of 283 recombinant inbred lines originated from the genetic cross between the near isogenic lines C 101 LAC (resistance gene Pi-1 and Pi-33) x C 101 A51 (resistance gene Pi-2) were developed for the identification of micro satellites (McCouch *et al.*, 2001) associated with these three resistance genes and the development of isogenic lines with different combinations of these genes to evaluate their resistance to *P. grisea* in the field and greenhouse inoculations. Our results demonstrated that the combination of the three resistance genes Pi-1, Pi-2 and Pi-33 is the most relevant and effective for breeding and development of rice cultivars with durable resistance to the blast pathogen in Colombia.

The virulence frequency on each one of the three resistance genes is high within some of the genetic lineages but no isolate is compatible with all three genes (Table 2, Table 3). We have shown that either in greenhouse inoculations or field evaluations over several growing seasons, the combination of the three resistance genes (Pi-1+Pi-2+Pi-33) in the isogenic lines CT 13432-34, CT 13432-107, CT 13432-189, and CT 13432-246 (Table 2, Table 3, Figure 3) confers complete resistance to the blast populations present in Colombia, which are compatible with any single gene or the combination of two of them (Table 3, Figure 3). According to our studies, the avirulence genes avr-Pi-1 and avr-Pi-33 are highly conserved in the blast population from Colombia, suggesting that these genes have a lower frequency of mutation towards virulence and probably are important genes in the fitness of the fungus. On the other hand, the avirulence gene avr-Pi-2 is conserved only in lineage SRL-5, suggesting that this avirulence gene has a larger rate of mutation towards virulence and that it would be less durable by itself.

The plant-pathogen interaction in the complex system rice-blast has been shown to follow the gene-for-gene hypothesis (Flor, 1971), where the resistance results from the genetic interaction between the resistance genes in the plant and the avirulence genes in the pathogen. The presence of only one of these corresponding genes in each organism (plant and pathogen) results in a resistant reaction. The mutation or loss of one of these genes in any of the two organisms is required for a susceptible reaction.

Many avirulence genes may act as "fitness" factors highly related to the survival capacity of the fungus (Leach *et al.*, 2001). It has been suggested that avirulence genes of high rate of mutation in the pathogen are probably less critical on the fitness processes than those that rarely mutate (Flor, 1971), and has been pointed out that the fitness of a pathogen is the factor or main force in the evolution and stability of a pathosystem in agriculture (Van der Plank, 1968). In this way, it can be considered that the quality and durability of a resistance gene is a direct function of the effects imposed to the pathogen after losing the function of

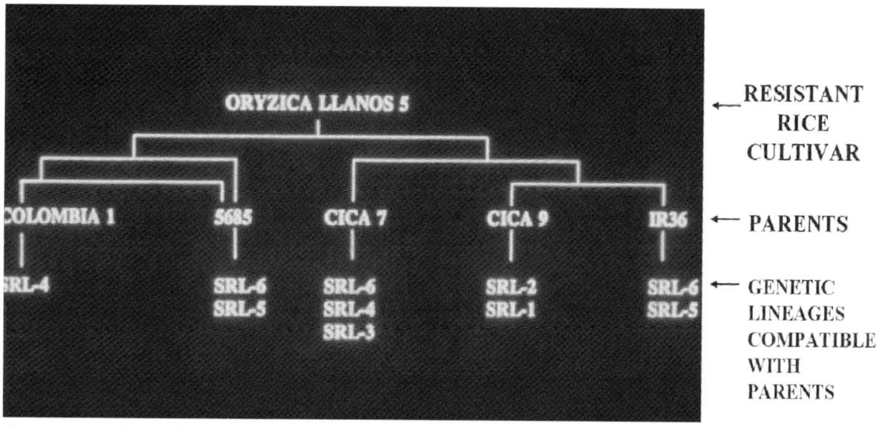

Figure 2 *Genealogy of the rice cultivar Oryzica Llanos 5 with durable resistance to P. grisea and the complementary blast reaction of the progenitors to the different genetic lineages of the pathogen*

Table 3 *Leaf and panicle reaction to Pyricularia grisea in isogenic lines with different combinations of the resistance genes Pi-1, Pi-2 and Pi-33*

Isogenic line	Resistance gene	Leaf blast		Neck blast
		BL4[1]	BL5[1]	Incidence (%)
CT 13432-95	None	9	9	Dead
CT 13432-110	None	9	9	Dead
CT 13432-68	1	9	9	Dead
CT 13432-54	2	9	9	Dead
CT 13432-6	33	9	9	Dead
CT 13432-230	1 + 2	9	9	Dead
CT 13432-26	1 + 33	9	9	Dead
CT 13432-193	2 + 33	8	5	15
CT 13432-34	1 + 2 + 33	2	2	0
CT 13432-107	1 + 2 + 33	2	1	0
CT 13432-189	1 + 2 + 33	2	1	0
CT 13432-246	1 + 2 + 33	2	1	0

[1] Fourth and fifth evaluations of leaf blast (BL).

the corresponding avirulence gene to overcome the corresponding resistance gene (Leach *et al.*, 2001). Therefore, the frequency of the avirulence genes in the pathogen can be used as an indicator to predict the durability of a resistance gene (Leach *et al.*, 2001) or the combination of several resistance genes. In other words, those resistance genes imposing to the pathogen a high value or deleterious effect after losing the corresponding avirulence gene to breakdown the resistance of the gene, will probably be a durable resistance gene.

Our studies at CIAT have been focused on developing the capacity to predict the durability of the resistance genes with the objective of using them efficiently in our breeding program. Isogenic lines developed at IRRI and CIAT with different resistance genes and combinations have been useful to determine the frequency of different pathogen genotypes present during different crop cycles both in commercial fields as well as in the experiment station. Even though, there are many studies demonstrating that individual resistance genes may confer durable resistance, our studies indicate that it is not the case for the control of rice blast in Colombia. It can be observed in Table 3 that for the eight possible genotypes using three resistance genes, the pathogen has developed probably seven genotypes, except the one losing the function of the three avirulence genes corresponding to the three resistance genes Pi-1, Pi-2, and Pi-33, combination which exhibits a complete blast resistance (Figure 3). The results suggest that the pathogen can lose the function of any of the three avirulence genes, or the combination of any two of them, but not the three avirulence genes (Table 2). It is possible that after losing one or two of the avirulence genes the other gene confers the needed fitness component to survive. This capacity would be lost in the case of losing the three avirulence genes.

Table 4 *Possible resistance genes present in commercial rice cultivars from Colombia inferred from inoculations with Pyricularia grisea isolates carrying the corresponding avirulence genes*

	Resistance gene									
Rice cultivar	*Pi-1*	*Pi-2*	*Pi-33*	*Pi-z*	*Pi-zt*	*Pi-ta2*	*Pi-sh*	*Pi-kh*	*Pi-k*	*Pi-b*
Oryzica 2	X[1]	X				X	X	X	X	X
Oryzica 3						X			X	
Cica 8	X					X			X	X
Cica 9		X				X				
IR 22						X	X		X	
Linea 2		X								
Oryzica Llanos		X					X		X	
Oryzica Caribe		X					X		X	
Oryzica Yacu 9		X								
Oryzica Llanos		X	X	X	X	X	X		X	X
Fedearroz 50		X	X	X	X	X	X		X	X

1 X = Presence of resistance gene

It is important to note that even for isolates with broad spectrum of virulence (many avirulence genes lost) as it is the case of isolate 1 of lineage L6B (SRL-6) in Table 2, this isolate has lost the avirulence genes for the resistance genes Pi-1 and Pi-2, but has conserved the avirulence gene for Pi-33. Similarly, isolate 4 identified as L4 (SRL-4), which has lost the avirulence genes for Pi-2 and Pi-33, has conserved the avirulence gene for Pi-1, and isolate 9 identified with lineage L5 (SRL-5) has lost the avirulence genes avr-1 and avr-33, but has conserved avr-2. Isogenic lines with individual resistance genes are also useful for the identification of avirulence genes present in different isolates of the blast pathogen, which can be used to predict which resistance genes could be present in a rice cultivar. The characterization of the avirulence genes in several hundred isolates of the fungus in Colombia has allowed us to infer the possible resistance genes present in the commercial cultivars of rice in Colombia (Table 4) after inoculating them with those isolates in the greenhouse. The rice cultivars Oryzica Llanos 5 and Fedearroz 50 that have exhibited a durable blast resistance (Table 1) carry the greatest number and similar resistance genes (Table 4). These two cultivars seem to carry the resistance genes Pi-2 and Pi-33 but not the gene Pi-1. The durability of the resistance of these two cultivars is probably due to the action of these two genes plus the action of the other genes shown in Table 4, and probably the presence of other unidentified genes.

With the objective to predict a future breakdown of the durable resistance exhibited by the cultivars Oryzica Llanos 5 and Fedearroz 50, we have been collecting leaf and panicle samples of single lesions occasionally developing on these two cultivars. These samples have been analyzed in order to advance a breeding program by identifying and incorporating resistance genes against the genetic lineages found to be in an evolution process to breakdown the resistance of these cultivars. The results of the molecular analysis of the isolates collected from these two cultivars in the last few years indicate that more than 90% belong to lineage SRL-4 and a few to lineage SRL-6. These isolates however, do not re-infect the two cultivars in greenhouse inoculations and do not develop typical lesions of the disease. Occasionally, some of these SRL-4 isolates induce under nutritional stress the formation of small non-sporulating lesions, or in a few cases typical blast lesions. However, these results suggest that lineage SRL-4 is in the process of evolution with higher probabilities than the lineages SRL-5 and SRL-6 of breaking down the resistance of the cultivars Oryzica Llanos 5 and Fedearroz 50. This observation is happening despite the fact that lineage SRL-6 has predominated in the entire pathogen population for many years. Our question of why lineage SRL-4 is found in more frequency than the other two lineages on these two cultivars, and why it is evolving with potentiality of breaking down their resistance, can be logically deduced from the data presented in Table 2. As it can be observed, isolate 4 in Table 2 exhibits the most recent evolution of lineage SRL-4 with virulence towards all the resistance genes present in the cultivars Oryzica Llanos 5 and Fedearroz 50 according to Table 4. This lineage however has not lost the avirulence gene to the resistance gene Pi-1, and it does not need to lose it, as this resistance gene is not present in any of the two cultivars (Table 4). On the other hand, those isolates with broader spectrum of virulence within lineages SRL-6 (isolate 1, Table 2) and SRL-5 (isolate 9, Table 2) would have to mutate or lose the avirulence genes to the resistance genes Pi-33 and Pi-2, respectively, which are present in the cultivars Oryzica Llanos 5 and Fedearroz 50. These two lineages have already lost the avirulence gene to Pi-1, and therefore to breakdown the resistance of the two cultivars would need to lose the three avirulence genes, avr-Pi-1, avr-Pi-2, and avr-Pi-33. According to our previous discussion, losing the three avirulence genes would have a deleterious effect on the pathogen.

The lineage SRL-4 has lost the avirulence genes for the resistance genes Pi-ta^2, Pi-k, and Pi-b (Table 2), which are present in those cultivars with durable resistance (Table 4). This lineage has not lost the avirulence gene for the resistance gene Pi-k^h, which is not present in the cultivars Oryzica Llanos

***Figure 3** Complete resistance to Pyricularia grisea in the isogenic line CT 13432-34 conferred by the combination of the resistance genes Pi-1, Pi-2, and Pi-33*

5 and Fedearroz 50. Therefore, this gene together with Pi-1, could be used in a breeding program to be incorporated in the two cultivars complementing their durable resistance, and moving ahead to the evolution that is apparently going on in this lineage of the pathogen that would lead them to the breakdown of the durable resistance present in the two commercial cultivars.

Our results also suggest that the two cultivars with durable resistance might carry other unknown resistance genes that need to be identified. We are in collaboration with Kansas State University working on the dissection of all the resistance genes present in the cultivar Oryzica Llanos 5.

4. CONCLUSIONS

- The genetic structure of *P.* grisea in Colombia ifs formed by few genetic lineages whose frequency depend on the resistance genes in the planted cultivars.
- The use of isogenic lines with different resistance genes for determining the frequency of avirulence genes in the pathogen allow us to identify relevant combinations of resistance genes for the control of rice blast.
- The rice blast pathogen in Colombia exhibits virulence to all known resistance genes. Durable resistance then can only be developed through combinations of resistance genes.
- Based on the relative contribution of the avirulence genes to fitness of *P. grisea*, plant genotypes can be designed and generated carrying the most effective combinations of resistance genes.
- A pathogen such as *P. grisea* which is mainly clonal, do not seem to be able in tolerating a continuous increase of mutations in its avirulence genes even in the case that those genes have minimal contributions to fitness.
- The combination of the resistance genes Pi-1, Pi-2, and Pi-33 confers resistance to the blast populations in Colombia. The use of molecular markers and greenhouse inoculations are useful for following the incorporation of resistance genes in breeding populations.
- Rice populations developed in breeding programs with the objective of accumulating several blast resistance genes should be evaluated and selected using field methodologies that maintain a high disease pressure and pathogen diversity.

REFERENCES

Correa-Victoria FJ and Martinez C (1994) Genetic structure and virulence diversity of Pyricularia grisea in breeding for rice blast resistance, in Induced Mutations and Molecular Techniques for Crop Improvement, IAEA, Vienna, pp. 133-145.

Correa-Victoria FJ and Zeigler RS (1993a) Pathogenic variability in Pyricularia grisea at a rice blast "hot spot" breeding site in eastern Colombia. Plant Disease 77: 1029-1035.

Correa-Victoria FJ and Zeigler RS (1993b) Field breeding for durable rice blast resistance in the presence of diverse pathogen populations, in Th Jacobs and JE

Parlevliet (eds.), Durability of Disease Resistance, Kluwer Academic Publishers, Dordretch, pp. 215-218.

Correa-Victoria FJ and Zeigler RS (1995) Stability of complete and partial resistance in rice to Pyricularia grisea under rainfed upland conditions in eastern Colombia. Phytopathology 85: 977-982.

Correa-Victoria FJ, Zeigler RS and Levy M (1994) Virulence characteristics of genetic families of Pyricularia grisea in Colombia, in RS Zeigler, SA Leong and PS Teng (eds.), Rice Blast Disease, CAB International, UK, pp. 211-229.

Correa-Victoria FJ, Escobar F, Prado G and Aricapa G (2000) Population dynamics of the rice blast pathogen in a screening site in Colombia and characterization of resistance, in D Tharreau, MH Lebrun, NJ Talbot and JL Notteghem (eds.), Advances in Rice Blast Research, Kluwer Academic Publishers, Dordreth, pp. 214-220.

Flor HH (1971) Current status of the gene-for-gene concept. Annual Review of Phytopathology 9: 275-296.

George MLC, Nelson RJ, Zeigler RS and Leung H (1998) Rapid population analysis of Magnaporthe grisea by using rep-PCR and endogenous repetitive DNA sequences. Phytopathology 88: 223-229.

Johnson R (1984) A critical analysis of durable resistance. Annual Review of Phytopathology 22: 309-330.

Leach JE, Vera-Cruz C, Bai J and Leung H (2001) Pathogen fitness penalty as a predictor of durability of disease resistance genes. Annual Review of Phytopathology 39: 187-224.

Levy M, Correa-Victoria FJ, Zeigler RS, Xu S, and Hamer JE (1993) Genetic diversity of the rice blast fungus in a disease nursery in Colombia. Phytopathology 83: 1427-1433.

McCouch SR, Temnykh S, Lukashova A et al (2001) Microsatellite markers in rice: abundance, diversity, and applications, in GS Kush, Brar DS and Hardy B (eds.), Rice Genetics IV, IRRI, Los Baños, Philippines, pp. 117-135.

Van der Plank JE (1968) Disease Resistance in Plants. London-New York: Academic. 206 pp.

DEVELOPMENT OF DIFFERENTIAL VARIETIES FOR BLAST RESISTANCE IN IRRI-JAPAN COLLABORATIVE RESEARCH PROJECT

Y. Fukuta[1], E. Araki[1], M. J. T. Yanoria[1], *T. Imbe[2], H. Tsunematsu[3], H. Kato[4], L. A. Ebron[1], D. Mercado-Escueta[1] and G. S. Khush[1]

[1]*International Rice Research Institute, DAPO Box 7777, Metro Manila, The Philippines*
[2]*National institute of Crop Science, 2-1-18, Kannondai, Tsukuba, Ibaraki, 305-8517 Japan*
[3]*Japan International Research Center for Agricultural Sciences, 1-2-1, Ohwashi, Tsukuba, Ibaraki, 305-8686 Japan*
[4]*Miyazaki Agricultural Experiment Station, Minemae 5851, Shimonaka, Sadowara, Miyazaki, 880-0212 Japan*
**Corresponding author e-mail to: timbe@affrc.go.jp*

Abstract: In the IRRI-Japan Collaborative Research Project, four kinds of differential varieties, monogenic lines and near isogenic lines (NILs) with three kinds of genetic backgrounds have been developed for the following 24 resistance genes *Pia, Pib, Pii, Pik, Pik-h, Pik-m, Pik-p, Pik-s, Pish, Pita, Pita-2, Pit, Piz, Piz-5* (*Pi2*)*, Piz-t, Pi1, Pi3, Pi5*(t)*, Pi7, Pi9* (t)*, Pi11* (t), *Pi12* (t), *Pi19* (t), and *Pi20*. Monogenic lines that have only a single resistance gene in each background, were developed from one to three time's backcrosses between donor varieties and a Japonica-type variety Lijianxintuanheigu (LTH) as a recurrent parent, and were distributed as first universal differential varieties and gene sources from the International Rice Research Institute (IRRI). Two kinds of NILs with Lijianxintuanheigu (LTH) and US2 genetic backgrounds have also only one resistance gene each. The NILs of an Indica-type variety CO39 are including an additional gene *Pia* in the genetic background. These NILs are confirming fixations of target resistance gene and morphological characters in the advanced generations BC_6F_7 or more, and will be released in the near future.

Key words: differential variety, monogenic line, near isogenic line, blast (*Pyricularia grisea* Sacc), rice (*Oryza sativa* L.)

S. Kawasaki (ed.), Rice Blast: Interaction with Rice and Control, 229–233.
© 2004 *Kluwer Academic Publishers. Printed in the Netherlands.*

1. INTRODUCTION

Blast, caused by *Pyricularia grisea* Sacc., is a serious disease of rice (*Oryza sativa* L.). In the tropics, it is particularly severe in upland and rainfed lowland environments. Susceptible varieties can also be severely infected under irrigated conditions. A set of near-isogenic lines (NILs) with single genes for blast resistance is useful as differential varieties and as a gene source in rice breeding programs. However, the development of NILs for blast resistance has been limited to a small number of genes (Mackill et al 1985. The IRRI-Japan Collaborative Research Project have been trying to develop and characterize monogenic lines and NILs with single blast resistant genes since 1994.

2. MONOGENIC LINES AS A DIFFERNTIAL VARIETIES (DV)

2.1. Monogenic lines

Thirty-one monogenic lines with 24 kinds of resistance genes -*Pia, Pib, Pii, Pik, Pik-h, Pik-m, Pik-p, Pik-s, Pish, Pita, Pita-2, Pit, Piz, Piz-5* (*Pi-2*), *Piz-t, Pi1, Pi3, Pi5* (t), *Pi7, Pi9* (t), *Pi11* (t), *Pi12* (t), *Pi19* (t), and *Pi20*- were developed (Tsunematsu et al 2000, Table 1). These genes originated from 25 resistant donors. These lines were bred by the backcross method to transfer the single gene into the recurrent parent, a Chinese Japonica variety Lijiangxintuanheigu (LTH), without any known resistance genes. Backcross generations were inoculated with isolates avirulent to the target gene and homozygous resistant lines were selected.

Twenty representative isolates from the Philippines were used to confirm the reaction pattern of each line. Monogenic lines showed distinct reaction patterns. However, *Pik* alleles could not be differentiated and there were no incompatible isolates for *Pit* and *Pi19* (t) (Table 2). These monogenic lines have been distributed from the International Network for Genetic Evaluation of Rice (INGER, IRRI) and the IRRI-Japan Collaborative Research Project.

3. NEAR ISOGENIC LINES (NILS) AS A DV

Three kinds of NILs with different genetic backgrounds, LTH, CO39 (Indica), and US-2 (Indica-Japonica hybrid) have been developed. In each background, backcrosses have been carried out at least four to six times to generate the NILs (Table 1). A total of 20 resistance genes *Pia, Pib, Pii, Pik, Pik-h, Pik-m, Pik-p, Pik-s, Pish, Pita, Pita-2, Piz, Piz-5* (*Pi2*), *Piz-t, Pi1, Pi3, Pi5* (t), *Pi7, Pi9* (t), and *Pi12* (t)- have been introduced into each genetic background.

3.1. LTH NILs

LTH NILs have been developed for the targets of 15 resistance genes *Pia, Pib, Pik, Pik-h, Pik-s, Pita, Pita-2, Piz, Piz-5* (*Pi2*)*, Piz-t, Pil, Pi3, Pi5* (t)*, Pi7, Pi9* (t), *Pi19* (t), and *Pi20*. It is necessary to generate and select the fixed lines in each combination, because of the segregation for some morphological traits.

Table 1. Development of differential varieties and near-isogenic lines with blast resistance genes in 2002

	Genetic background (resistance gene)			
Target gene	Monogenic line (LTH as recurrent parent)	LTH (Lijianxintuanheigu) (nothing)[a]	CO39 (*Pia*)	US2 (nothing)
Pia	BC_1F_{15}	BC_6F_{10}	-	BC_6F_2
Pib	BC_1F_{13}	BC_6F_9	BC_6F_8	BC_6F_2
Pii	BC_1F_{15}	-	-	BC_4F_1
Pik	BC_1F_{14}	BC_6F_{10}	BC_6F_8	BC_6F_2
Pik-h	BC_1F_{10}	BC_6F_9	BC_6F_8	BC_6F_2
Pik-m	BC_1F_{11}	-	BC_6F_8	-
Pik-p	BC_1F_{13}	-	BC_6F_8	BC_6F_2
Pik-s	BC_1F_{15}	BC_6F_9	BC_6F_8	BC_6F_2
Pish	BC_1F_{13}, BC_1F_{15}	-	BC_6F_8	BC_5F_1
Pita	BC_2F_{13}, BC_3F_{13}, BC_5F_{11}	BC_6F_{11}	BC_6F_8	BC_6F_2
Pita-2	BC_1F_9, BC_1F_{11}	BC_6F_{10}	BC_7F_6, BC_6F_8	-
Pit	BC_2F_{10}	-	-	-
Piz	BC_1F_{15}	BC_6F_8	BC_6F_8	-
Piz-5	BC_3F_{13}, BC_5F_{11}	BC_6F_{11}	BC_6F_8	-
Piz-t	BC_1F_{15}	BC_6F_{10}	BC_7F_6	BC_6F_2
Pil	BC_3F_{13}	BC_6F_{11}	BC_6F_8	BC_6F_2
Pi3	BC_2F_{13}	BC_6F_{11}	-	-
Pi5(t)	BC_3F_{13}	BC_6F_{11}	-	BC_6F_2
Pi7	BC_3F_{13}	BC_6F_{11}	BC_6F_8	BC_6F_2
Pi9(t)	BC_3F_{13}	BC_6F_{11}	-	BC_6F_2
Pi11(t)	BC_3F_{13}	-	-	-
Pi12(t)	BC_3F_{13}	-	BC_6F_8	BC_6F_1
Pi19(t)	BC_1F_{12}	-	-	-
Pi20	BC_1F_{11}	-	-	-

[a]Nothing = no known resistance genes.

3.2. CO39 NILs

In CO39 NILs, fifteen resistance genes *Pib, Pik, Pik-h, Pik-m, Pik-p, Pik-s, Pish, Pita, Pita-2, Piz, Piz-5* (*Pi2*)*, Piz-t, Pi3, Pi5* (t), *Pi7,* and *Pi12* (t) are targeted, and these reactions to isolates and morphological characteristics were nearly fixed based on field observation. Detailed evaluation of morphological

^characters will be needed for each line. CO39 is an Indica-type variety and has a resistance gene *Pia* in its genetic background.

Table 2 *Reaction to Philippine isolate of monogenic lines.*

		Isolate								
Target gene[a]	Line name[b]	PO6-6	CA 89	IK81-25	IK81-3	BN209	BN111	M36-1-3-10-1	B90002	C923-49
Pia	IRBLa-A	S	S	S	MS	S	S	S	R	R
	IRBLa-C	S	S	S	S	S	S	S	R	R
Pib	IRBLb-B	S	S	S	S	R-M	S	S	R	R
Pi11 (t)	IRBL11-Zh	S	MS-S	MS-S	M	R-M	S	MS	R	R
Pi12(t)	IRBL12-M	S	S	S	S	R	S	S	R	R
Pii	IRBLi-F5	R	S	R	R-M	S	R	S	MS-S	S
Pi3	IRBL3-CP4	R	S	R	M	S	R-M	S	S	S
Pik-s	IRBLks-F5	S	S	S	S	S	S	S	S	S
	IRBLks-S	M	M	M	R-M	M	M	M	M-MS	M
Pik, Pik-p,	IRBLk-Ka	R	R	S	R	R	R	S	R	R
Pik-h, Pik-m	K60	R	R	S	R	R	R	S	R	R
	IRBLkh-K3	R	R	M	R	R	R	M-MS	R	R
	IRBLkm-Ts	R	R	M-MS	R	R	R	M	R	R
Pi1	IRBL1-CL	R	R	S	R	R	R	S	R	R
Pi7 (t)	IRBL7-M	R	R	S	R	R	R	S	R	R
Pish	IRBLsh-S	R-M	M	R	R-M	M	R-M	R	R-M	M
	IRBLsh-B	R-M	R-M	R	R-M	R-M	R-M	R	M	M
Pit	IRBLt-K59	MS-S	S	MS-S	MS	S	S	S	S	S
Pita	IRBLta-K1	MS	S	R	R	S	S	M-MS	R	S
	IRBLta-CT2	S	S	R	R	S	S	MS	R	S
	IRBLta-CP1	S	S	R	R	S	S	S	M	S
Pita-2	IRBLta2-Pi	S	S	R	R	R	R	R	R	S
	IRBLta2-Re	S	S	R	R	R	R	R	R	S
Piz	IRBLz-Fu	R-M	M	R	R	R-M	S	R	M	R
Piz-5	IRBLz5-CA	R	M	R	R	M	R	R	M	M
	IRBL25-CA(R)	R	R	R-M	R	M	R	R	R-M	R
Piz-t	IRBLzt-T	S	S	S	S	S	S	S	R	R
Pi5 (t)	IRBL5-M	R	S	R	R-M	MS	R-M	R-M	M	MS
Pi9 (t)	IRBL9-W	R	R	R	R	R	R	R	R	R
Pi19 (t)	IRBL19-A	S	S	S	S	S	S	MS	S	S
Pi20 (t)	IR24	S	S	MS-S	MS-S	S	R	R	R	R

[a]Target resistance gene introduced from donor variety.

[b]Monogenic lines are designated as IRBL, followed by the resistance gene, then designated abbreviation of the resistance donor.

Donor varieties: A: Aichi asahi, C: CO 39, F5: Fujisaka 5, S: Shin 2, Ka: Kanto 51, K60: K60, K3: K3, Fu: Fukunishiki, CA: C101A51, T: Toride1, K1: K1, CT2: C105TTP2LP, B: BL1, CL: C101LAC,

CP4: C104PKT, M: Moroberekan (RIL 10, RIL29 and RIL 249), W: WHD-1S-75-1-127. Ts: Tsuyuake, IR24: ARL 24, Pi: Pi No. 4, Re: Reiho, CP1: C101 PKT, Zh: Zhaiyeqing.

*: *Pish* from Shin 2 might be included in the background.

**: Unknown gene(s) might be included in the background.

3.3. US-2 NILs

In US-2 NILs, resistance genes *Pia, Pib, Pii, Pik, Pik-h, Pik-p, Pik-s, Pish, Pita, Piz-t, Pi1, Pi3, Pi5, Pi7*, and *Pi9* are currently used for development. Some lines will be backcrossed one or two more times and continued until the F_9 generation or more. US-2 is a progeny from Indica-Japonica hybrid and does not contain any resistance genes in the genetic background.

Although it is still required to confirm the fixation in each line and to cover the vacant combination for target resistance genes in CO39 and LTH NILs, almost all these lines will be released with the information of the reaction patterns to blast isolates and morphological traits in the near future.

The seed yielding of monogenic lines, CO39 NILs, and LTH NILs is quite low in tropical conditions and multiplication of seeds is one of the cultivation problems in 'these genetic backgrounds' lines. US-2 shows good seed yielding and using this specific genetic background will solve problems in multiplication.

Developed differential varieties will be useful tools to clarify the pathogenesis of blast races based on the gene-for-gene theory (Flor 1956) and gene source in breeding programs.

REFERENCES

Flor H. H. (1956) The complementary genetic systems in flax and flax rust. Adv. Genet. 8: 29-54.

MacKill, D.J., J.M. Bonman, H.S. Suh and R. Srilingam. 1985. Genes for resistance to Philippine isolates of the rice blast pathogen. Rice Genet. Newsl. 2: 80-81.

Tsunematsu H., M. J. T. Yanoria, L. A. Ebron, N. Hayashi, I. Ando, H. Kato, T. Imbe, and G. S. Khush (2000) Development of monogenic lines of rice for blast resistance. Breed Sci. 50: 229-234.

IDENTIFICATION OF BLAST RESISTANCE GENES IN ELITE INDICA-TYPE VARIETIES OF RICE (*ORYZA SATIVA* L.)

**L.A. Ebron,*[1] *Y. Fukuta,*[1] *T. Imbe,*[2] *H. Kato,*[3] *M. J. T. Yanoria,*[1] *H. Tsunematsu,*[4] *G. S. Khush and*[1] *M. Yokoo*[5]

[1]*International Rice Research Institute, DAPO Box 7777, Metro Manila, The Philippines*
[2]*National Institute of Crop Science, 2-1-18, Kannondai, Tsukuba, Ibaraki, 305-8517, Japan*
[3]*Miyazaki Agricultural Experiment Station, Minemae 5851, Shimonaka, Sadowara, Miyazaki, 880-0212 Japan*
[4]*Japan International Research Center for Agricultural Sciences, 1-2-1, Ohwashi, Tsukuba, Ibaraki, 305-8686 Japan*
[5]*Institute of Agriculture and Forestry, University of Tsukuba, Tennodai, Tsukuba, Ibaraki, 305-8572 Japan*

**E-mail: lebron@cgiar.org*

Abstract: Eight blast resistance genes --*Pia, Pib, Pik-s, Pik-h, Pita, Piz-t, Pi20,* and one of three (*Pii, Pi3, or Pi5*(t)) were identified in 42 IRRI-bred varieties following a differential system using Philippine blast isolates. Among these genes, two genes *Pib* and one of *Pik* alleles (*Pik-s* or *Pik-h*), were found in almost all IRRI varieties. The number of resistance genes in each variety varied from two to five, and the kinds of gene combination were also limited. This limitation might be due to the gene sources such that IR8, IR24, IR36, or their hybrid progenies were often included in the breeding of these varieties. Using the gene-for-gene theory, this study was carried out in three steps: 1) estimation of genes following the reaction patterns of monogenic lines to blast isolates, 2) genetic analysis using BC_1F_2 populations with a susceptible variety as recurrent parent, and 3) allelism test with Japanese differential varieties. Through genetic analysis, each investigation's ability to identify resistance genes was demonstrated.

Key words: resistance gene, IRRI-bred variety, gene-for gene theory, blast fungus (*Pyricularia grisea* Sacc.), rice (*Oriza sativa* L.)

S. Kawasaki (ed.), Rice Blast: Interaction with Rice and Control, 235–240.
© 2004 *Kluwer Academic Publishers. Printed in the Netherlands.*

1. INTRODUCTION

In the tropics, blast is particularly severe in upland and rainfed lowland environments and even under irrigated conditions where susceptible varieties are growing. The control of the disease can be achieved with the use of blast resistant varieties.

IRRI-bred varieties such as IR8, IR24, and IR36 have been widely distributed and grown in many countries. However, little is known regarding the blast resistance genes in these varieties and lines. This is partly attributed to the lack of a suitable differential system for efficient identification of these genes. The presence of several resistance genes in Indica-type varieties makes genetic studies of blast resistance genes difficult (Mackill et al 1985, Yu et al 1987). Based on the gene-for-gene theory (Flor 1956), the present study aimed to identify the blast resistance genes in IRRI rice varieties using Philippine isolates of the blast fungus *Pyricularia grisea* Sacc.

2. MATERIALS AND METHODS

2.1. Estimation of blast resistance genes

Forty-two Indica-type rice varieties bred in IRRI were used in this study. The resistance genes were estimated based on the reaction patterns of 10 kinds of monogenic lines to 14 isolates whose pathogenicities were previously described (Tsunematsu et al 2000, Yanoria et al 2000). Each monogenic line carrying different single resistance genes was developed in the genetic background of Lijiangxintuanheigu (LTH) by backcross breeding method. The reaction to blast was evaluated using a slightly modified visual scoring method (0-5) with slight described by Mackill and Bonman (1992).

The IRRI varieties were classified into seven groups based on the presence of *Pi20, Pita*, and one of the alleles *Pik, Pik-h, Pik-m,* or *Pik-p* (the gene was tentatively designated as *Pik** in this study). These genes showed a relatively wide spectrum of resistance and incompatibility with only one of these three isolates --M36-1-3-10-1, IK81-25, and PO6-6.

2.2. Genetic analysis of resistance in IRRI varieties

The number and kind of genes were analyzed from the segregation of BC_1F_2 lines derived from the crosses between six IRRI varieties (IR34, IR36, IR46, IR60, IR64, and IR74) and a susceptible Indica-type variety, CO39, as recurrent parent. Several blast isolates avirulent to resistance genes were used in the analysis.

Allelism tests were carried out in nine varieties (IR34, IR24, IR36, IR60, PSBRc1, IR74, IR56, IR70, and IR64). The F_2 populations derived from crosses between IRRI and the differential varieties were used to confirm the estimated genes.

3. RESULTS

The blast resistance genes identified in IRRI varieties by reaction pattern and genetic analyses are shown in Table 1. On the basis of the presence of three genes (*Pi20, Pita*, and *Pik**), the 42 IRRI varieties were classified into seven variety groups (VG1-7).

In VG1, the three genes, *Pi20, Pita,* and one of *Pik** were not identified; seven varieties were included. This group was divided into two subgroups, VG1a and VG1b, and the *Piz-t* in VG1b was different from that in VG1a. Additionally, two genes, *Pib* and *Pik-s,* were observed to be common in both subgroups. *Pia, Pib, Pik-s, Piz-t,* and an unknown gene were also detected in IR34 by BC_1F_2 analysis. *Piz-t* and *Pib* in IR34 were also confirmed in allelism test. From these findings, IR34 was confirmed to have five kinds of genes--*Pia, Pib, Pik-s, Piz-t*, and at least one unknown one.

VG2 was characterized by only one gene *Pi20*, and seven varieties including IR8 and IR24 were classified as belonging to this group. In addition, *Pib* was common in this group. There were three subgroups: VG2a, VG2b, and VG2c. Varieties in VG2a and VG2c carried *Pik-s* additionally. *Piz-t* and an unknown resistance gene were also estimated in VG2c and VG2b, respectively. Imbe et al (1997) has previously reported that IR24 in VG2a included genes *Pia, Pib, Pik-s*, and *Pi20*. Our allelism test also confirmed *Pib* and *Pik-s*.

VG3 was characterized by *Pita*, and 17 varieties, including IR36 and its derived progenies, belonged to this group. Additionally, these varieties had *Pib*. In IR36, four genes (*Pita*, *Pib, Pik-s* and an unknown one), and three genes (*Pib, Pita, Pik-s,* and *Piz-t*) were confirmed using BC_1F_2 analyses and allelism tests with differential varieties, respectively. Three genes --*Pib*, *Pik-s,* and *Piz-t*-- were confirmed in the F_2 population derived from crosses between IR36 and two varieties IR24 and IR34 for the allelism test. Using BC_1F_2 analyses and allelism test, respectively, six genes (*Pib, Pita, Piz-t, Pik-s,* and two kinds of unknown ones), and two genes (*Pib* and *Pik-s* were found) in IR60.

VG4 characterized by *Pik** included only one variety PSBRc1. Because PSBRc1 showed a moderate resistance to one isolate, which was different from the susceptible reaction of *Pik*,* it was estimated that additional unknown gene conferred the reaction. The allelism test of F_2 population from a cross between IR56 including *Pik** and PSBRc1 confirmed the presence of the gene.

Two genes *Pik** and *Pi20* were found in VG5, and this group included only IR74. Four kinds of resistance genes, *Pia, Pib, Pik*,* and *Pi20*, were identified in BC_1F_2 analyses, and *Pik** was confirmed based on the allelism test using IR56 with *Pik** as a differential variety.

Table 1. *Blast resistance genes estimated and identified in IRRI rice varieties.*

Variety group[a]		Variety	Resistance genes		
			Reaction pattern[b]	BC1F2 analysis	Allelism test
VG 1	a	IR20, IR28, IR30, IR45, IR66	*Pib*, *Pik-s*	-	-
	b	IR29	*Pib*, *Pik-s*, *Piz-t*	-	-
		IR34		*Pia*, *Pib*, *Pik-s*, *Piz-t*, one unknown	*Pib*, *Piz-t*
VG 2	a	IR8, IR22, IR26, PSBRc30	*Pib*, *Pik-s*, *Pi20*	-	-
		IR24		*Pia*[c], *Pib*[c], *Pik-s*[c], *Pi20*[c]	*Pib*, *Pik-s*, *Pi20*[c]
	b	IR43	*Pib*, *Pi20*, one unknown	-	-
	c	PSBRc2	*Pib*, *Pik-s*, *Pi20*, *Piz-t*	-	-
VG 3		IR5, IR32, IR38, IR40, IR42, IR44, IR50, IR52, IR54, IR58, IR62, IR65, IR68, IR72, PSBRc4	*Pib*, *Pita*	-	-
		IR36		*Pib*, *Pik-s*, *Pita*, one unknown	*Pib*, *Pik-s*, *Pita*, *Piz-t*
		IR60		*Pib*, *Pik-s*, *Pita*, *Piz-t*, two unknown	*Pib*, *Piz-t*
VG 4		PSBRc1	*Pik-h*, one unknown	-	*Pik** (*Pik-h*)
VG 5		IR74	*Pi20*, *Pik** (*Pik-h*)	*Pia*, *Pib*, *Pik** (*Pik-h*), *Pi20*	*Pik** (*Pik-h*)
VG 6		IR56	*Pita*, *Pik** (*Pik-h*)	-	*Pita*, *Pik** (*Pik-h*)
		IR70		-	-
VG 7	a	IR48, PSBRc28	*Pi20*, *Pita*, *Pib*	-	-
		IR46		*Pib*, *Pik-s*, *Pi20*, *Pita*, one unknown	-
		IR60		*Pib*, *Pi20*, *Pita*, *Piz-t*, one unknown	-
		IR64		-	*Pib*, *Pik-s*, *Pita*, *Piz-t*
	b	PSBRc10, PSBRc18, PSBRc20	*Pi20*, *Pita*, *Pib*, *Pii*, *Pi3* or *Pi5(t)*	-	-

[a]Classification based on the presence of three genes, *Pi20*, *Pita* and *Pik** (*Pik-h*)
[b]Resistance genes were estimated based on the reactions to 14 blast isolates from the Philippines
[c]Imbe et al. 1997

Two genes *Pita* and *Pik** characterized VG6, and two varieties IR56 and IR70 were included in reaction pattern analysis. The presence of *Pita* and *Pik* was confirmed by allelism tests with differential varieties, too.

VG7 was characterized by *Pi20* and *Pita,* and consisted of two subgroups VG7a and VG7b based on the reaction pattern analysis. Seven varieties were included in VG7 and *Pib* was also included in all varieties. VG7a had four varieties and these varieties had only the *Pi20* and *Pita*. In VG7b, the three varieties were estimated to contain one additional gene, either *Pii, Pi3,* or *Pi5* (t). In VG7a, five genes, *Pib, Pik*, Pita, Pi20* and one unknown in IR46, and five genes *Pita, Pi20, Piz-t, Pib* and one unknown in IR64 were identified by BC_1F_2 analyses. In IR64, *Pita* and *Piz-t* were confirmed by allelism tests with differential varieties and IR56. The allelism test by F_2 segregation of IR64 with IR24 also indicated that these two varieties have same genes *Pib* and *Pik-s*. From these results, IR64 was confirmed to carry four genes *Pita*, *Piz-t, Pib,* and *Pik-s*.

4. DISCUSSION

The blast resistance genes in IRRI bred rice varieties were identified following a differential system using Philippine isolates of the blast fungus *Pyricularia grisea* (Cooke) Sacc. On the basis of the presence of the three resistance genes, *Pi20, Pita* and one of the *Pik* alleles (*Pik**, except for *Pik-s*), the 42 IRRI varieties were classified into seven variety groups. Seven genes, *Pib, Piz-t, Pi20, Pita, Pik-s*, and *Pik** and one of *Pii, Pi3, or Pi5*(t) were able to identify based on the reaction patterns of the monogenic lines to the blast isolates, at least. The seven genes, *Pia, Pib, Pik-s, Piz-t, Pi20, Pita,* and *Pik**, were confirmed by the genetic analyses, BC_1F_2 and allelism test. A total of eight kinds of genes were identified from these analyses.

The two genes *Pib* and one of the *Pik* alleles (*Pik-s* or *Pik**) among them were included in almost all of the varieties. Kiyosawa (1972), and Kiyosawa and Ando (1997) reported that five alleles, *Pik, Pik-h, Pik-m, Pik-p,* and *Pik-s,* were found in the *Pik* locus. In this study, *Pik** was detected in PSBRc1 of VG4, IR74 of VG5, and IR56 and IR70 of VG6. In those varieties, three varieties Dawn, Tadukan, and Tetep were included in their pedigrees, and Kiyosawa (1981) reported that *Pik-h* was identified in these varieties. On the basis of this information, the gene of *Pik** in VGs4, 5, and 6 was estimated to be *Pik-h*.

We could identify a limited number and kind of genes compared with the number of over 40 blast resistance genes reported previously (RGC 1998). Because IR8, IR24, and IR36 or their progenies were often used in IRRI bred varieties, the limitations might be due to the near relation among the IRRI varieties.

This study was carried out based on the gene for gene theory (Flor 1956) in the three steps, reaction pattern analysis, BC_1F_2 analysis, and allelism test. The reaction patterns to blast isolates were particularly useful in identifying the blast resistance genes in the varieties. In some cases, however, genes like *Pia, Pib* and *Pik-s* were not easy to estimate by reaction pattern due to the masking effect of other genes like *Pi20*, *Pita* or the *Pik-h*. To solve this problem, genetic analyses were conducted to detect and clarify these genes.

This information generated in this study will be important and useful in rice blast resistance breeding.

REFERENCES

Flor, H.H. 1956. The complementary genetic systems in flax and flax rust. Advan. Genet. 8: 29-54.

Imbe, T., S. Oba, M.J.T. Yanoria and H. Tsunematsu. 1997. A new gene for blast resistance in rice cultivar, IR24. Rice Genet. Newsl. 14:60-62.

Kiyosawa, S. 1972. Genetics of blast resistance. In: Rice breeding. IRRI, Manila, Philippines: 203-225.

Kiyosawa S. 1981. Gene analysis for blast resistance. Oryza 18: 196-203.

Kiyosawa S. and I. Ando. 1997. Chapter 4: Inheritance of resistance. In: Matsuo, T., Y.Futsuhara, F. Kikuchi, and H. Yamaguchi (eds.). *Science of the rice plant, vol. 3 Genetics*. Tokyo, Japan: Nosan Gyoson Bunka Kyokai (Nobunkyo), 470-506.

MacKill, D.J., J.M. Bonman, H.S. Suh and R. Srilingam. 1985. Genes for resistance to Philippine isolates of the rice blast pathogen. Rice Genet. Newsl. 2: 80-81.

RGC (Rice Genetics Cooperative). 1998. B. Report of the committee on gene symbolization, nomenclature and linkage groups. Rice Genet. Newslet. 15: 13-76.

Standard evaluation system for rice. 1996. IRRI, Manila, Philippines.

Tsunematsu H., M. J. T. Yanoria., L. A. Ebron, N. Hayashi, I. Ando, H. Kato, T. Imbe, and G. S. Khush 2000. Development of monogenic lines of rice for blast resistance. Breeding Science 50: 229-234.

Yanoria, M.J.T., T. Imbe, H. Tsunematsu, L.A.Ebron, D. Mercado, H. Kato and Y. Fukuta 2000. Phathogenicity of IRRI blast isolate to rice blast resistant genes. 4th International Rice Genetic Symposium. (Abstract 397).

Yu, Z.H., D.J. MacKill and J.M. Bonman.1987. Inheritance of resistance to blast in some traditional and improved rice cultivars. Genetics 77(2):323-326.

RESISTANCE OF SOME CHINESE HYBRID RICE, CONVENTIONAL EARLY *INDICA* AND LATE *JAPONICA* RICE TO *MAGNAPORTHE GRISEA*

*Shen Y[1,2], Adreit H[2], Zhu Xd[1], Milazzo J[2], Chen Hq[1] and Tharreau D[2]

[1]*China National Rice Research Institute, Hangzhou 310006, P R China;*
[2] *Centre de Cooperation Internationale en Recherche Agronomique pour le Developpement, UMR BGPI, TA 73 / 09, 34398 Montpellier, Cedex 05, France*

**E-mail: shenying@mail.hz.zj.cn*

Abstract: Thirty isolates of *Magnaporthe grisea* collected from 18 provinces/cities representing 21 pathotypes and 9 different lineages were inoculated to rice varieties with known resistance genes, some hybrid rice, and conventional early *indica* and late *japonica* varieties cultivated in recent years in China. The results indicated that some of them were resistant to most of the isolates tested and that they could be of interest as resistance sources for hybrid donors/parents or to be planted in the field directly.

1. INTRODUCTION

Rice blast [*Magnaporthe grisea* (Hebert) Barr], one of the main fungal diseases of rice [1,2], spreads all over the rice growing areas of China. It affects rice yield and decreases the quality of rice [3,4]. People build up integrated strategies to control the disease by using resistant varieties and fungicides. These strategies played an important role in high and stable yields and in decreasing the fungicide pollution to environment in the end of the 1970s. Farmers used to plant one or several varieties with the same or close genetic background. Thus, the frequency of virulent pathotypes

S. Kawasaki (ed.), Rice Blast: Interaction with Rice and Control, 241–250.
© 2004 *Kluwer Academic Publishers. Printed in the Netherlands.*

increased continuously in some areas, resulting in important losses under suitable climatic conditions and to resistance breakdown 3 to 5 years after large-scale release. To control the disease, fungicide use was increased resulting in higher cost of production, and pollution of soil and water resources [5].

Resistance breakdown is partly caused by incomplete knowledge of the resistance genes present in one particular variety. Exploiting the genetic diversity of the pathogen is one mean to better exploring the diversity of resistance genes. Thirty isolates of *M.grisea* were chosen to inoculat on 159 varieties/lines to better understand the genetic relationships between blast pathogens and rice varieties. The objective of this study was to explore the genetic diversity of resistance genes in Chinese varieties through inoculation with chosen isolates and to compare the observed reaction patterns of rice differential varieties with known resistance genes. The final goal of this work is to provide information on resistance genes in varieties to improve breeding for resistance by developing new varieties with broad resistance spectrum and durable resistance to blast.

2. MATERIALS AND METHODS

2-1 Test Isolates

Thirty isolates were chosen for their genetic diversity. Among them, 26 Chinese rice isolates were collected from 18 provinces/cities and represented 9 different lineages [6~8] and 21 Chinese pathotypes.

Two isolates, TH12 and TH16 were isolated from barley in Thailand. The last two isolates (101/1/1 and 101/4/8) are progenies from a cross between a Chinese isolate from rice (CH72) and a Thailand isolate from barley (TH12) (Table 1).

Table 1 *M. grisea Isolates*

Group	Pathotype	Lineage	Isolate No
ZA	A1, A5		101/1/1, 101/4/8, TH16, CH43
	A3, A9. A33, A59	PRC-4, 16	CH93, CH131, CH193
ZB	B1, B13, B15		CH45, CH46, CH49, CH52, CH53,
	B3, B5, B25	PRC-7, 10, 45	CH102, CH155, CH199, CH227
ZC	C9, C13	PRC-7	CH51, CH109
ZD	D3, D7	PRC-36	CH184, CH260
ZE	E1, E3	PRC-7	CH72, CH255, CH333
ZF	F1	PRC-7	CH209
ZG	G1	PRC-6, 8	CH63, CH77, CH174, CH188
ZH	H1		TH12, CH97
Total	21	9	30

2-2 Rice Varieties and Lines

One-hundred-fifty-nine varieties including 13 three-line and 3 two-line hybrids, 98 conventional indica and japonica varieties, 43 varieties or lines with known resistance genes to blast, and two susceptible check were tested in the study.

2-3 Pathogenicity Test and Disease Scoring

Rice plant were grown in trays in a greenhouse and inoculated by spraying 30 ml of a 50,000 conidia/ml suspension with 0.5% gelatin per tray at 5~6 leaf stage. After 16 hours in dew chamber at T 26°C/RH 95% and 6 days more in the greenhouse, lesion types on rice leaves were observed and scored 1 to 6 [9].

3. RESULTS

3.1. Classification of Varieties

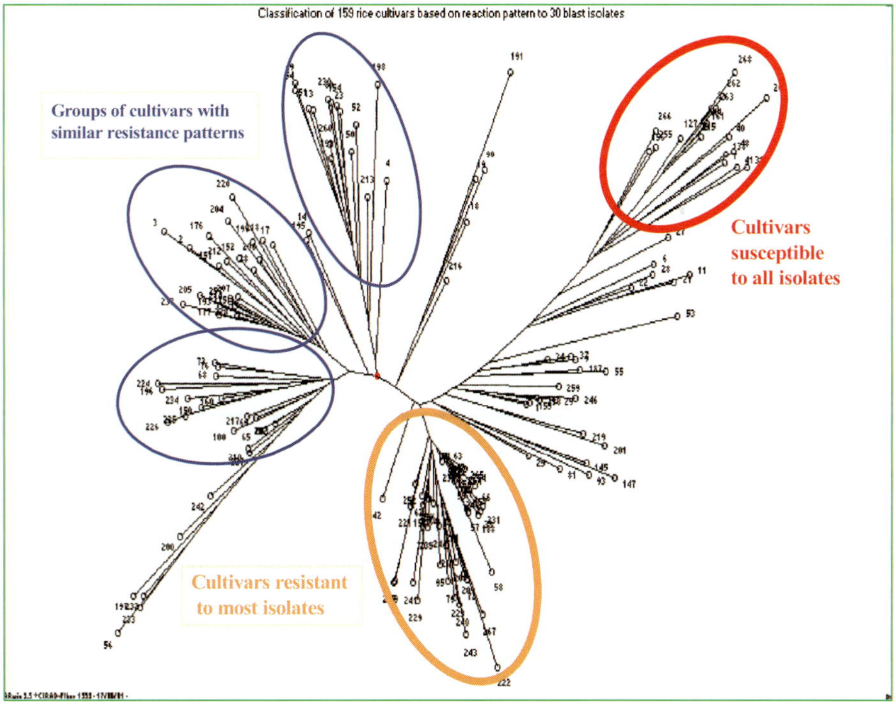

Figure 1. *Classification of rice varieties based on their reaction pattern to 30 M. grisea isolates.*

Symptom scoring of 30 isolates on 159 cultivars, were classified in three classes (resistant, moderately resistant and susceptible) and data were transformed in a similarity matrix by comparing virulence patterns of isolates over varieties or lines, or resistance patterns of varieties to isolates. An Ultrametric Pairwise Grouping Method Using Arithmetic Mean (UPGMA) tree was conducted to compare the virulence spectrum of isolates. Data analysis showed that all the isolates had a different virulence pattern, confirming that the 30 chosen isolates represented a wide range of diversity. Using the same method, varieties were classified into 5 main groups (Fig.1). In these groups, varieties have similar resistance spectrum and are likely to share resistance genes. Some of these groups contain a variety with already identified resistance genes. In these cases, it is possible to suppose that these varieties might carry the same resistance gene as the one described in the differential varieties.

3-2 Reactions of Varieties/Lines with Known Resistance Genes to M. grisea Isolates

Thirty isolates of *M.grisea* were inoculated on varieties with known resistance genes, near isogenic lines and new *indica* and *japonica* varieties cultivated mainly in China. The results indicated contrasting resistance

Table 2 *Resistance of varieties with known resistance genes M. grisea isolates*

Test	Known	A[+]		B		C		D		E		F		G		H		Total		Resistance
V / L	gene	R	S	R	S	R	S	R	S	R	S	R	S	R	S	R	S	R	S	frequency (%)
Aichi Asahi	a		7	1	8		2		2	2	1		1		4	1	1	4	26	15.4
BL1	b	5	2	3	6	1	1	1	1	3		1		3	1	1	1	18	12	60
Fukuhikari	z	3	4	6	3	1	1	1	1	2	1	1		4		2		20	10	66.7
Fukunishiki	z	3	4	6	3	1	1	1	1	2	1	1		4		2		20	10	66.7
Kanto 51	k		7	4	5	1	1		2		3	1		4		2		12	18	40
Kusabue	k	3	4	6	3	1	1	1	1	1	2	1		4		2		19	11	63.3
K1	ta	1	5	8		1	1		2		3	1		2	2	2		15	13	53.6
K3	kh	3	4	8	1	2		2		3		1		4		1	1	24	6	80
K59	t	1	6	3	6	1	1	2			3		1	1	3		2	8	22	26.7
K60	kp	2	5	4	5	1	1	1	1		3	1		2	2	2		13	17	43.3
Pi 4	ta2		5	6		2		1	1	1	2	1		4		2		17	8	68
Shin 2	ks		7	2	7		2		2	1	2		1		4		2	3	27	10
Toride 1	zt	5	2	9		2		2		3		1		4		1	1	27	3	90
Tsuyuake	km	1	6	9		2	1	1	1	3		1		3	1	1	1	21	9	70
Chokoto	k	1	6	7	2	2		1	1	3		1		3	1	1	1	19	11	63.3
BL8	b	2	3	1	5	1	1	2		3		1		3	1		1	13	11	54.2
Fujisaka No.5	i	2	5	3	6		2	2			3		1		4		2	7	23	23.3

[+] A to H = Group A to H according to Chinese classification of pathotypes.

between varieties (Tables 2, 3 and 4).

The most resistant variety was Moroberekan (from Africa) with complete resistance to 29 of 30 isolates. This variety is known to carry at least 3 resistance genes (*Pi-5*, *Pi-7*, *Pi-44*). ZYQ 8 (*Pi-11*, *Pi-33*), C101LAC (*Pi-1* + *Pi-33*), Bala (*Pi-33*) and Toride1 (*Pi-zt*) to more than 90 % of the isolates tested. Gu-mei No.2 with at least 2 genes, had also wide resistance spectrum. These varieties may carry other resistance genes. These lines could be used as donors/parents for breeding programs to blast resistance, but which resistance genes or resistance gene combinations are responsible for wide spectrum remains to be determined.

The most susceptible varieties to 30 isolates were CO39 (*Pi-a*), Aichi Asahi (*Pi-a*), Shin 2 (*Pi-a*), Nato (*Pi-i*), Caloro (*Pi-ks*), and Kassalath (*Pi-8*). These materials were resistant to 3 ~ 4 out of 30 isolates and have narrow resistance-spectrum to blast, showing that the corresponding resistance genes are of weak interest for the development of resistant lines.

3-3 Reaction of Some Hybrid Parents

The reactions of 16 parents of hybrids cultivated in recent years in China are listed in Table 5. Check varieties Wei-you 402 (early season hybrid rice parent) and Shan-you 63 (mid-late season hybrid rice parent) were resistant to 53.3% and 50.0% of the 30 isolates tested respectively. The 14 other hybrid parents were resistant to 36.7 % to 95.5 % of the blast isolates tested with an average of 63 %.

Table 3 *Resistance of near-isogenic lines with known resistant genes to M. grisea*

Test V / L	Known Gene	A[+]		B		C		D		E		F		G		H		Total		Resistant frequency (%)
		R	S	R	S	R	S	R	S	R	S	R	S	R	S	R	S	R	S	
C101 LAC	1 + 33	5	2	9		2		1	1	3		1		4		2		27	3	90
C104 LAC	1	2	5	7	2	1	1		2	3		1		4		1	1	19	11	63.3
C103 TTP	1 + 1b	3	4	8	1	2		1	1	3		1		3	1	1	1	22	8	73.3
C101 A51	2	3	4	5	4	1	1	1	1	3		1		1	3	1	1	16	14	53.3
C104 PKT	3	1	6	2	7		2	2		3			1	1	3		2	9	21	30
C101 PKT	4		7	5	4	1	1		2	3		1		1	3	1	1	12	18	40
C102 TTP	4		7	7	2	1	1		2	3		1		1	3	1	1	14	16	46.7
C105 TTP-1	4	1	6	6	3	1	1	1	1	3		1		2	2	1	1	16	14	53.3
C105 TTP-4-L23	4		7	7	2	1	1		2	3		1		2	2	1	1	15	15	50

[+] A to H = Group A to H according to Chinese classification of pathotypes

In another assay, five other parents of hybrid rice were tested for blast resistance. For two-line parents, blast resistance level of the photo-thermo sensitive genetic male sterile Pei-Ai 64S was better than An-Xiang S. For the three-line system parents, the level order of blast resistance was Long-te-fu A > Zhen-shan 97A > Jin 23A. The growing duration of Jin 23A was longer than Long-te-fu A and Zhen-shan 97A for hybrid rice production. But the grain quality of Jin23A was better than Long-te-fu A and Zhen-shan 97A.

3-4 Reactions of Some Japonica Varieties

Table 4 *Resistance to M. grisea isolates of varieties/lines with known resistance genes*

Test V / L	Geno type	Origin	A+		B		C		D		E		F		G		H		Total		Resistance frequency (%)
			R	S	R	S	R	S	R	S	R	S	R	S	R	S	R	S	R	S	
Tetep	k	Vietnam	1	6	9		2			2	3		1		4		2		21	8	72.4
ZYQ 8	11 + 33	China	5	2	9		2		1	1	3		1		3	1	2		27	3	90
Zhong	24	China	1	5	3	6	1	1	1	1	3			1	3	1	2		15	14	51.7
Gumei No.2	25	China	6	1	7	2	1	1	1	1	3		1		3	1	2		25	5	83.3
Morobere kan	5 + 7 + 44	Africa	7		9		2		2		3		1		4		1	1	29	1	96.7
Usen	a + z	CH-Taiwan		7	6	3	1	1		2	3			1	2	2	2		14	16	46.7
Caloro	ks	USA	1	6	1	8		2		2		3		1		4	1	1	3	27	10
Zenith	a + z	USA	9	2	4	3	1	1	1	1	3		1		3	1	1	1	18	9	66.7
Nato	i	Japan	1	6	1	8		2		2	3			1		4		2	5	25	16.7
Kassalath	8	Japan	1	6	1	8		2	1	1		3		1		4		2	3	27	10
Rico 1	ks	IRRI		4	1	7		2		2		3	1		1	3	1	1	5	22	18.5
Ou 244	z	IRRI	3	3	6	1	1	1	1	1	3		1		4		2		21	6	77.8
ST1	f	IRRI	2	5	3	6		2		2	1	2	1		3	1	2		12	18	40
CO39	a	IRRI		6		9		2		2	3			1		4		2	3	26	11.5

+ A to H = Group A to H according to Chinese classification of pathotypes

Some japonica type varieties from South China were chosen to evaluate blast resistance in this study (Table 6). The results indicated that 11 japonica varieties expressed broad-spectrum resistance to 30 isolates. For example, new lines Chun-jiang 25 and Chun-jiang 683, developed by China National Rice Research Institute, were resistant to the 30 isolates tested, and other lines such as Chun-jiang 20, Chun-jiang 26, Bing 95-13, Xiu-shui 63 were resistant to more than 90% of the isolates .The resistance spectrum was obviously larger than that of the Northern japonica varieties tested (Chang-wang-jing and Bei-jing-nuo). The results also showed that the new cultivated late-season japonica varieties had wider resistance spectrum than old and widely released variety Nong-hu No.6. The resistance spectrum of these japonica varieties was generally better than that of hybrid rice parents (Table 5 and 6).

Table 5 *Resistance of new hybrid parents to M.grisea isolates*

Season Type	Variety	Origin	2 / 3 - System	A⁺ R	A⁺ S	B R	B S	C R	C S	D R	D S	E R	E S	F R	F S	G R	G S	H R	H S	Total R	Total S	Resistance frequency(%)
Early	Xiangliangyou 68	Hu-nan	2	2	5	3	6	2		1	1	3			1	2	2	1	1	14	16	46.7
Hybrid	Shanyou 36 fu	Guang-xi	3	3	4	3	6		2	2		3			1	3	1	2		16	14	53.3
Rice	Jinyou 402	Hu-nan	3	2	5	1	7	1	1	1		3		1		2	2	1	1	12	16	42.9
	Weiyou 402 (CK)	Hu-nan	3	2	5	2	7	1	1	2		3		1		2	1	2		16	14	53.3
	Peizashuangqi	Guangdong	2	6	1	8		2				1				2		2		21	1	95.5
	Peizashanqing	Guangdong	2	5	2	6	3	2		2		3		1		3	1	2		24	6	80
	Teyou 63	Fu-jian	3	6	1	4	4	2		2		3		1		3	1	2		24	6	80
	Teyou 18	Guang-xi	3	5	2	1	8	1	1	2		3		1		4		1	1	18	12	60
Mid-Late	Teyou 524	Guangdong	3	6	1	2	7	1	1	2		3		1		3	1	1	1	19	11	63.3
Hybrid	Jinyou 974	Hu-nan	3	1	6	0	9		2	2		3		1		3	1	1	1	11	19	36.7
Rice	Jinyougui 99	Hu-nan	3	2	5	1	8	1	1	2		3		1		3	1	2		15	15	50
	Gangyou 22	Si-chuan	3	5	2	4	5	2		1	1	3		1		3	1	2		21	9	70
	Zhengyou No.3	Guangdong	3	3	4	2	7	1	1	2		3		1		3	1	2		17	13	56.7
	Zhengyou No.9	Guangdong	3	6	1	5	4	1	1	2		3		1		4		2		24	6	80
	Shanyougui	Guang-xi	3	4	3	3	6	1	1	2		3		1		3	1	2		19	11	63.3
	Shanyou 63(CK)	Fu-jian	3	2	5	2	6	1	1	1		3		1		2	2	2		14	14	50

⁺ A to H = Group A to H according to Chinese classification of pathotypes.

Table 6 *Resistance of some new japonica varieties to M. grisea isolates*

Test V / L	Re - sources	A⁺ R	A⁺ S	B R	B S	C R	C S	D R	D S	E R	E S	F R	F S	G R	G S	H R	H S	Total R	Total S	Resistance frequency (%)
Chun Jiang 11	CNRRI	4	3	9		2		2		1	2	1		3	1	2		24	6	80
Chun Jiang 20	CNRRI	6	1	9		2		2		3		1		4		2		29	1	96.7
Chun Jiang 25	CNRRI	7		9		2		2		3		1		4		2		30	0	100
Chun Jiang 26	CNRRI	5	2	9		2		2		2	1	1		4		2		27	3	90
Chun Jiang 683	CNRRI	7		9		2		2		3		1		4		2		30	0	100
Chun Jiang Nuo 2	CNRRI	4	3	9		2		2		1	2	1		3	1	2		24	6	80
Bing 95-13	ZJCAA	6	1	9		2		2		3		1		4		2		29	1	96.7
Xiu Shui 63	ZJCAA	5	2	9		2		1	1	3		1		4		2		27	3	90
Chang Wan Jing	HNAU		7	9		2		1	1	1	2	1		2	2	2		18	12	60
Bei jing nuo	CAAS	4	3	1	8	1	1	2		3		1		3	1	2		17	13	56.7
Nong hu No.6	CNRRI		7	9		1	1		2		3	1		1	3	2		14	16	46.7

⁺ A to H = Group A to H according to Chinese classification of pathotypes.

3-5 Reactions of Some Early Indica Varieties

The main early-season indica varieties grown in the South of China and some indica reference varieties from all over China were studied. According to the results of 21 new varieties or lines developed by CNRRI, HNAAS and JXAAS (Table 7), the variety with the large resistance spectrum was Zhong 20 jian 8 which was resistant to 90% of the tested isolates and was resistant to the Chinese races ZC to ZH. Zhong 20 Zao 8, 20 Zhong-8 and Zhong 20 Zao 55 were resistant to more than 80 % of the isolates. These varieties are considered of good grain quality. Resistance spectrum of Xiang-Zao-Xian 13, Xiang-Zao-Xian 19 and Zhong 98–4 ranged from 66.7% ~ 70%. These latter varieties are appreciated for their grain quality and started spreading in farmer fields.

4. DISCUSSION

Blast is one of the most widespread and serious diseases of rice in the world. Farmers control the disease by growing resistant varieties and spraying fungicide. Developing varieties with complete and durable resistance would be of low production cost and less damaging to the environment. This remains a challenge for pathologists, geneticists and breeders all over the world [10].

More than 60 resistance genes to blast have been reported. But very few were well characterized, in particular identity between genes described in different varieties and alleles were seldom tested. This makes it more difficult to detect a particular gene in a particular variety. Our knowledge of resistance genes to blast probably remains very poor compared to the potential number of genes to be discovered in such a vary diverse species as rice. Exploiting the diversity of the pathogen to characterize resistance genes and identify new ones is a first step towards a better understanding of this diversity. Identifying the resistance genes or combinations of resistance genes that will lead to durable resistance will require much more efforts.

In this study, some varieties such as Chun jiang 25, Chun jiang 683, Qing jing 11, Pin 9596 and Yuan zhen 116 showed very broad resistance spectrum to 30 rice blast isolates from China. China is a presumed center of diversity of the rice blast fungus. Thus, the diversity of the isolates used in this study is very high. Understanding which resistance genes are responsible for this large spectrum is of interest for breeding durable resistance to blast.

The resistance genes *Pi1*, *Piz*^{t}, and *Pi33* alone revealed a wide resistance spectrum to the 30 isolates of *M.grisea* tested in this study. Using them for hybrid rice breeding could help improving the resistance level of Chinese hybrid rice. But more interestingly, these genes could be used in combinations. Development of marker-assisted selection (MAS) could be of interest to develop rapidly such resistant lines.

Table *7 Resistance of some new indica varieties to M.grisea isolates*

Test	Resource	A+		B		C		D		E		F		G		H		Total		Resistanc
V / L		R	S	R	S	R	S	R	S	R	S	R	S	R	S	R	S	R	S	frequency (%)
Zhong 156	CNRRI	1	5	3	6	1	1	2		3			1	3	1	2		15	14	51.7
Zhou 903	Zhejiang	3	4	1	8	1	1	2		3		1		3	1	1	1	15	15	50
Xiangzaoxian 17	Hunan	2	5	2	7		2	2		3			1		4	1	1	10	20	33.3
Zhe 9248	Zhejiang	2	5	2	7	1	1	1	1	3		1		3	1	2		15	15	50
Xiangzaoxian 13	Hunan	6	1	2	7	1	1	2		3		1		3	1	2		20	10	66.7
Xiangzaoxian 19	Hunan	6	1	2	7	1	1	2		3		1		3	1	2		20	10	66.7
Zhongyouzao 81	Hunan	2	5		9	2		2		3		1		4		1	1	15	15	50
Zhongjian 100	CNRRI	6	1	1	8	1	1	2		3		1		3	1	1	1	18	12	60
Ganzaoxian No.26	Jiangxi	5	2	2	7	1	1	2		3			1	3	1	1	1	17	13	56.7
Ganzaoxian No.37	Jiangxi	3	4	8	1	1	1	1	1	3		1		2	2	2		21	9	70
Zhe 733	Zhejiang	4	3	5	3	2		2		3		1		3	1	1	1	22	8	73.3
Zhong 20 Zao 8	CNRRI	7		6	3	2		2		3		1		4		1	1	26	4	86.7
Zhong 98-15	CNRRI	2	5		9	1	1	2		3		1		3	1	1	1	13	17	43.3
Zhong 98-18	CNRRI	1	6	1	8	1	1	1	1	3			1	3	1	2		12	18	40
Zhong 98-19	CNRRI	1	6	1	8	1	1	1	1	3		1		3	1	2		12	18	40
20 Zao Jian 6	CNRRI	6	1	3	6	2		2		3		1		3	1	1	1	21	9	70
Zhong You Zao 12	CNRRI	5	2		9	1	1	2		3		1		3	1	2		17	13	56.7
Zhong 20 Zao 55	CNRRI	5	2	6	3	2		2		3		1		3	1	2		24	6	80
20 Zhong-8	CNRRI	4	3	9		2		2		2	1	1		3	1	2		25	5	83.3
Zhong 98-4	CNRRI	5	2	4	5	2		2		3		1		3	1	1	1	21	9	70
Zhong 20 Jian 8	CNRRI	6	1	7	2	2		2		3		1		4		2		27	3	90

+A to H = Group A to H according to Chinese classification of pathotypes

REFERENCES:

[1].Barr ME. 1977, *Magnaporthe*, *Telimenella* and *Hyponectria* (Physosporellaceae). Mycologia, 69: 952 –956.

[2] Giatong P and Frederiksen RA. 1969, Pathogenic variability and cytology of monoconidial subcultures of *Piricularia oryzae*. Phytopathology, 59: 1152 – 1157.

[3] Shen Y, et al. 1994,Studies on the perfect stage of *Pyricularia oryzae* in China. Scientia Agricultura Sinica, 27(1): 25-29 (in Chinese).
[4] Shen Y, et al. 1998, Molecular probes for epidemiological studies of the rice blast fungus. Journal of Southweat Agricultural University, 20(5): 401 - 408 (in Chinese).
[5] Sun GC, et al. 1999, The directional selection and stabilizing selection from the interaction between rice variety and blast fungus, *Magnaporthe grisea*. Acta Phytopathologica Sinica, 29(1): 45 - 49 (in Chinese).
[6] Shen Y, et al. 1996, The genetic diversity and geographic distribution of *Magnaporthe grisea* in China. Scientia Agricultura Sinica, 29(4): 39 - 46 (in Chinese).
[7] Shen Y, et al. 2002,Geographic distribution of mating type in *Magnaporthe grisea* and its relationship between fertile isolates in China. Scientia Agricultura Sinica, 1(6): 648 – 656.
[8] Levy M, et al. 1991, DNA fingerprinting resolves pathotype diversity in a plant pathogenic fungus. The Plant Cell, 3: 95 - 102.
[9] Tharreau D, et al. 2000,New tools for resistance gene characterization in rice. *In* Advances in rice blast research. Proceedings of the 2nd International Rice Blast Conference, August 4-8, 1998, Montpellier, France. D Tharreau, MH Lebrun, N Talbot and JL Notteghem (eds), Kluwer Academic Publishers, Wageningen, The Netherlands, pp54-62.
[10] Liu EM, et al. 1999, Preliminary study on inheritance of blast resistance to three durable resistant rice varieties. Plant Protection, (2):10–11(in Chinese).

TRITICALE AND BARLEY: NEW HOSTS OF *MAGNAPORTHE GRISEA* IN SÃO PAULO, BRAZIL - RELATIONSHIP WITH BLAST OF RICE AND WHEAT

*A.S. Urashima[12], T. Dias Martins[12], C.R.N.C. Bueno[12], D.B. Favaro[1], M.A. Arruda[12] *and* Y.R. Mehta[3]

[1]*Centro de Ciências Agrárias, Universidade Federal de São Carlos, P.O. Box 153, 13600-000 Araras, SP, Brazil*

[2]*FAPESP scholars:* [3]*Instituto Agronômico do Paraná, P.O. Box 481, 86047-902 Londrina, PR, Brazil*

**E-mail: alfredo@dbv.cca.ufscar.br*

Abstract: Triticale (*X. triticosecale*) and barley (*Hordeum vulgare*) are the most recent crops to be infected by *Magnaporthe grisea,* the causal agent of blast disease. This is a newly emerging problem in Sao Paulo state where frequent yield losses due to rice and wheat blast already occur. The present work is aimed at examining the relationship among isolates from these four important crops through host range, cross-infectivity and DNA fingerprinting. Data from controlled-environment inoculations revealed that *M. grisea* from triticale and barley could infect triticale, wheat, barley, oat, and rye but not rice, sorghum, maize, common millet, sugarcane, and *Brachiaria brizantha*. A positive cross-infection of triticale by wheat and barley isolates was observed. Positive cross-infection was also detected between barley and triticale and wheat. Cross-infection was not possible with rice but rice isolates infected barley, triticale and wheat. Southern hybridization of DNA digested with *Eco*RI, probed with *Pot2* showed similar DNA fingerprints among wheat, triticale, and barley isolates suggesting that the disease on triticale and barley originated from wheat blast pathogen.

S. Kawasaki (ed.), Rice Blast: Interaction with Rice and Control, 251–260.

© 2004 *Kluwer Academic Publishers. Printed in the Netherlands.*

1. INTRODUCTION

Magnaporthe grisea (Hebert) Barr [anamorph *Pyricularia grisea* (Cooke) Sacc.] attacks a wide range of grasses. Its most important crop host is rice (*Oryza sativa* L.) for which this pathogen causes major losses in many rice growing countries of the world. In Brazil, wheat (*Triticum aestivum* L.) is another crop to suffer economic impact due to this fungus. It was first reported in 1985 in Northern Parana state and soon became one of the major diseases of wheat in the country (Igarashi et al., 1986). Besides these two crops, two other hosts of economic importance were noticed in 2001 with severe blast infection during survey of winter crop diseases in Southern parts of São Paulo. Many fields of triticale (*X. triticosecale* Wittmack) showed high percentage of neck blast, the most destructive form of the disease, with considerable yield loss. Blast on barley (*Hordeum vulgare* L.) grown at the experimental station of Instituto Agronômico de Campinas, located in Capão Bonito was also detected. This is the first report of blast on both crops in the state of São Paulo.

The first blast epidemic on triticale occurred in 1995 in the state of Parana where 60% of yield loss was reported. Its causal organism was considered to be different from the rice and wheat blast pathogens because of the difference in pathogenicity (Mehta & Baier, 1998). This had been the first and the only report of blast on triticale in Brazil or elsewhere.

Occurrence of blast on barley has been documented in other parts of the world (Asuyama, 1965, Sarma et al., 1971). In Brazil, the first report of this disease is as recent as 2000 when infection of barley by *M. grisea* in natural condition was observed in Brasília (Anjos & Charchar, 2000). Lately, this disease seems to be widespread in the country as reports of its occurrence in the southern state of Rio Grande do Sul and the Brazilian Cerrado have been documented (Casa et al., 2002, Goulart et al., 2002). No details on the relationship with other blast pathogens are available.

An important step toward a successful disease management of these diseases is to clarify the relationship among isolates from these four important crops. Thus, this work aimed at examining the relationship of *M. grisea* detected on triticale and barley with blast of rice and wheat based on host range, cross-infectivity and DNA fingerprinting.

2. MATERIALS AND METHODS

2.1. Pathogen

Plants of triticale and barley with neck blast were collected from various fields in the Southern part of São Paulo. The method of single spore isolation from diseased plants and storage of isolates were described previously (Urashima et al., 1993).

2.2. Host

A set of seedlings of 11 gramineous plants was inoculated to examine the host range of *M. grisea* from triticale and barley. Seeds employed were: triticale cultivars BRS148, BRS203; wheat Frontana, CNT8, BR40; barley BRS195, Antartica5, Embrapa127; oat IAC107, UFRGS14, Embrapa29; rye BR1; rice Epagri109, IAC201, Urucuia; sorghum BR700, BR800; maize P30K75, P30F33; common millet BRS1501; sugarcane BR72454 and *Brachiaria brizantha.* For cross-infection assay were employed rice cultivars IAC101, IAC202, Caiapó, IAC242; wheat IAC350, IAC1003, Anahuac, Embrapa10, BR18; triticale BRS203, BRS148, Embrapa53; barley Embrapa127, BRS180, BRS195, Antartica5. Seeds were soaked for three days and 3-5 germinated seeds were sown in plastic pots in uniformly fertilized soil. They were kept at 23-28 C until three- to four- leaf stage when they were inoculated. Four replicates were prepared for each isolate.

2.3. Inoculation and Disease Assessment

Inoculation and disease assessment were performed according to Urashima et al. (1993).

2.4. DNA manipulation

Isolates were grown in CM media on a rotary shaker for 5-6 days at room temperature. Subsequently, mycelia were harvested and stored at –20C. DNA was extracted as previously described (Urashima et al., 1999).

3. RESULTS

The reaction of various gramineous plants to *M. grisea* from triticale is described in Table 1. All isolates were pathogenic to triticale, completing the Koch's postulate. Besides triticale, wheat, barley, rye, and oat were also susceptible. Nevertheless, pathogenicity reactions were not uniform among them, 7 out 18 isolates from triticale could not infect any of the three cultivars of wheat and 3 isolates were not pathogenic to rye. On the other hand, barley showed to be extremely susceptible to blast pathogen from triticale as the only host on which compatible reaction occurred on all cultivars with all isolates. As a whole triticale formed by two cultivars and oat formed by three, were the most susceptible hosts because of the predominance of lesion type 4. Hosts with incompatible reaction to *M. grisea* from triticale included rice, maize, sorghum, common millet, sugarcane and *B. brizantha.* From this set, reaction on sorghum was characterized by formation of many lesions classified as type 2, brown color with no discernible centers that never sporulated. Same type of reactions were also observed on sugarcane and rice but in much lesser frequency. *B. brizantha* stood out for reacting with immune reaction (lesion type 0) to all isolates from triticale.

Table 1 also presents the reaction of various hosts to *M. grisea* from barley. Triticale, wheat, barley, rye, and oat were susceptible to the barley isolates. Among them, barley and rye were extremely susceptible with lesions classified as type 4 towards all barley isolates. Triticale and oat also had the majority of reaction classified as lesion 4. Wheat characterized by reacting differently according to isolates and one of them was not pathogenic to any of the three wheat cultivars. Other barley isolates infected at least one cultivar but none infected all cultivars. Blast isolates from barley were not pathogenic to rice, maize, sorghum, common millet, sugarcane and *B. brizantha*.

Table 1. Reaction of gramineous plants to *Magnaporthe grisea* from *X. triticosecale* and *Hordeum vulgare*

		*Isolate - Original host**	
Host	*Cultivar*	*Triticale*	*Barley*
Triticale	*BRS148*	*S(18)*	*S(9)*
	BRS203	*S(18)*	*S(9)*
Rice	*Epagri109*	*R(18)*	*R(9)*
	IAC201	*R(18)*	*R(9)*
	Urucuia	*R(18)*	*R(9)*
Wheat	*Frontana*	*S(5)*	*S(5)*
	CNT8	*R(18)*	*R(9)*
	BR40	*S(11)*	*S(8)*
Maize	*P30K75*	*R(18)*	*R(9)*
	P30F33	*R(18)*	*R(9)*
Barley	*BRS195*	*S(18)*	*S(9)*
	Antartica5	*S(18)*	*S(9)*
	Embrapa127	*S(18)*	*S(9)*
Rye	*BR1*	*S(15)*	*S(9)*
Oat	*IAC107*	*S(18)*	*S(9)*
	UFRGS14	*S(16)*	*S(5)*
	Embrapa29	*S(16)*	*S(8)*
Sorgum	*BR700*	*R(18)*	*R(9)*
	BR800	*R(18)*	*R(9)*
Common millet	*BRS1501*	*R(18)*	*R(9)*
Brachiaria brizantha	*Common*	*R(18)*	*R(9)*
Sugarcane	*RB72454*	*R(18)*	*R(9)*

*18 isolates from triticale and 9 from barley were tested. Number in the parentheses indicates the number of fungal isolates which gave the scored reaction.

Table 2 shows results of cross-infection assay of rice, wheat, triticale, and barley inoculated with *M. grisea* from these four hosts. Isolates from triticale were pathogenic to triticale, barley, and wheat. All the 9 isolates of triticale were pathogenic to the four cultivars of barley with high predominance of typical blast lesions (lesion type 4). On its own host, all isolates from triticale provoked typical blast symptoms (lesion type 4) on the three cultivars. On wheat, triticale isolates caused all types of reaction, from immune to typical blast lesions; 2 out of 9 isolates were not pathogenic to any of 5 wheat cultivars and none was pathogenic to all wheat cultivars.

Table 2. Cross-inoculation of *Magnaporthe grisea* from rice, wheat, triticale, and barley.

		Isolate - Original host*			
Host	Cultivar	Rice	Wheat	Triticale	Barley
Rice	IAC101	S(2)	R(8)	R(9)	R(7)
	IAC202	S(3)	R(8)	R(9)	R(7)
	IAC242	R(6)	R(8)	R(9)	R(7)
	Caiapo	S(2)	R(8)	R(9)	R(7)
Wheat	IAC350	R(6)	S(2)	S(1)	S(1)
	IAC1003	S(2)	S(6)	S(6)	S(5)
	Anahuac	S(1)	S(5)	S(4)	S(6)
	Embrapa10	S(1)	S(8)	S(6)	S(6)
	BR18	R(6)	S(3)	S(0)	S(0)
Triticale	BRS148	S(3)	S(7)	S(9)	S(7)
	BRS203	S(4)	S(8)	S(9)	S(7)
	Embrapa53	S(3)	S(7)	S(9)	S(7)
Barley	BRS180	S(6)	S(8)	S(9)	S(7)
	BRS195	S(5)	S(3)	S(9)	S(6)
	Antartica5	S(5)	S(7)	S(9)	S(7)
	Embrapa127	S(6)	S(8)	S(9)	S(6)

*6 isolates from rice, 8 from wheat, 9 from triticale, and 7 from barley were tested. Number in parentheses indicates the number of fungal isolates which gave the scored reaction.

Barley isolates of *M. grisea* had the same pattern of pathogenicity as the triticale isolates. The seven isolates employed were able to infect wheat, triticale, and barley but not rice. The lesion type provoked on triticale and barley cultivars was the same as the one caused by the isolates from triticale, with predominance of lesion type 4. The same variation observed in wheat towards triticale isolates was also detected with barley isolates. Cross-inoculation was successful between triticale, barley and wheat since these isolates were pathogenic to their own host and the two other cereals.

None of the blast isolates from these three hosts were pathogenic to rice. The great majority of rice cultivars exhibited an immune reaction but some isolates from triticale and barley caused minute or small brown lesions on a few rice cultivars that were classified as non-pathogenic because no spores were formed when lesions were submitted to suitable conditions. On the other hand, rice isolates were capable of infecting wheat, triticale, and barley. Barley was the most susceptible, showing typical blast lesions to all rice isolates. Interestingly, more cultivars of barley and triticale were susceptible to rice blast fungus than rice cultivars themselves. However, cross-inoculation was not complete as no isolates from other hosts could infect rice.

Figure 1 shows the DNA fingerprinting of *M. grisea* from rice, wheat, triticale, and barley when digested with *Eco*RI, eletrophoresed in 0.8% agarose gels, hybridized with *Pot2,* and detected in chemilluminiscence. A distinctive DNA pattern was produced by *M. grisea* isolates from rice (lanes

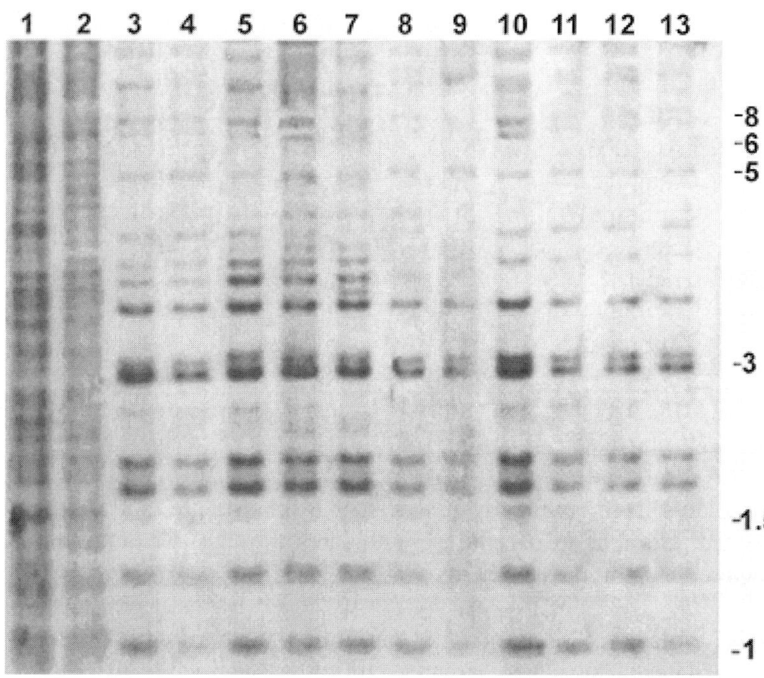

Figure 1. DNA fingerprints of *Magnaporthe grisea* from rice, triticale, barley and wheat. DNA was digested with *Eco*RI and hybridized with *Pot2.* The lanes contain genomic DNA from: 1-2: rice, 3-7: triticale, 8-10: barley, 11-13: wheat. Kilobase markers (kb) are indicated on the right side.

1 and 2). These have high copy number of *Pot2,* typical of rice-infecting isolates which distinguished them from pathogens of other hosts. Isolates

from barley, triticale and wheat carried this element in a moderate copy number, approximately 30 copies between 10kb-1kb. DNA fingerprints of triticale (lanes 3 to 7), barley (lanes 8 to 10) and wheat (lanes 11 to 13) were similar. The *Pot2* probe hybridized strongly with these isolates and produced 6-8 intense hybridization signals (Figure 1). All barley isolates were very uniform with little diversity, triticale isolates showed diversity according to date of the sampling, samples collected in 2001 presented high similarity. Wheat isolates from the same field were uniform but diverse from other fields.

DISCUSSION

This is the first report of blast disease on triticale and barley in the state of Sao Paulo. Barley blast was detected in field trials at the IAC Experimental Station and not at commercial scale. On the other hand, blast on triticale was identified in commercial fields spread over a large area in the region. Many of these triticale fields showed high percentage of head infection with severe yield loss. In those fields blast symptoms were easily identified because of fungal colonization that caused the spikelets to turn white above the infection point on the rachis. Another characteristics visible during sampling of those fields was that wheat growing in triticale fields due to seed contamination presented heads with no symptoms. This phenomenon suggested that wheat blast pathogen was not the origin of the disease on triticale. In fact, the first report of blast on triticale suggested its causal organism to be different from *M. grisea* of wheat (Mehta & Baier, 1998). However, data of the present work demonstrated a close similarity between *M. grisea* from wheat and triticale. Results of host range on a set of gramineous plants showed same pattern of pathogenicity between wheat and triticale isolates (Table 1) and cross-infection was positive between isolates of these two hosts (Table 2).

A definite proof of the close relationship between wheat and triticale isolates came from the DNA fingerprints where similar profile was observed between triticale isolates and isolates of wheat. Six intensely staining bands were shared by isolates from barley, triticale and wheat at the same molecular sizes indicating that they are closely related. Therefore, data of the present work clearly demonstrated that *M. grisea* that caused blast epidemics on triticale in São Paulo state originated from the wheat blast pathogen.

The fact of wheat heads remained symptom-free in a heavily infected triticale fields can be explained by the cultivar specificity reaction. Among the gramineous plants examined wheat characterized by the highest variation in host reaction with response varying from immune (lesion type 0) to typical blast symptom (lesion type 4) depending on the isolate and/or host cultivar (Table 1 and Table 2). Differential reaction of wheat to different blast isolate was due to the presence of different races, characteristic of *M. grisea* populations.

Isolates of *M. grisea* from triticale and barley were similar. This conclusion is supported by host range which showed the same results (Table 1), cross-inoculation assay with successful infection on both hosts (Table 2), and similar DNA fingerprints (Figure 1). The high fingerprint similarity among isolates of *M. grisea* from triticale and barley provide evidence of one single source of primary inoculum for the blast outbreak on both hosts and this source was from wheat pathogen from a field located in Paraná. Because wheat blast was not observed in São Paulo during outbreak of blast on triticale and barley it is still unclear how this pathogen reached these new hosts.

The fact that wheat blast pathogen gave origin to blast epidemics on triticale and barley is a matter of extreme concern because they are cultivated in the same geographical area and during the same growing season. It is not unusual farmers having fields of wheat, triticale and barley cultivated nearby. Another aspect to be pointed out is the high susceptibility of other winter crops like rye and oat demonstrating that wheat blast disease poses as potential threat to many winter crops. Recent reports have already documented severe blast infection on black oat (*Avena strigosa*) and *Lolium multiflorum* although their relationship with wheat blast were not examined (Prestes et al.,2002, Nunes et al., 2002).

Rice was not the source of the blast disease on triticale and barley even though rice isolates could infect cultivars of these two hosts (Table 3). The capacity of rice isolates to infect different winter crops has already been documented (Prabhu et al., 1992, Urashima et al., 1993, Kato et al., 2000). Thus, it would be possible to identify rice-infecting isolate from fields of triticale and barley specially where rice has been cultivated as summer crop.

Triticale has become popular among growers because high yield can be obtained without any application of fungicides. Thus, the occurrence of blast can jeopardize the expansion of this crop in the region since methods of controlling the disease will demand a much higher cost, and thus making it less cost-effective. A disease management program to be accepted by farmers will require methods to keep the disease in low level without the use of chemicals products. The monogenic resistant cultivar, the most employed method of controlling rice blast disease all over the world, cannot be recommended straightforward on these crops because of the high susceptibility of different cultivars demonstrated in this work. Thus, breeding for useful cultivars will demand a long process since no genes from commercial cultivars seem to be useful and a search for other sources of resistance is urgently needed.

Barley and other winter crops that could be useful for crop rotation in a disease management program also seems not promising because of their susceptibility, in fact blast on barley has already been documented in commercial fields in other regions of Brazil (Casa et al., 2002, Goulart et al, 2002).

ACKNOWLEDGEMENTS

We are thankful to Y. Tosa and S. Mayama for their assistances. To L.F. Rasera, F. Giacomini, S. Cleto, B.C. Barros, A. Linhares, and L. Federizzi for providing research materials.

REFERENCES

Anjos, J. R. N., & Charchar, M. J. A. (2000). Natural infection of barley by *Pyricularia grisea* in Brazil. *Fitopatologia Brasileira, 25,* 205.

Asuyama, H. (1965). Morphology, taxonomy, host range, and life cycle of *Pyricularia oryzae*. In: S. H. Ou (Ed.), The rice blast disease (pp. 9-22). Baltimore: Johns Hopkins Press.

Casa, R. T., Reis, E. M., & Panisson, E. (2002). Occurrence of the head blight on barley in planalto region of the Rio Grande do Sul State caused by *Pyricularia grisea*. *Fitopatologia Brasileira, 27,* 92.

Goulart, A. C. P., Amabili, R. F., Nasser, L. C. B., & Freitas, M. A. (2002). Occurrence of fungi on barley seeds produced under central pivot irrigation in the Brazilian Cerrado. *Fitopatologia Brasileira, 27,* 116.

Igarashi, S, Utiamada, C. M., Igarashi, L. C., Kazuma, A. H., & Lopes, R. S. (1986). *Pyricularia* em trigo. 1. Occurrence of *Pyricularia* sp. in Paraná state. *Fitopatologia Brasileira, 11,* 351-352.

Kato, H., Yamamoto, M., Yamaguchi-Ozaki, T., Kadouchi, H., Iwamoto, Y., Nakayashiki, H., et al. (2000). Pathogenicity, mating type and DNA Restriction Fragment Length Polymorphisms of *Pyricularia* populations isolated from gramineae, bambusideae and zingiberaceae plants. *J. Gen. Plant Pathol., 66*, 30-47.

Mehta, Y. R., & Baier, A. (1998). Variação patogenica entre isolados de *Magnaporthe grisea* atacando triticale e trigo no estado do Paraná. *Summa Phytopathologica, 24,* 119-125.

Nunes, C. D. M., Brancão, N., Rodrigues, R. C., & Reis, J. C. (2002). Blast occurrence on ryegrass (*Lolium multiflorum*) in different sites from RS, Brazil. *Fitopatologia Brasileira, 27,* 231.

Prabhu, A. S., Filippi, M. C., & Castro, N. (1992). Pathogenic variation among isolates of *Pyricularia oryzae* infecting rice, wheat, and grasses in Brazil. *Tropical Pest Management, 38,* 367-371.

Prestes, A. M., Rigo, L. L., Bevilaqua, G. A. P., & Muhl, F. R. (2002). Occurrence of leaf blast on black oat in Passo Fundo/RS. *Fitopatologia Brasileira, 27,* 153.

Sarma, Y. R., Nayudu, M. V., & Suryanarayanan, S. (1971). *Pyricularia* on barley. *Current Science, 24,* 670-671.

Urashima, A. S., Igarashi, S, Kato, H. (1993). Host range, mating type and fertility of *Pyricularia grisea* from wheat in Brazil. *Plant Disease, 77,* 1211-1216.

Urashima, A. S., Hashimoto, Y., Don, L.D., Kusaba, M., Tosa, Y., Nakayashiki, H., et al. (1999). Molecular analysis of the wheat blast population in Brazil with a homologue of retrotransposon MGR583. *Annals of the Phytopathological Society of Japan, 65,* 429-436.

GENETICS OF FERTILITY AND MATING TYPE IN *MAGNAPORTHE GRISEA*

*Anil S. Kotasthane[1], J. Kumar[2], U.S. Singh[2] & R.S. Zeigler[3]

[1]*Department of Biotechnology, Indira Gandhi Agricultural University, Raipur (C. G.), INDI*
[2]*Department of Plant Pathology, G. B. P. U. A. & T, Pantnagar. INDIA*
[3]*KSU-Plant Pathology, Manhattan, KS 6650-65502 USA*

**E-mail:Jkotasthane@yahoo.com*

Abstract: Blast disease of rice (*Oryza sativa* L.) caused by *Magnaporthe grisea* (anamorph = *Pyricularia grisea*) is one of the principal disease of rice. *M. grisea* includes a large number of monocots among its host range, however individual isolates are specific to only one or few hosts species. Sexual fertility has repeatedly been reported in rice pathogen from mountain regions of South and East Asia. In the Indian Himalayas where the rice and non-rice crops are grown in adjoining fields or some times mixed together the population of *M. grisea* was observed to be highly diverse, suggesting sexual recombination and random mating in pathogen population. The likelihood of sexual recombination occurring in a population is a function of coexisting biological competence (fertility) and potential, and can be assessed by the presence of both the mating types, and fertility. Our present observation regarding genetics of fertility and host range in *M. grisea* indicates that, the isolates from Foxtail millet*(Setaria italica)* are highly compatible and hermaphrodite and mated with most of the field isolates tested. High compatibility and fertility was observed in non-rice isolates but was low in non-rice x rice crosses. Rice isolates did not mate within themselves. The extent of gene flow between resident *M. grisea* from rice and non-rice host can be resolved by the distribution and frequencies of these rare alleles. The pattern of fertility and mating type distribution in the two typical rice-derived and setaria-derived isolates from the collection of Himalayan isolates revealed that Foxtail millet*(Setaria italica)* isolate III 202 was identified as highly compatible, hermaphrodite isolate. Progenies from non-rice x rice crosses produced very few fertile combinations on backcrossing, while the frequency was higher with the non-rice x non-rice derived progenies. The behavior of III 202 both as male and female depending upon opposite strengthens the view that *M. grisea* is self-sterile hermaphrodite. Genetic ratio for fertility suggests that a single gene did not control it, and a large number of genes functions to affect the mating competence. Fertility segregated independent of mating type. All the progenies did not cross with standard tester isolate III 202, although some of them were compatible within themselves. Some progenies of non-rice crosses were able to show compatible reactions with both the parents, thus suggesting the presence of both mating types, which in normal case any one of them is present in the isolates. The mating type identity for such progenies remains obscure in the absence for the evidence of dual maters, which has not so far been reported in nature. Higher frequency of male progenies indicates the dominance of male fertile parent. Asci with abnormal or aborted ascospores were frequently observed in non-rice x rice crosses. Two-point inoculation technique in test tubes was evaluated as more efficient over widely used three point inoculation technique and a more efficient technique of single ascospore isolation was standardised.

S. Kawasaki (ed.), Rice Blast: Interaction with Rice and Control, 261–270.
© 2004 *Kluwer Academic Publishers. Printed in the Netherlands.*

1. INTRODUCTION

The important rice (*Oryza sativa* L.) fungal pathogen, *Magnaporthe grisea* (Hebert) Barr (anamorph *Pyricularia grisea*) Zeigler *et al.* 1994, offers a pertinent case study of the multidisciplinary approach required to clarify the significance of recombination in agricultural pathosystem. The heterothallic ascomycetes has a wide host range, causing sometimes extensive necrotic lesions (blast) on aerial structures of approximately 50 graminaceous spp., including economically important crops such as rice, wheat, and millets (*Elusine spp., Echinochola spp., Panicum spp., and Setaria spp*) (Asuyama 1965; Kato *et al.*. 1977). The blast pathogen from all host are now considered to be one species (Rossman *et al.*. 1990), the extent of gene flow among pathogens of different host is still unclear.

It is generally considered that hybridization is an important means of pathogenic variation in fungi having sexual state as a part of their life cycle. The event of meiosis transpires within a fruiting body known as perithecium, was induced as early as in 1971 by Hebert in cultures and demonstrated heterothallism in this fungus. Like most filamentous ascomycetes, the *M. grisea* mating system is controlled by a single master locus (MAT 1 Coppin *et al.*, 1997; Yoder *et al.* 1986), with two different mating types (MAT 1-1 and MAT 1-2) being idiomorphs required for sexual recombination. These two types are found in natural population in many rice growing areas, but teleomorph stage has not been observed in nature yet (Kato 1974; Yaegashi 1977). Using genomic subtraction approach Kang *et al.* (1994) cloned and sequenced the 2.5-kb MAT 1 and 3.5-kb MAT 2 idiomorphs from *M. grisea*. The PCR primers subsequently developed (Xu & Hammer 1995) now permit convenient mating type assessment that is independent of other incompatibility factors.

The likely-hood of sexual recombination occurring in a population is a function of coexisting biological competence (fertility) and potential, which can be judged through test crosses with known fertile strains and by the presence of both mating types in a population. The perfect stage, provides a good basis to analyze genetically the distribution of mating type alleles, fertility and host range in the field isolates as well as in the progenies derived from them and also provides an evaluation of genetic variability.

2. MATERIALS AND METHODS

Fifty seven single spore field isolates were obtained from Dr. J. Kumar, Plant Pathology Laboratory, Ranichauri (Hill Campus, GBPUA&T, Pantnagar). For use the purified cultures were multiplied on OMA (Oat-Meal Agar) slants and the original were preserved. A female fertile, MAT 1-1 standard tester III 202 a Foxtail millet (*Setaria spp.*) isolate was used whose fertility and mating type was identified through repeated in-vitro mating with compatible isolate, and also reconfirmed as MAT 1-1 (J Kumar personal communication) with MAT primers using PCR.

Single spore cultures of field isolates of *M. grisea* were multiplied on OMA medium at 27-28° C. Isolates were paired against each other in every possible combination in petri-plate following the three point inoculation technique (Ito *et al.,* 1983) and in the test tube following two point inoculation technique (Kotasthane and Singh 2000) and were incubated at about 19 $\pm$ 1° C with continuous fluorescent light. All combinations were observed regularly w.e.f. 10 days of inoculation till one month for the formation perithecia. The fertile combinations were again dual cultured on the OMA plates to confirm the fertility and mating type of the isolates and for the isolation of single ascospores from the selected crosses.

Single ascospore progenies were isolated by the technique of Kotasthane and Singh 2000 and were back-crossed with their respective parents (thrice with progenies of rice x non-rice crosses) and (twice with progenies of non-rice x non-rice crosses. The fertility and mating type was finally confirmed by pairing the field isolates with a standard tester isolate III 202 (female fertile, MAT 1-1) and the progenies with respective parents were assessed as follows.

1. If perithecia formed on both the sides of the area of confluence, both isolates were considered to be hermaphroditic.
2. If perithecia were formed only one side, the isolates towards which perithecia were formed was considered as female while the opposite isolate was considered as female sterile and functioning as male in the mating.
3. If perithecia formed only in the region of confluence, no judgement of sexuality was made for that mating.

The mating type and fertility was designated as of the opposite mating type with which they were found compatible and produced perithecia.

3. RESULTS AND DISCUSSION

3.1. Fertility and Mating Type of the Field Isolates

Fertility and mating type of the isolate was confirmed twice. The combination, which showed the production of perfect stages, was selected for further studies. Of, 30 fertile combination with III 202, and the isolates V35, V292 (rice infecting), and V 63 and V 67 (Ragi *(Elusine coracana)* infecting), were selected for single ascospore isolation and genetic analysis (Table 1). Foxtail millet*(Setaria italica)* isolate III 202 was identified as highly fertile hermaphrodite isolate and was able to mate with most of the field isolate tested. Non-rice x non-rice crosses were highly fertile. Non-rice x rice crosses exhibited low fertility. Crosses between isolates with different host specificity's were possible and were reported by several workers (Yaegashi and Nishihara 1976; Kato *et al.* 1976; Valent *et al.* 1986; Yaegashi and Udagawa 1978; Ueyama and Tsuda 1975 ;Yaegashi 1977). Failure to obtain rice x rice fertile cross in our present indicates lack of female fertility in rice isolates (Yaegashi 1977; Yaegashi and Nishihara 1976; Itoi *et al.* 1983; Kolmar and Ellingboe 1988; Leung and Williams 1985; Valent *et al.* 1986).

Table 1. *Fertility, Mating Type, of selected parent isolates for genetic analysis*

S. No.	Selected Field Isolates Fertility and Mating Type		Designation to the Progenies of Crosses
	MAT 1-1	**MAT 1-2**	
1	III 202 (F)	V 35A (M) Rice	**AK** 1-102
2	Foxtail	V 292 (M) Rice	**CK** 1-119
3	Millet	V 67A (H) Ragi	**BK** 1-105
4	Isolate	V 63A (M) Ragi	**DK** 1-100

Foxtail Millet = *Setaria italica;* Ragi = *Eleusine coracana*

with hermaphrodite isolates from other hosts.The fertile combinations differed in their ability and time to produce perithecia. Crosses producing high number of perithecia in a short period (approx. 10 days) were considered highly compatible while those producing less number in longer time (14 to 20 days) were rated as poorly compatible are in agreement with the observations reported by Shen et al. (1994).

3.2. Fertility and Mating Type of Progenies

The back-crossing of the progenies with their respective parents was done thrice to ensure the fertility and mating type (Table 2). Scoring was usually easy, as there was either an abundance of perithecia or none. Sexuality of an isolate was designated, as Male (M) / Female (F) / Hermaphrodite (H). The progenies of the crosses involving non-rice x rice parental isolates were observed to produce few fertile combinations, while the progenies of the crosses involving non-rice parental isolates were observed to produce fertile combinations in higher frequencies. The variable behaviour of III 202 with the

Table 2 *. Segregation of progenies in four crosses*

With Female Parent	Fertility & Mating Type of Progenies: Male (MAT 1-2)	Female (MAT 1-2)
Segregation of progenies when back-crossed with III 202 (F) MAT 1-1	**AK**,102*	**AK**,51*,68,71
	CK,15*,21,32,36,38,52,58,61,62,65,84,88,	**CK**,30*,35,57,79,100
	BK5*,6,7,8,9,11,12,14,19,22,23,24,26,29,31,32,33,34,35,36,39,46,48,49,50,51,53,55,56,58,60,63,66,67,68,69,73,75,79,80,81,83,85,88,90,91,93,94,97,98,99,100,101,103,105	**BK**,17*,38,43,54,65
	DK,1*,2,11,22,25,26,32,37,47,54,57,58,66,67,69,70,71,77,78,79,84,86,92,95,96	**DK**,3*,6,19,23,24,41,49,64,85,94,98

With Male Parents	Fertility & Mating Type of Progenies: Male (MAT 1-1)	Female (MAT 1-1)
V 35 (MAT 1-2)	Nil	**AK**,1*,5,14,21,28,29,84,85,95,96,102
V 292 (MAT 1-2)	Nil	**CK**,6*,11,31,83,94,105,114
V 67(MAT 1-2)	**BK**,10*,30,69	**BK**,3*,18,41,59,62,76,77,89,96
V 63 (MAT 1-2)	Nil	**DK**,4*,7,8,10,13,16,36,46,50,52,55,61,62,68,73,81,90,91,93,97

progenies of non-rice x non-rice and non-rice x rice crosses is suggestive of the preference to behave as Female/Male or as Hermaphrodite with the mated isolate. Chao and Ellingboe (1991) while selecting for mating competence in *M. grisea* pathogenic to rice, led to the selection of isolates where the mating type alleles were the only restriction on mating competence. Isolates can function as male, female, or hermaphrodite in mating depending on the isolate with which they are mated suggesting that mating competence was controlled by the genes in both mated isolates. They gave the concept of a perithecial allele, which enables an isolate to function as a female in cross.

3.3. Non-Rice x Rice Crosses

III 202 X V 35 Very few progenies of the crosses III 202 X V 35 showed fertility when they were back-crossed (Table 2). The progenies which showed fertile combinations with the male parent V 35 were all female fertile. With III 202, AK 51, 68, and 71 behaved as female. Looking to the few fertile combinations obtained by back-crossing, all the progenies derived from III 202 X V 35 were also crossed with two selected progenies AK 1 and AK 28 (female MAT 1-1) and comparatively more fertile combinations were formed than with the parental isolate. The M:F fertility segregation ratio was approximated to 5:3 irrespective of the isolate tested with which they formed the perfect stages (Table 3) and 1:3 on back-crossing with III 202 (Table 3).

***Table 3**. Segregation for M : F fertility in progenies of III 202 X V 35 and III 202 X V 292*

Cross	Parents	Ratio M : F	Genetic ratio & expected X^2 1:1	3:1	1:3	5:3	9:7	15:1
III 202 X V 35	III 202	1:3	1.00	----	0.0	2.4	1.59	----
III 202 X V 292		12:5	2.88	0.17	----	0.47	1.42	----
III 202 X V 35	V 35	0:11	----	----	----	----	----	----
III 202 X V 292	V 292	7:0	----	----	----	----	----	----
Back-crossing with parent isolate								
Parents	Male	Female						
III 202 X V 35*	1	14	----	----	----	----	----	----
III 202X V 292	12	12	0.00	----	----	1.6	0.38	----
Irrespective of the standard isolate tested								
Parents	Male	Female						
III 202 X V 35*	28	17	2.68	----	----	0.001	0.65	----
III 202X V 292	19	19	0.00	----	----	2.53	0.60	----

*Excluding hermaphrodite reactions of cross III 202 X V 35

III 202 X V 292 Progenies of the cross III 202 X V 292 also produced few fertile combinations after back-crossing (Table 2). The progenies, which produced fertile combinations with the male, parent V 292 behaved as female. When backcrossed with III 202 twelve progenies behaved as male and 5 as female. The segregation ratio for M:F fertility was approximated to 1:1 irrespective of the isolate tested with which they formed the perfect stages (Table 3) and 3:1 on back-crossing with III 202 (Table 3). The genetic ratio suggests that fertility was not controlled by a single gene and a large number of gene(s) functions to affect the mating competence.

3.2. Non-Rice x Non-Rice Crosses

III 202 X V 67 Most of the progenies derived from III 202 X V 67, showed fertility irrespective of the parents with which they crossed. The progenies when mated with the female parent (III 202) and male parent (V67) segregated for M:F fertility which approximated to 15:1 and 1:3 respectively (Table 5). Irrespective of the isolate tested 66 progenies were male and 22 were female, and the segregation ratio was approximated to 3:1 (Table 5). With III 202 two rows of perithecia were formed by eight progenies and were designated as hermaphrodite (Table 4).

Table 4. *Hermaphrodite ascospores progenies of III 202 X V 67*

Progeny No.	Mating Type Reaction with		Fertility
	III 202 (F MAT 1-1)	V 67 (M MAT 1-2)	
	Fertility & Mating Type Reaction of the Progenies		
	MAT 1-2	MAT 1-1	
BK20	F/M	---	H
BK25	H	---	H
BK28	F/M	---	H
BK47	H/M	---	H
BK52	F/M	---	H
BK86	H/M/M	---	H
BK87	F/H/H	---	H
BK104	F/H	---	H

M= Male; F= Female; H= Hermaphrodite

15:1 suggest for duplicate gene action due to the presence of completely two dominant genes, and produced the same phenotype whether they are alone or together. The contrasting results are observed when both the genes were in homozygous recessive state, the point to be noted here is both the parents were hermaphrodite, thus indicating the possibility of such ratio. It was also observed that some of the progenies were showing compatible reaction with both the parents, all the progenies were designated as hermaphrodite, but the mating type to them could not be designated. The higher frequency of male progenies indicates the dominance of male parent.

III 202 X V 63 In another set of 100 progenies derived from non-rice parent, III 202 X V 63, when mated with the female (III 202) the M:F segregation was approximated to 3:1 (Table 5). Irrespective of the isolate tested with which they formed the perfect stages the M:F segregation ratio was approximated to 1:1 (Table 5).The genetic ratio so obtained in the progenies of non-rice cross also suggest for fertility to be controlled by more then one gene.

Table 5. *Segregation for M : F fertility in progenies of* III 202 X V 67 and III 202 X V 63

Cross	Parents	Ratio M : F	Genetic ratio & expected X^2					
			1:1	3:1	1:3	5:3	9:7	15:1
III 202 X V 67	III 202	55:5	----	----	----	----	----	0.44
III 202 X V 63		25:11	----	0.59	----	0.74	2.54	----
III 202 X V 67	V 67	3:9	----	----	0.00	----	----	----
III 202 X V 63	V 63	0:20	----	----	----	----	----	----
Back-crossing with parent isolate								
Parents	Male	Female						
III 202 X V 67*	58	14	----	1.19	----	----	----	----
III 202 X V 63	25	31	0.64	----	----	----	----	----
Irrespective of the standard isolate tested								
	Male	Female						
III 202 X V 67*	66	22	----	0.00	----	----	----	----
III 202 X V 63	25	31	0.64	----	----	----	3.06	----

*Excluding hermaphrodite reactions of cross III 202 X V 67.

Leung and Williams (1985) in their genetic studies of hermaphroditism of two Japanese rice isolates observed that it was controlled by a single gene and progenies segregated 1:1 for hermaphroditism and maleness. Kolmar and Ellingboe (1988) observed that there was no clear segregation for single hermaphroditism gene in the first generation but was observed in the second generation. They suggested that a single gene did not control female fertility.

3.4. Mating Type

The progenies of the cross III 202 X V 35 segregated for MAT 1-1 : MAT 1-2 approximating to the genetic ratio of 3:1, 5:3 , 9:7 or 13:3 while III 202 X V 292 it approximated to 1:3. No approximation of genetic ratio could be made for the progenies derived from the crosses involving non-rice parents but a higher number of MAT 1-2 mating type were observed (Table 6). The mating type reaction suggests that mating type genes segregated independently of fertility. The 9:7 and 13:3 ratio suggest for the complementary and inhibitory gene action. The presence of high frequency of MAT 1-2 which seems to be a dominant feature of non-rice isolates support the reports of several workers. Kolmar and Ellingboe (1988) observed that mating type, female fertility and pathogenicity to rice segregated randomly. Ellingboe *et al.* (1988) observed the female fertile and rice pathogenic progenies in the third generation and many of these progenies to be inter-fertile and reported that a large number of gene functions to affect mating competence.

Table 6. *Segregation for mating type reaction (MAT 1-1 : MAT 1-2) in progenies of four crosses*

Cross	MAT 1-1	MAT 1-2	Genetic ratio & Expected X^2						
			1:1	**3:1**	**1:3**	**5:3**	**9:7**	**13:3**	**15:1**
III 202 X V 35	11	4	----	0.02	----	0.75	1.78	0.62	----
III 202X V 292	7	17	----	----	0.22	----	----	----	----
III 202 X V 67	12	68	----	----	----	----	----	----	----
III 202 X V 63	20	36	----	----	----	----	----	----	----

Table 7. *Hermaphrodite ascospores progenies of III 202 X V 67 and III 202 X V 63*

Progeny No	Fertility reaction with III 202 F (mating type=?)	Fertility reaction with V 67 M (mating type=?)	Fertility	Progeny No	Fertility Reaction with III 202 F (mating type=?)	Fertility Reaction with V 63 M (mating type=?)	Fertility
III 202 x V 67				**III 202 x V 63**			
BK16	F	F	H	DK9	F	F	H
BK27	M	F	H	DK 20	H	F	H
BK42	F	F	H	DK 30	M	F	H
BK57	M	F	H	DK 53	M	F	H
BK64	M	F	H	DK 87	F	F	H
BK70	M	F	H	DK 27	H/M	F	H
BK71	F/M	F	H	DK 99	F/M	F	H
BK72	F	F	H	DK 88	F/M	F	H
BK74	M	F	H				
BK84	F	F	H				

Progenies derived from III 202 X V 63 showed compatible reaction with both the parents (Table 7). Does it indicate that the recombinants were dual maters? Mating type status could not be designated, as the isolates with dual mating capabilities have not been reported so far. Through transformation of *M. grisea* by mating type genes Kang *et al.* (1994) produced several dual maters containing both mating types, that successfully mated with both MAT 1-1 and MAT 1-2 testers. They further reported that they appear to lose one of the mating type genes during vegetative growth. Ascospores isolated from these homothallic crosses were either MAT 1-1 or MAT 1-2, but with no dual maters.

Following the three point inoculation technique, i.e. one tester and two test isolates, it was observed that none of the progenies showed compatible reaction with their respective parent, but the two test isolates formed compatible reaction between themselves. Thus the fertility of the progenies could be worked out but the mating type to these isolates could not be designated (Table 8). The resulting recombinants from the crosses, do they indicate that during the recombination process, there was a change in the mating competence, which resulted in incompatibility with the parents?

Table 8. *Fertile combinations formed within the progenies in different crosses*

Progeny No.	Fertility		Progeny No.	Fertility	Progeny No.	Fertility		Progeny No.	Fertility
III 202 X V 35					**III 202 X V 292**				
AK51	F	Vs.	AK52	M	CK09	F	Vs.	CK10	M
AK63	F	Vs.	AK64	M	CK31	F	Vs.	CK32	M
AK69	F	Vs.	AK27	M	CK35	F	Vs.	CK37	M
AK85	F	Vs.	AK86	M	CK83	F	Vs.	CK 84	M
AK95	F	Vs.	AK96	M	CK13	M	Vs.	CK14	F
AK96	F	Vs.	AK98	M	CK27	M	Vs.	CK28	F
AK67	M	Vs.	AK68	F	CK99	M	Vs.	CK100	F
AK101	M	Vs.	AK102	F					
III 202 X V 67					**III 202 X V 67**				
BK17	F	Vs.	BK18	M	BK37	M	Vs.	BK38	F
BK57	F	Vs.	BK58	M	BK67	M	Vs.	BK68	F
BK71	F	Vs.	BK75	M	BK77	M	Vs.	BK78	F
BK87	F	Vs.	BK88	M	BK85	M	Vs.	BK86	F

4. CONCLUSIONS

The pattern of fertility and mating type distribution in the two typical rice-derived and setaria-derived isolates from the collection of Himalayan isolates revealed that

- Foxtail millet *(Setaria italica)* isolate III 202 was highly compatible, and the behaviour of III 202 both as male and female depending upon opposite strengthens the view that *M. grisea* is a self-sterile hermaphrodite.
- Non-rice x non-rice crosses was highly fertile, while non-rice x rice crosses exhibited low fertility and there was a failure to obtain a rice x rice fertile cross. Higher frequency of male progenies indicated the dominance of male fertile parent.
- Fertility segregated independent of mating type. Genetic ratio of fertility suggests that it was not controlled by a single gene, but a large number of gene functions to affect the mating competence.
- All the progenies did not cross with standard tester isolate III 202, although some of them were compatible within themselves.

- Compatibility of the non-rice x non-rice derived progenies with both the parents, suggest the presence of both mating types, which in normal case any one of them is present in the isolate. The mating type identity of such progenies remains obscure in the absence for the evidence of dual maters, which has not been reported in nature so far.

REFERENCES

Asuyama H. (1965). Morphology taxonomy, host range and life cycle of *Piricularia oryzae*. In: Symposium on the Rice Blast Disease, Los Banos, Jul 1965. The rice blast disease. Baltimore Johns Hopkins Press. pp. 9-22.

Chao C.T. and Ellingboe A.H. (1991). Selection for mating competence in *Magnaporthe grisea* pathogenic to rice *Can. J. Bot.* 69: 2130-2134.

Coppin E., Debuchy R., Arnaise S. and Picard M. (1997). Mating type and sexual development in filamentous fungi. *Microbiol. Mol. Biol. Rev.* 61:411-428.

Ellingboe A.H., Wu B.C. and Robertson W (1988) Fertility and pathogenicity to rice in *Magnaporthe grisea*. In: Palacios, R. and Verma, D. P. S. eds., Molecular genetics of Plant-microbe Interaction. APS. Press St. Paul, Minnesota. Pp.235-240.

Hebert T.T.(1971). The perfect stage of *Pyricularia grisea. Phytopathology*, 61: 83-87.

Itoi S. T., Mishima S. Arase., and Nozu M (1983) Mating behaviour of Japanese isolates of *Pyricularia oryzae*. *Phytopathology*. 73:155-158.

Kang S., Chumley F.G. and Valent B. (1994). Isolation of the mating type genes of the phytopathogenic fungus *Magnaporthe grisea* using genome subtraction. *Genetics*. 138 (2): 289-296.

Kato H. (1974). Epidemology of rice blast disease. *Rev. Pl. Protec. Res.* 7: 1-20.

Kato H., Yamaguchi T. and Nishihara N. (1976). The perfect state of *Pyricularia* Cav. in culture. *Ann. Phytopath. Soc. Japan* 42: 507-510.

Kato H., Yamaguchi T. and Nishihara, N. (1977). Seed transmission, pathogenicity and control of ragi blast fungus and susceptibility of ragi to *Pyricularia* spp. from grasses. Cereals and migo. *Ann. Phytopath. Soc. Japan* 43:392-401.

Kolmar J.A. and Ellingboe A.H. (1988). Genetic relationship between fertility, pathogenicity and virulence in *Magnaporthe grisea*. *Can. J. Bot.* 66(5): 891-897.

Kotasthane A. S. and Singh. U. S. (2000). Simple technique for single ascospore isolation and a modified two point inoculation technique of mating in test tubes. *Journal of Mycology and Plant Pathology* 30(3): 172-179

Leung H. and Williams P.H. (1985). Genetic analysis of electrophoretic enzyme variants, mating types, and hermaphroditism in *Pyricularia oryzae* Cavara. *Can. J. Gen. & Cyto.* 27:697-704.

Rossman Y.A., Howard J.R. and Valent B. (1990). *Pyricularia grisea,* the correct name for the rice blast fungus. *Mycologia*. 82: 509-512.

Shen Y., Zhu P.L., Yuan X.P., Jin M.Z., Chai R.Y. and Zang Q.S. (1994). Studies on the perfect stage of *Pyricularia oryzae* in China. *Scientia Agricultura Sinica*. 27(1): 25-29.

Ueyama A. and Tsude M. (1975). Formation of the perfect state in culture of *Pyricularia* sp. from some graminaceous plants. *Trans. Myso. Soc. Japan*. 16: 420-422.

Valent B., Crawford S.M., Weaves G.C. and Chumley G. F. (1986). Genetic studies of fertility and pathogenicity in *Magnaporthe grisea (Pyricularia oryzae). Iowa State Journal of Research*. 60: 569-594.

Xu J-R.and Hamer J.E. (1995). Assessment of *Magnaporthe grisea* mating type by spore PCR. *Fungal Genet. Newsl.* 42:80.

Yaegashi H. (1977). On the sexuality of blast fungi, *Pyricularia* spp. *Ann. Phytopath. Soc. Japan*. 43: 432-439.

Yaegashi H. and Nishihara N. (1976). Production of the perfect stage in *Pyricularia* from cereals and grasses. *Ann. Phytopath. Soc. Japan* 42:511-515.

Yaegashi H. and Udagawa G. (1978). The taxonomical identity of the perfect stage of *Pyricularia grisea* and its allies. *Can. J. Bot.* 56: 180-183.

Yoder O.C., Valent B., and Chumley F. (1986). Genetic nomenclature and practice for plant pathogenic fungi. *Phytopathology.* 76:383-385.

Zeigler R.S., Teng P.S., and Leong S.A. eds (1994). Rice Blast Disease. Commonwealth Agricultural Bureaux International, Wilmington, England. pp.287-29.

MATING TYPE ALLELES, FEMALE FERTILITY AND GENETIC DIVERSITY OF *MAGNAPORTHE GRISEA* POPULATIONS PATHOGENIC TO RICE FROM FIVE ASIAN COUNTRIES

*Shen Y [1], Milazzo J [2], Yuan XP [1], Adreit H [2], Wang YL[1], Notteghem JL [3] and Tharreau D [2]

[1] *China National Rice Research Institute, Hangzhou 310006, China*

[2] *Centre de Cooperation Internationale en Recherche Agronomique pour le Développement, UMR BGPI, TA 73 / 09, 34398 Montpellier, Cedex 05, France*

[3] *Institut National de la Recherche Agronomique, UMR BGPI, 2 Place Viala, 34060 Montpellier, France*

**E-mail: shenying@mail.hz.zj.cn*

Abstract: Five hundred and twenty-two isolates of *Magnaporthe grisea* isolated from rice in 5 Asian countries were characterized for their mating type by crossing them with 4 hermaphroditic isolates (KA3 and TH12: MAT1.1; Guy11 and TH16: MAT1.2). Among them, 41% were MAT1.1 and 25 % were MAT1.2. The remaining 34% did not produce perithecia with any of the 4 hermaphroditic testers. In Bangladesh, India, Nepal, Vietnam and in most provinces of China, both mating types were present. Only one mating type was found in 3 provinces and 1 city of China. Almost all the isolates had very low fertility, as they were in general female sterile and sometimes also male sterile. Hermaphroditic isolates were recovered from the 5 countries. In these countries, they represented between 13% and 75% of the isolates. In Zhejiang, Guizhou, Guangdong, Hunnan, Yunnan and Hubei provinces of China, hermaphroditic isolates represented between 6% and 67%. The genetic diversity of 143 isolates from these countries and provinces, where hermaphroditic isolates had been collected, was analyzed using SCAR markers. Genetic diversity was high and population structure did not resemble classical clonal structure described in most rice growing regions. The existence of sexual reproduction in the field, localization of a center of diversity in China, and migration between countries are discussed in this paper.

S. Kawasaki (ed.), Rice Blast: Interaction with Rice and Control, 271–279.
© 2004 *Kluwer Academic Publishers. Printed in the Netherlands.*

1. INTRODUCTION

Magnaporthe grisea (Hebert) Barr.[1, 2] is a heterothallic Ascomycete pathogenic to rice, causing the rice blast disease, but is also pathogenic to other grass hosts. Analyzing the genetic structure of *M. grisea* populations pathogenic to rice clearly showed that clonal reproduction is the rule in most rice growing areas [3-8]. Despite these clear results, the existence of hermaphroditic isolates in some restricted areas of the world [9, 10] leads to the assumption that sexual reproduction could occur in the field in some limited areas. This idea is supported by the fact that the complete sexual reproduction could be obtained *in vitro*, on culture medium, or in dead tissue [11-16] in different laboratories and with different isolates. The observation of the sexual stage of the fungus in the field would allow to clearly demonstrate this assumption. It was never observed in the field but it might difficult to find. Population study is another possible way of demonstrating the existence of recombination in natural populations. But which populations must be analyzed? Obviously, sexual reproduction requires isolates of opposite mating types and hermaphroditic isolates. The purpose of this study was to further investigate mating type and hermaphroditism of isolates from some Asian countries to identify populations suitable for testing the hypothesis of sexual reproduction by population genetics analysis.

2. MATERIALS AND METHODS

2.1 Mating Type Determination

Five-hundred and twenty-two *M.grisea* isolates were isolated from disease symptoms sampled on different rice types in Bangladesh, China, India, Nepal, and Vietnam. Most of the isolations were performed at the Plant Pathology laboratory at CNRRI and Rice Pathology Laboratory of CIRAD. Some isolates were received from colleagues. As already described[17], mating type was determined by crossing with 4 hermaphroditic testers (KA3 and TH12: MAT 1.1, Guy11 and TH16: MAT 1.2) on Petri dishes containing rice flour medium (rice flour 20 g; yeast extract 2.5 g; agar 15 g; water 1 L; and penicillin 500,000 UI added after autoclaving) and incubated at 28°C for 3 days. Then, the Petri dishes were placed during 18 days under continuous fluorescent light at 20°C. Formation of perithecia was observed 21 days after co-inoculation of the medium.

2.2 Genomic DNA Extraction

One-hundred and forty-three *M. grisea* isolates were cultured for extraction of genomic DNA. Mycelium mass of the fungus was grown in liquid culture in 2YEG medium (yeast extract 2 g/L, glucose 10 g/L, KNO_3 3 g/L, KH_2PO_4 2 g/l, penicillin G 500,000 I.U./L, pH 6.0), lyophilized and ground into powder. Total genomic DNA was extracted following the methods previously described in Dioh *et al.*[18] and was diluted in TE buffer (10 mM Tris-HCl, pH 7.5, 1 mM EDTA) at 5 ng/µl.

2.3 SCAR Primers and PCR Amplification Conditions

SCAR markers were developed and used as described by Soubabere *et al.*[19]. Seventeen pairs of SCAR primers (C7-1.4, Cut, E10, ERG2, H6-0.8-4, H6-0.8-5, H6-1.1-5, H6-1.1-7, J13-1.4, J16-0.7, J16-0.9, P9-0.4, P9-1.9, PWL.2, S9, Y16, G18) were used to amplify the DNA of 143 isolates collected in the countries or Chinese provinces where hermaphroditic isolates were identified. Amplified products were separated by electrophoresis on 1% agarose gel using 1 x TBE buffer (0.09 M/L Tris borate, 0.002 M/L EDTA) and stained with ethidium bromide (Figure 1).

2.4 Cluster Analysis

PCR products obtained with a pair of SCAR primers were scored as alleles: 0 for the absence of amplification, 1, 2, 3, ..., n for amplification products of different size. Similarity was calculated using the Simple Matching Index, and the resulting matrix was used to construct a dendrogram by Ultrametic Pairwise Grouping Method using Arithmetic mean (UPGMA). Calculations were performed using Darwin 4 (CIRAD).

3. RESULTS

3.1 Mating Type Frequency in some Asian Rice Blast Populations

Among the 522 Asian isolates of *M. grisea* characterized for their mating type, 216 isolates (41%) were determined as *MAT1.1*, 130 (25%) as *MAT1.2*, and the remaining 176 isolates (34%) did not mate with any of the 4 hermaphroditic fertile testers and mating type was not determined. These frequencies varied greatly between countries or Chinese provinces but sample size might be responsible for such variations. In all the countries studied and in most Chinese provinces, both mating types were sampled.

3.2 Geographic Distribution of Hermaphroditic Isolates

Hermaphroditic isolates of *M.grisea* were identified in Bangladesh, India, Nepal, Vietnam and Guangdong, Guizhou, Hubei, Hunan, Yunnan and Zhejiang provinces of China. The frequency of hermaphroditic isolates in these countries or provinces varied between 6 % and 75 %. Highest frequencies were recorded in Vietnam and Hubei province of China (75 % and 67 % respectively). But, again sample size and sampling method may be responsible for these high frequencies. In Yunnan and Hunan provinces, the frequency of hermaphroditic (10 % and 11 % respectively) is probably high enough to influence population structure. For these populations, sample size

(91 and 95 isolates respectively) is probably sufficient to avoid overestimation of frequencies. MAT 1.2 hermaphroditic isolates were sampled more frequently than MAT1.1 hermaphroditic isolates. Hermaphroditic isolates of both mating types were identified in Nepal and Guizhou, Hubei, Hunan and Zhejiang provinces of China (Table 1).

Table 1 Distribution of mating type alleles and hermaphroditic isolates of *Magnaporthe grisea* from 5 Asian countries

	Isolates Number						Frequency				
	Mat1.1		Mat1.2		ND	Total	Mat1 .1	Mat1 .2	Nd	Herma-phroditi c	Mat1.1/ Mat1.1 +
Country	Male	Herm a.	Male	Herm a.							
BANGLA DESH		**1**	**3**		**3**	**7**	**0.14**	**0.43**	**0.43**	**0.14**	**0.25**
INDIA	**15**	**4**	**4**		**7**	**30**	**0.63**	**0.13**	**0.23**	**0.13**	**0.83**
NEPAL	**5**	**1**		**1**	**2**	**9**	**0.67**	**0.11**	**0.22**	**0.22**	**0.86**
VIETNA	**1**			**6**	**1**	**8**	**0.13**	**0.75**	**0.13**	**0.75**	**0.14**
CHINA	**178**	**11**	**96**	**20**	**163**	**468**	**0.4**	**0.25**	**0.35**	**0.07**	**0.62**
Fujian	13		9		12	34	0.38	0.26	0.35	0	0.59
Guangdong	6	2	8		6	22	0.36	0.36	0.27	0.09	0.5
Guangxi			5		5	10	0	0.5	0.5	0	0
Guizhou	14	1	5	1	13	34	0.44	0.18	0.38	0.06	0.71
Hainan	2		2		2	6	0.33	0.33	0.33	0	0.5
Hebei/Beijing/ Tianjin	10		2		2	14	0.71	0.14	0.14	0	0.83
Hubei	1	1	1	3		6	0.33	0.67	0	0.67	0.33
Hunan	24	5	18	5	43	95	0.31	0.24	0.45	0.11	0.56
Jiangxi	1		2		5	8	0.13	0.25	0.63	0	0.33
Liaoning	4				7	11	0.36	0	0.64	0	1
Shanghai	8				3	11	0.73	0	0.27	0	1
Shanxi	5		1		10	16	0.31	0.06	0.63	0	0.83
Sichuan	16		14		15	45	0.36	0.31	0.33	0	0.53
Yunnan	25		20	9	37	91	0.27	0.32	0.41	0.1	0.46
Zhejiang	37	2	9	2	20	70	0.56	0.16	0.29	0.06	0.78
Jiangsu	12				6	18	0.67	0	0.33	0	1
Total	***199***	***17***	***103***	***27***	***176***	***522***	***0.41***	***0.25***	***0.34***	***0.08***	***0.62***

3.3 Rice Blast Genetic diversity

In an attempt to evaluate the diversity of isolates from the five Asian countries studied here, 143 *M. grisea* isolates from Bangladesh (12 isolates), China (109 isolates), India (6 isolates), Nepal (6 isolates), and Vietnam (10 isolates) were characterized with SCAR markers. An example of SCAR amplification of some of these isolates is shown in Figure 1. For this study, the sample included all the hermaphroditic isolates and was completed by additional isolates from the same country or province of China. For this latter country, some reference isolates of known lineage were also included.

Fig 1. Separation on agarose gel of amplification products generated with the P9-1.9 SCAR primers

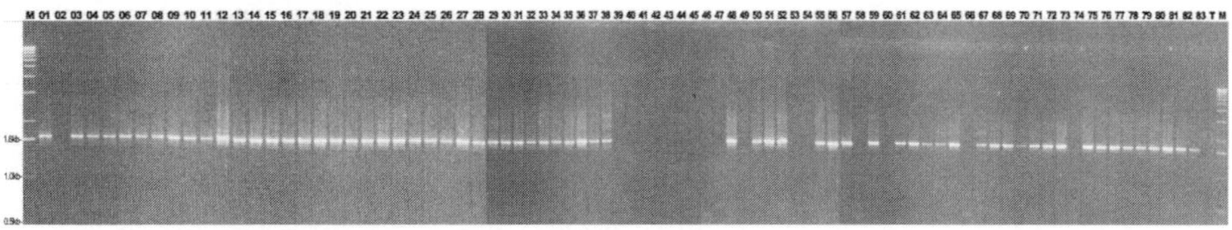

Isolates from China: 1~37; Bangladesh: 38~45; India: 46~49; Nepal: 50; Vietnam: 51~60; Thailand: 61~83; T: Control (Empty); M: 1 kb Ladder.

Similarity between tested isolates ranged from 100% to 50% (Figure 2). This range of variation is similar to that of a worldwide sample analyzed with the same markers (data not shown). With the exception of isolates from Bangladesh and some isolates from Hunan and Yunnan (probably clones, because they were 100% identical), isolates from the same geographic origin were distributed all over the tree. On the contrary some isolates from Vietnam, China and Nepal were identical. Unlike in classical clonal populations of the blast fungus, no obvious clustering could be seen. Hermaphroditic isolates were also distributed all over the tree and did not tend to cluster.

4. DISCUSSION

Population genetics can bring some evidence supporting the occurrence of mating in *M. grisea* populations, in the absence of observation of the organs required for sexual reproduction (perithecia) in the field. Population studies require appropriate sampling in specific geographic areas where the sexual reproduction is possible (presence of hermaphroditic isolates of opposite mating type). The collection we analysed gathers isolates from diverse geographic origin within Asia. Such collection is valuable for the estimation of mating type and hermaphroditism frequencies in large area. We identified a significant number of hermaphroditic isolates in this sample (8% of the isolates tested). The genetic characterisation of these hermaphroditic isolates with SCAR markers showed that their genetic diversity is relatively high. For the moment, the laboratory crosses used for the genetics of *M. grisea* rely mostly on the fertile isolate Guy11. The high diversity we identified in these new hermaphroditic isolates from Asia is likely to provide a large number of parental isolates that could be used for genetic studies of the rice blast fungus.

When considering the geographic distribution of the *M. grisea* hermaphroditic isolates, we found that these isolates were mostly collected in regions of the Himalayan foothills, either from the Indian side or the Chinese side. This localised geographic distribution of the hermaphroditic isolates suggests that they could originate from an ancestral *M. grisea* mating population that was located in the Himalayan foothills. The observed diversity in these regions strengthens this hypothesis. Our results confirm previous diversity analysis conducted in China[7, 20, 21] and India[22,23]. According to the Red Queen hypothesis, sexual reproduction is likely to be maintained in areas where the genetic diversity of the host is high[24]. Southern China and Northern India are the two centres of diversification and domestication of cultivated rice. The traditional growing practices used in these regions have maintained a high level of genetic diversity among populations of cultivated rice. Therefore, according to the Red Queen hypothesis, the populations of the rice blast fungus that has developed on the rice cultivars from these regions, in order to maintain a high number of different genotypes, should be highly diversified genetically and are likely to be mating populations. Indeed, we observed that the rice blast fungus populations from the Himalayan foothills have a high genetic diversity, that both mating types and hermaphroditic isolates are present, all things required for mating.

But, all these results are not enough to demonstrate the existence of recombination. The distribution of mating types and hermaphroditic isolates are clues to identify populations to be studied in detail. The geographic distribution of hermaphroditic isolates indicates where samples of the rice blast fungus should be collected. These populations will then be analyzed with molecular markers and the hypothesis of linkage disequilibrium will be tested. This work is in progress for some provinces of China.

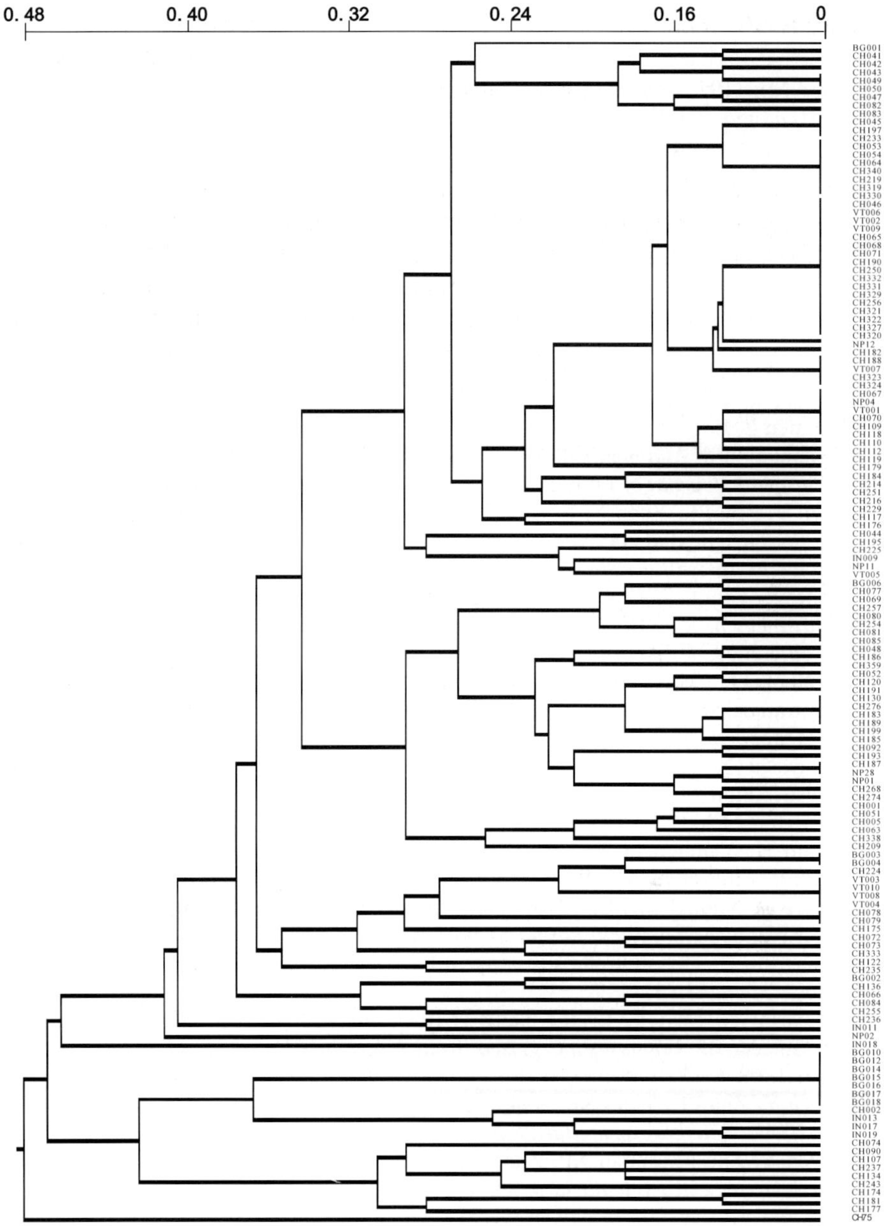

Figure 2. Dendrogram showing the genetic relationships measured with SCAR markers between 143 *M. grisea* isolates from 5 Asian countries
CH: China; BG: Bangladesh; IN: India; NP: Nepal; VT: Vietnam.

REFERENCES

[1] Barr ME. 1977. *Magnaporthe, Telimenella* and *Hyponectria (Physosporellaceae). Mycologia,* 69:952-956.[2] Hebert TT. 1971. The perfect stage of *Pyricularia grisea*. *Phytopathology*. 61:83-87.

[2] Hebert TT. 1971. The perfect stage of *Pyricularia grisea*. *Phytopathology*. 61:83-87.

[3] Levy M, Romao J, Marchetti MA, Hamer JE .1991. DNA fingerprinting with a dispersed repeated sequence resolve pathotype diversity in the Rice Blast fungus. *The Plant Cell*. 3:95-102.

[4] Levy M, Correa-Victoria FJ, Zeigler RS, Xu S, Hamer JE. 1993. Genetic diversity of the Rice Blast fungus in disease nursery in Colombia. *Phytopathology*. 83:1427-1433.

[5] Chen D, Zeigler R, Leung H, Nelson RJ. 1995. Population structure of *Pyricularia grisea* at two screening sites in the Philippines. *Phytopathology*, 85:1011-1020.

[6] Roumen E, Levy M, Notteghem JL. 1997. Characterization of the European pathogen population of *Magnaporthe grisea* by DNA fingerprinting and pathotype analysis. *European Journal of Plant Pathology*, 103, 363-371.

[7] Shen Y, Zhu PL, Yuan XP, Zhao XH, Manry J, Rojas C, Shahjahan AKM, Levy M. 1996. Genetic diversity and geographic distribution of *Pyricularia grisea* in China. *Scientia Agricultura Sinica,*. 29(4):39-46 (in Chinese).

[8] Sone T, Tomita F (2000) Clonal lineage of Japenese blast fungus population. In *Advances in Rice Blast Research*. Proceedings of the 2nd International Rice Blast Conference. 4-7 August 1998, Montpellier, France. D. Tharreau, M.H. Lebrun, N.J. Talbot, J.L. Notteghem (ed). Kluwer Academic Press. P221-233.

[9] Mekwatanakarn P, Kositratana W, Phromraksa T, Zeigler RS. 1999. Sexually fertile *Magnaporthe grisea* rice pathogens in Thailand. *Plant Disease*. 83:939-943.

[10] Hayashi N, Li, CY, Li JR, Iwano M, Naito H, Yoshino R, Kato H. 1997. Distribution of fertile *Magnaporthe grisea* fungus pathogenic to rice in Yunnan Province, China. *Annals of the Phytopathological Society of Japan*. 63:316-323.

[11] Kato H, Yamaguchi T. 1982. The perfect stage of *Pyricularia oryzae* CAV. in culture. *Ann. Pytopath. Soc. Japan*. 48:607-612.

[12] Silue D, Notteghem JL; 1990. Production of perithecia of *Magnaporthe grisea* on rice plants. *Mycological Research*. 94:1151-1152.

[13] Yaegashi H. 1977. On the sexuality of blast fungi, *Pyricularia* sp. *Ann. Phytopath. Soc. Japan*. 43:432-439.

[14] Hayashi N, Li, CY, Li JL, Naito H.1997. *In vitro* production on rice plants of perithecia of *Magnaporthe grisea* from Yunnan, China. *Mycological Research*. 101:1308-1310.

[15] Shen Y, Zhu PL, Yuan XP.1994, Studies on the perfect stage of *Pyricularia oryzae* in China. *Scientia Agricultura Sinica,* 27(1):25–29 (in Chinese).

[16] Shen Y, Zhu PL, Yuan XP, Zhao XH, Peng YL, Chen GH. 1995, Inducing of the perfect stage in *Pyricularia oryza* and the pathogenicity of its progeny. *Journal of Southwest Agricultural University*, 8(3):74–79 (in Chinese).

[17] Notteghem JL, Silué D. 1992. Distribution of mating type alleles in *Magnaporthe grisea* populations pathogenic to rice. *Phytopathology*. 82: 421-424.

[18] Dioh W, Tharreau D, Notteghem JL, Orbach M, Lebrun MH. 2000. Mapping of avirulence genes in the Rice Blast Fungus, *Magnaporthe grisea*, with RFLP and RAPD Markers. *Molecular Plant-Microbe Interactions*. 2:217-227.

[19] Soubabère O, Jorge V, Notteghem JL, Lebrun, MH, Tharreau D. 2001. Sequence characterized amplified region markers for the rice blast fungus, *Magnaporthe grisea. Molecular Ecology Notes*. 1:19-21.

[20] Shen Y, Yuan XP, Wang YL. 1998, Molecular probes for epidemiological studies of the rice blast fungus. *Journal of Southwest Agricultural University,* 20(5):401–408 (in Chinese) .

[21] Shen Y, Notteghem JL, Milazzo J, Yuan XP, Adreit H, Zhao XH, Wang YL, Tharreau D. 2002, Geographic distribution of mating type in *Magnaporthe grisea* and its relationship between fertile isolates in China. *Agricultural Sciences in China*. 1(6):648–656.

[22] Kumar J, Nelson RJ, Zeigler RS. 1999. Population structure and dynamics of *Magnaporthe grisea* in the Indian Himalayas. *Genetics*. 152:971-84.

[23] Zeigler RS. 1998. Recombination in *Magnaporthe grisea. Annual Review of Phytopathology*. 36:249-275.

[24] Ebert D, Hamilton WH. 1996. Sex against virulence: the coevolution of parasitic diseases. Tree. 11:7982.

EFFICACY OF CARPROPAMID AGAINST MUTANTS OF *MAGNAPORTHE GRISEA* AT CODON 75 ON SCYTALONE DEHYDRATASE

*T. Shigyo, Y. Kuchii, Y. Araki and H. Sawada

Nihon Bayer Agrochem K.K. (now Bayer CropScience K.K.), Yuki Research Center, 9511-4, Yuki, Yuki City, Ibaraki 307-0001 Japan

**E-mail: takuma.shigyo@bayercropscience.com*

Abstract: In 2001, *Pyricularia oryzae* isolates, showing reduced sensitivity to scytalone dehydratase inhibitors, were collected from the Matsuura river area in Saga prefecture, Japan. A single-point mutation causing substitution of one amino acid in scytalone dehydratase was found in those isolates. However, the blast control efficacy of carpropamid in the field has not been completely lost in the area where the mutant isolate is dominant. The results of inoculation tests suggest that enhancement of host defense responses, the other mode-of-action of carpropamid, may contribute to the residual efficacy of carpropamid against the mutants.

Key words: carpropamid, rice blast, mutant, scytalone dehydratase

1. INTRODUCTION

Rice blast is the most important rice disease from old days in Japan. Organomercurials, antibiotics, organophosphorus compounds, melanin biosynthesis inhibitors (MBI), and resistance inducers were widely used in Japan (Uesugi, 1998, Clark and Yamaguchi, 2002). But the control efficacy of antibiotics and organic phosphorus fungicides has decreased due to the

S. Kawasaki (ed.), Rice Blast: Interaction with Rice and Control, 281–287.
© 2004 *Kluwer Academic Publishers. Printed in the Netherlands.*

field resistance after some years of their launch to the market. On the contrary, MBIs and resistance inducers have not faced any resistance problems for more than 20 years.

Carpropamid [(1*RS*, 3*SR*)-2,2-dichloro-*N*-[1-(4-chlorophenyl) ethyl]-1-ethyl-3-methylcyclopropanecaroboxamide] is a systemic fungicide for rice blast control. It inhibits the scytalone dehydratase activity in the melanin biosynthesis pathway (Figure 1).

Carpropamid 4% granule formulation is the first long-lasting blasticide applicable in nursery box application, launched in 1998.

In 2001, insufficient efficacy of carpropamid against blast was reported in the Matsuura River area in Saga prefecture, the south-western part of Japan. It was suspected that the less sensitive isolates to carpropamid were the major reason. However, the control efficacy of carpropamid in the field has not been completely lost (Yamaguchi *et al.*, 2002). Takagaki *et al.* (2002) revealed that the rice blast isolates from Saga, showing reduced sensitivity to carpropamid, have a change at the 75th amino acid of scytalone dehydratase, valine to methionine, due to a single point mutation on the coding gene. This amino acid change causes transformation of the sterical structure, resulting in a reduced binding activity of carpropamid to this enzyme, which is about 20 times lower for the V75M mutants in comparison to wild type. However, inoculation tests results with the mutant isolates suggested that the control efficacy of carpropamid *in vivo* is not as low as 1/20 level to wild type isolates (So *et al.*, 2002).

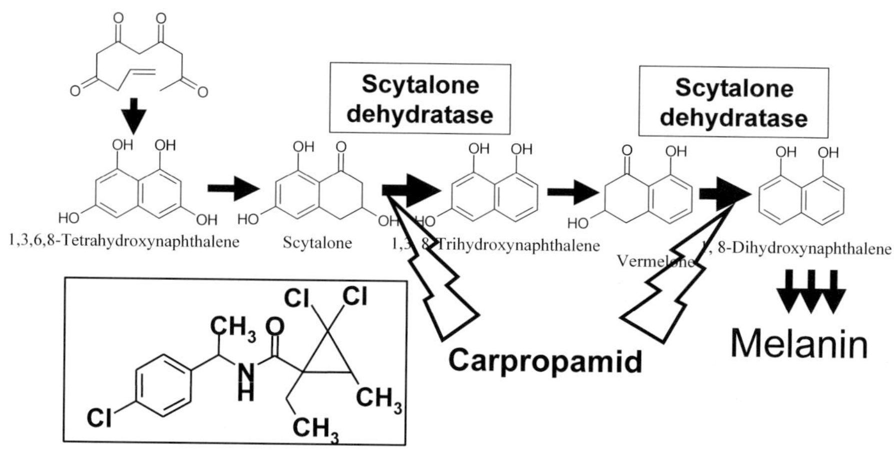

***Figure** 1. Carpropamid inhibition sites of melanin biosynthesis*

2. CORRELATION BETWEEN THE MUTATION AND SENSITIVITY TO CARPROPAMID

To investigate where V75M mutants exist in Saga prefecture, rice blast lesions on leaves or panicles were additionally collected in 28 cities, towns, or villages except the areas along the Matsuura River, in September 2001. Those lesions were supplied for single spore isolation followed by genetic analysis and inoculation tests.

2.1. PCR-SSCP analysis

Diagnosis of the single point mutation of scytalone dehydratase gene was carried out by PCR-SSCP (Sugihara *et al.*, 2002). From each isolate, mycelia were picked up from the media, for the DNA extraction. SSCP was carried out using the primer A, 5'-ATGGGTTCGCAAGTTCAAAAG-3', and the primer B, 5'-ACAAGCTCTGGGAGGCAATG-3'. Amplifying cycle, 94 degrees (30 seconds), 50 degrees (30 seconds), and 72 degrees (30 seconds) was repeated 30 times. From electrophoresis results, comparing with a wild type isolate and a V75M mutant of known sequence the 166 isolates were classified to 81 wild type and 75 V75M mutants.

2.2. Blast inoculation just after carpropamid spray

The 2nd leaf stage rice (cultivar Aichi-Asahi) was sprayed with carpropamid, at a concentration of 10, 50, or 250 ppm. The spray inoculation with conidia spore suspension (1.0×10^4 spores/ml) took place at the same day. The number of lesions was counted 6 days after inoculation to calculate the protective value with reference to untreated plants. The results (Figure 2) at 10 ppm indicated that efficacy of carpropamid against the V75M mutant isolates are reduced by more than 90%. Because of the 20 times lower binding activity of carpropamid to the V75M scytalone dehydratase, melanin biosynthesis should not be inhibited for V75M strains with concentrations up to 400 ppm (= 20 times of minimum effective dosage by spray application). However, some efficacy was observed already at 50 and 250ppm. The EC_{50} value was calculated for both wild and V75M strains. EC_{50} values were lower than 0.1mg/l for the wild-type isolates, which is considered to represent the baseline sensitivity. EC_{50} values for the mutants were clearly higher than those for the wild strains, but they were distributed across a wide range (0.85 to 3.1×10^2 mg/l) and occasionally could not be calculated due to an irregular dose response. This indicates that the other mode-of-action of carpropamid may contribute to some control efficacy at those higher application rates.

It has been reported previously, Araki *et al.* (1999), that accumulation of some phytoalexins after blast infection is much faster and higher in carpropamid treated rice than in the untreated control. It has also been reported that carpropamid enhances the activity of some known enzymes related to host defense reactions in rice (Kuchii *et al.*, 2002).

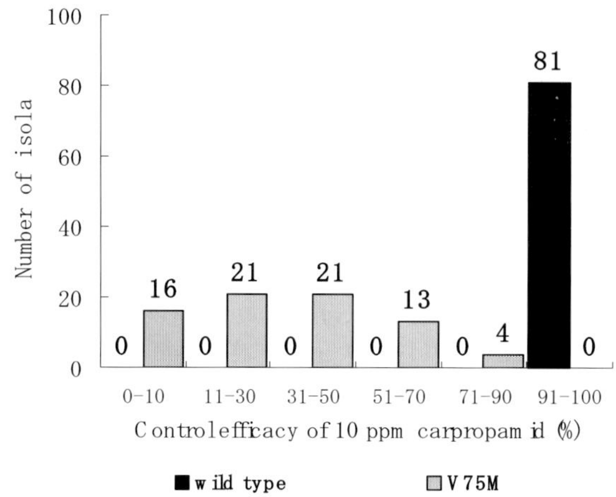

Figure 2. *Distribution of sensitivity to carpropamid in an inoculation test*

3. DISEASE CONTROL EFFICACY OF CARPROPAMID FIVE DAYS AFTER APPLIATION

To investigate whether the suggested other mode-of-action, the enhancement of rice plant defense mechanisms, is responsible for the remaining efficacy of carpropamid, two different inoculation tests were carried out.

3.1. Magnaporthe grisea, V75M mutant

Another inoculation test was carried out using same the method as previously described, except using another rice cultivar (Koshihikari) and a different interval between spraying and inoculation (5 days = 5DAT). For a V75M mutant isolate, W-1, control efficacy of carpropamid at 50 ppm in 5DAT test was significantly higher than that of 0DAT (inoculation on the day of spraying) (Figure 3). In this test, rice leaves were sampled just before each inoculation, to measure the amount of carpropamid residue on leaf surface. Based on the residue and control efficacy in the 0DAT test, the approximate equation was calculated. (Figure 4). The Control efficacy of carpropamid in 5DAT tests was much higher than the theoretical value obtained from the approximate equation.

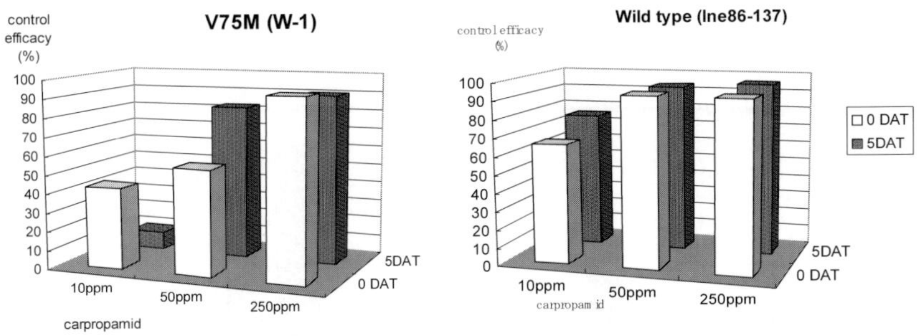

Figure 3. Carpropamid control efficacy against V75M mutant strain (W-1) and wild type strain (Ine86-137)

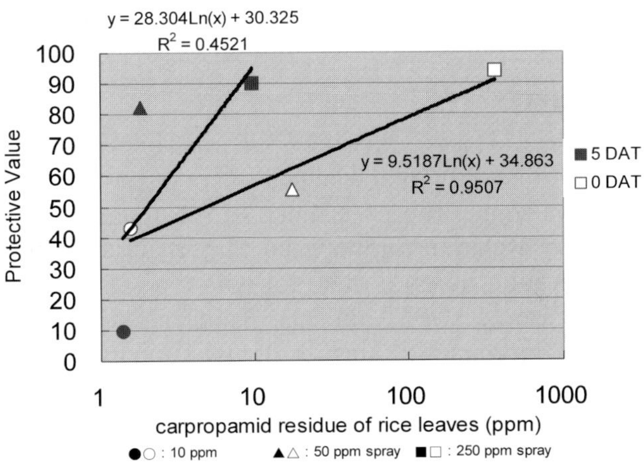

***Figure 4.** The correlation between residues of carpropamid on the rice leaf surface and control efficacy of V75M mutant (W-1)*

3.2. *Xanthomonas campestris* pv. *oryzae*

Inoculation test was conducted with *Xanthomonas campestris* pv. *oryzae*, which doesn't synthesis the melanin. Bacterial suspension was used for inoculation, by dipping the leaf end, which had been already cut off by scissors. The inoculation timing was either on the day of carpropamid application (0DAT) or 5 days after it (5DAT). Carpropamid controlled the disease in 5DAT tests better than 0DAT (Figure 5). The level of the control efficacy and the difference between 0DAT and 5DAT showed patterns similar to the results with V75M mutant blast isolates.

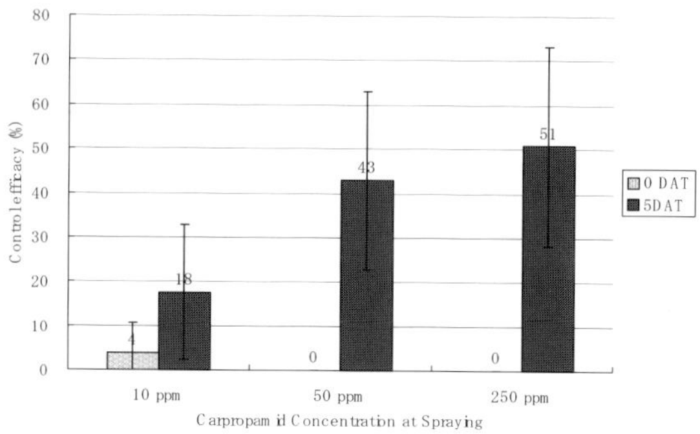

Figure 5. *Control efficacy of carpropamid against rice bacterial blight*

4. CONCLUSION

The results suggest that only by the enhancement of host defense mechanism, carpropamid shows some efficacy against V75M mutant isolates as well as bacterial pathogens. Relationship between carpropamid residue on rice leaf surface and its control efficacy is clear for wild type blast isolates, though it is not simple in case of V75M mutants, in other words, without the first mode of action of carpropamid. This result also support the reported control efficacy of carpropamid against several diseases which do not need melanin for their infection process to the host plants (Araki *et al.*, 1998; Tanaka *et al.*, 2000)

Under the practical conditions in Japan, carpropamid residues in rice are in general much lower than the level necessary for the second mode of action. However, in case of spray application outside Japan, where it is sprayed at higher concentration than 100 ppm, the enhancement of host defense mechanism by carpropamid may contribute to control efficacy in the field and play some role to protect rice from quick prevalence of V75M mutant isolates.

REFERENCES

Araki, Y., & Kurahashi, Y. (1999). Enhancement of phytoalexin synthesis during rice blast infection of leaves by pre-treatment with carpropamid. *J. Pesticide Sci.* 24, 369-374.

Araki, Y., Sakuma, H., Kinbara, T., Ohshima, A., & Sawada, H. (1998). Efficacy of Carpropamid (WIN) against some bacterial and fungal diseases. *Ann. Phytopathol. Soc. Jpn.* 64 (4), 391.

Clerk, J.M. & Yamaguchi, I. (2002). Scope and status of pesticide resistance. In J.M. Clark and I. Yamaguchi (Ed.), *Agrochemical Resistance*, (p.p. 1-22). Washington, DC: American Chemical Society.

Kuchii, Y., Araki, Y., & Sawada, H. (2002). Effect of carpropamid on enzymatic activities of rice plant that are related to the resistant reactions. *Abst. 27th Ann. Meet. J. Pesticide Sci.* 81.

So, K., Fuji, M., Iwabuchi, H., Kanayama, M., & Yamaguchi, J. (2002). Effects of various fungicides against less carpropamid-sensitive rice blast fungus isolated from the northwest area in Saga prefecture. *Jpn. J. Phytopathol.* 68, 262.

Sugihara, M., Tanaka, K., Shigyo, T., Kuchii, Y., Araki, Y., Kusaba, M., Takagaki, M., Nagayama, K., & Sawada, H. (2002). Distribution of carpropamid less-sensitive Pyricularia oryzae in Saga prefecture and countermeasure against them. *Jpn. J. Phytopathol.* 68, 262.

Takagaki, M., Shimizu, T., Miura, I., Araki, Y., Sawada H., So, K., & Nagayama, K. (2002). Mechanism of less-sensitive *Pyricularia oryzae* to carpropamid and the method for testing the sensitivity. *Jpn. J. Phytopathol.* 68, 262.

Tanaka, K., Araki, Y., Joo, W.J., Choi, Y.C., & Sawada, H. (2000). Control of bacterial diseases in rice nursery by carpropamid. *Jpn. J. Phytopathol.* 66, 182.

Uesugi, Y. (1998). Fungicide classes: Chemistry, uses and mode of action. In D. Hutson and J. Miyamoto (Ed.), *Fungicidal Activity,* (p.p. 23-56). Chichester: John Wiley & Sons Ltd.

Yamaguchi, J., Kuchiki, F., Hirayae, K., & of Saga prefecture in 2001. *Jpn. J. Phytopathol.* 68, 261.

COMPUTER SIMULATION APPROACHES FOR RICE BLAST DISEASE FORECASTING IN JAPAN

*Fumihiro NEMOTO[1] and Kiyoshi ISHIGURO[2]

[1]*Fukushima prefecture Agricultural Experiment Station, Koroyama, Fukushima 963-804, Japan*
[2]*National Agricultural Research Center for Tohoku Region, Morioka, Iwate 020-0198, Japan*

**E-mail: fnemoto@agri-exp.pref.fukushima.jp*

Abstract: Blast disease causes the most serious damage to rice production in Japan. Air temperature and rainfall are known as critical climatic factors that determine the favorable conditions for infection, and hence the severity of blast epidemics. The Japan Meteorological Agency (JMA) has developed an automated meteorological data acquisition system (AMeDAS) which retrieves data. A decision rule model, BLASTAM, has been developed to evaluate the favorable conditions for blast infection of leaves based on AMeDAS data. BLASTAM can identify when and where favorable infection conditions occur on a meso-scale. The purposes of these extension services are not only to provide information about the current and projected situation of local epidemics, but also to recommend topical countermeasures to growers. For the latter purpose, simulation model, BLASTL, of blast epidemics are very useful. With AMeDAS data automatically retrieved from JMA, the Internet-based national network system (JPP-NET) allows the latest information from BLASTAM and BLASTL to be displayed on a Web site.

Key words: Computer simulation, Blast disease forecasting, AMeDAS

Blast disease is one of the most serious diseases affecting rice production in Japan. Significant and frequent blast epidemics have occurred in recent decades, including those in 1988, 1991, 1993, 1995 and 1998 (Fig. 1) when rice growers suffered considerable yield losses. In 1993, a severe blast

S. Kawasaki (ed.), Rice Blast: Interaction with Rice and Control, 289–295.
© 2004 *Kluwer Academic Publishers. Printed in the Netherlands.*

epidemic occurred together with cool temperatures and caused yield losses enough to force the Japanese government to accept rice imports from foreign countries.

In Japan, the public sector on agricultural affairs includes an extension

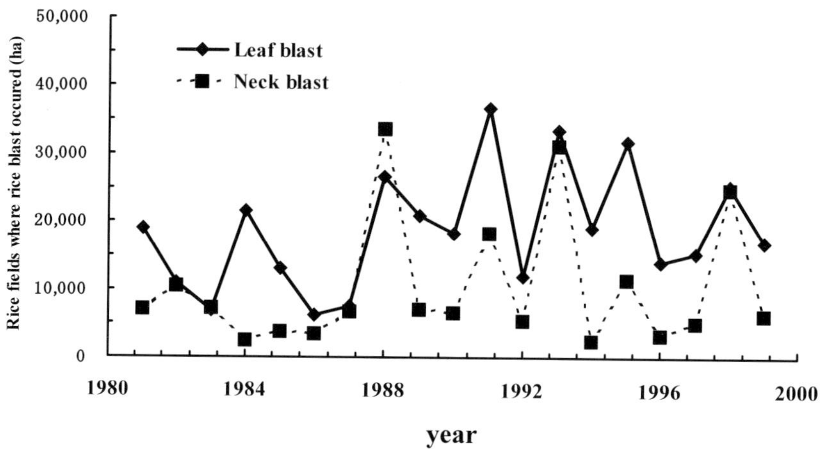

Fig.1. Occurrence of rice blast disease in Fukushima prefecture.

organization, which publishes information about crop pests and diseases. This information is comprised of observations forecasts alerts and risk assessments with regards to pest and disease outbreaks, and recommendations for their control. Figure 2 shows how the dissemination of information is organized within the extension service.

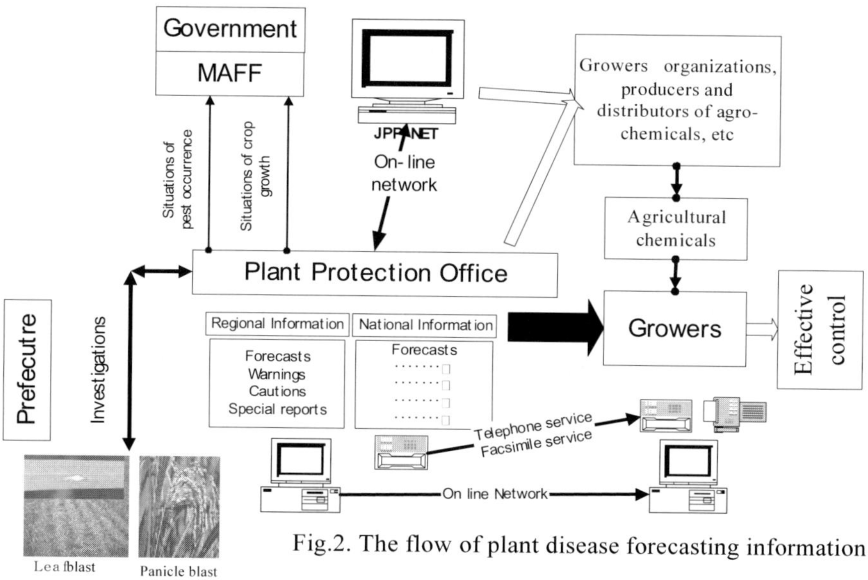

Fig.2. The flow of plant disease forecasting information.

Plant protection offices, which are managed locally by prefectural government authorities with partial support from the central government (MAFF, The Ministry of Agriculture, Forestry and Fisheries of Japan), have the responsibility to publish this information. We expect that clients using it will be growers, agricultural extension personnel, city and town officers, and producers and distributors of agrochemicals.

To make the information more accessible, MAFF has established JPP-NET (the Japan Plant Protection Network), an Internet web site for plant protection affairs. So far, this web site includes two computer simulation models, BLASTAM and BLASTL, which are able to forecast rice blast epidemics.

To run these models, input of meteorological data for the current season is essential. In particular, air temperature and precipitation are established as critical climatic factors that determine the favorableness of conditions for rice blast infection, and hence the severity of blast epidemics. Accordingly, JPP-NET was constructed to accommodate real-time weather data from the Japan Meteorological Agency (JMA) which uses an automated meteorological data acquisition system (AMeDAS) to retrieve data at 1-h intervals of air temperature, wind velocity/direction, rainfall and sunshine duration from 840 fixed observation stations distributed over all parts of the Japanese archipelago at spatial resolutions of 21 km. AMeDAS data are readily available on the Internet (Fig. 3). Plant protection offices can download AMeDAS data from JMA through JPP-NET (Fig. 4), and can also use the data to run BLASTAM and BLASTL on the same web site.

BLASTAM, a decision rule model, was developed to evaluate the favorableness of conditions for leaf blast infection based on AMeDAS data. BLASTAM can identify when and where favorable conditions for infection occur and display the information graphically (Fig. 4). Since favorable conditions for rice blast infection are thought to occur on a meso-scale level,

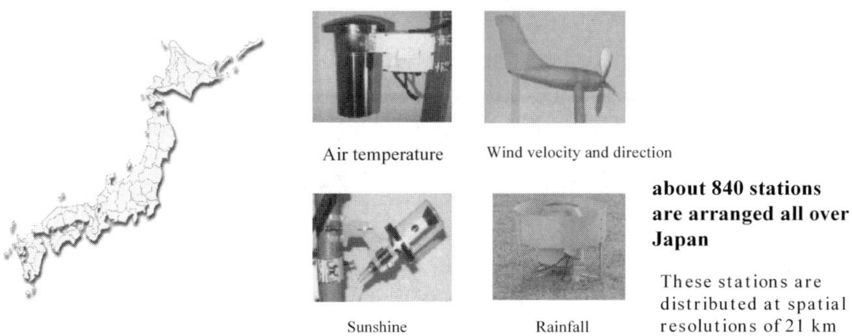

Fig.3. The Japan Meteorological Agency has developed an automated metrological data acquisition system, named "AMeDAS".

users of JPP-NET can easily distinguish which areas are at risk of infection. The information about intervals and frequencies of favorable conditions provided in the chronological table (Fig. 4) is also useful for understanding the current situation of local epidemics.

Many local plant protection offices have operated BLASTAM to fulfill their legal responsibilities for disease forecasting and public notification of the risk of blast epidemics. The purposes of these extension services are not

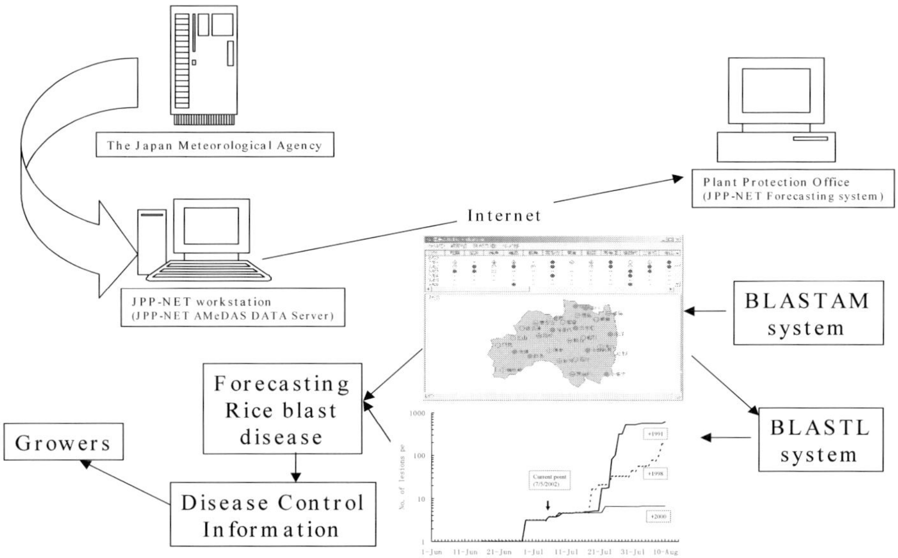

Fig.4. Computer system for public forecasting service for rice blast disease.

only to provide information about the current and projected situation of local epidemics, but also to advise growers on topical countermeasures. For the latter purpose, the BLASTL models are useful to the BLASTAM system, which simulates rice leaf blast epidemics, uses knowledge of each component of the disease cycle to predict the temporal progress of the disease according to meteorological input variables (Fig 4). Parameters of the model include morphological and phenological observations of single plants, the amount of initial inocula and the effect of fungicide on pathogen behavior of the pathogen.

As an example of how JPP-NET can be used to forecast blast epidemics, we now describe how JPP-NET was used as a decision–making tool by the Fukushima Prefecture Plant Protection Office in 2002. Leaf blast epidemics usually start as focal epidemics before developing into general epidemics. From a practical point of view, detecting the regional focus of an epidemic is critical for accurately forecasting the disease progress. Focal epidemics are known as phenomena occurring on a meso-scale level. Observations show that once BLASTAM detects the simultaneous occurrence of favorable conditions for infection at most AMeDAS stations on a meso-scale level, a general epidemic usually follows in that region. Consequently, plant protection offices using BLASTAM in JPP-NET closely monitor weather conditions during the early growing season of rice.

After detecting the local epidemic of leaf blast with BLASTAM and confirming it with field observations, the plant protection office went on to use BLASTL to investigate how the epidemic would increase and what recommendations to publish for the timing of fungicide spraying. In the 2002 season, the leaf blast epidemic became generalized at the end of June (Fig. 5). In JPP-NET, a window for BLASTL can be accessed by clicking on an AMeDAS site in the BLATSAM window (Fig. 4). A meteorological

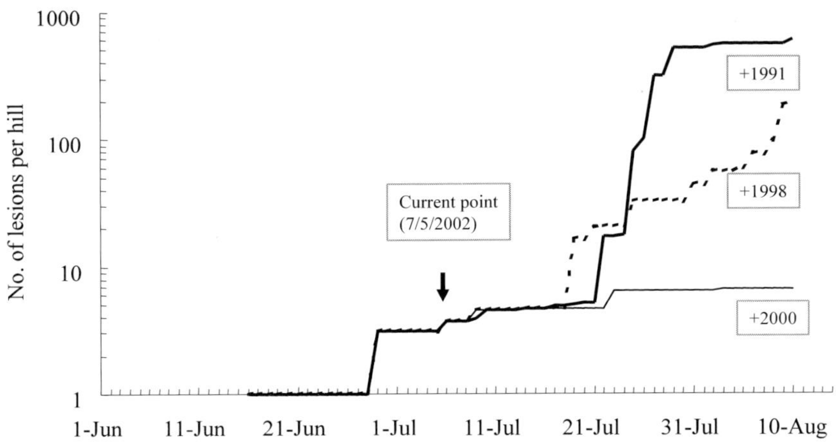

Fig 5. Forecast of infestation leaf blast.

data set for input is established by combining data of the current season prior to the current data with data for a historical season after the same date of that year. Based on historical meteorological data sets, the plant protection office used BLASTL to predict three possible disease progression curves for 1991, a season with low temperatures and high rainfall, 2000, a season with high temperatures and low rainfall, and 1998, a season with moderate temperatures and rainfall.

Figure 5 shows the three possible disease progressive curves predicted by the BLASTL model. Mid-term weather forecasts indicating cool and rainy conditions similar to those in 1991 predict a rapid disease increase and would prompt a recommendation for almost immediate fungicide spraying. On the other hand, weather forecasts indicating hot and dry conditions similar to those in 2000 predict a slower disease increase (Fig. 5) and would prompt a recommendation that no fungicide splaying is necessary. In fact, the weather forecast in 2002 was closest to conditions in 1998, which predicted a relatively rapid disease increase in mid July (Fig. 5). On the basis of this result, the Fukushima Prefecture Plant Protection Office published a watching report.

As a final point, we discuss contemporary advisory policies for the effective and efficient control of blast disease. Most policies take a cost benefit analysis approach (Fig. 6). If measures for disease control are not sufficient, the economical loss due to the disease is considerable. In contrast, over use of control measures means that the cost of control will exceed the cost of losses due to disease. In either case, growers would not make a profit.

However, with the disease progress simulations provided by the BLASTAM and BLASTL programs on JPP-NET, plant protection offices can publish more accurate and precise disease forecasts and hence provide more appropriate recommendations for blast disease control. Consequently, growers can readily implement control measures to ensure reasonable profit for themselves and also to bring down costs for consumers.

Recently, emphasis has been placed on the importance of enhancing the ability of growers to make decisions about disease control based on disease forecasting and economic management analysis rather than on information obtained from ready-made spraying calendars. Clearly, simulation approaches will continue to be very useful in this respect.

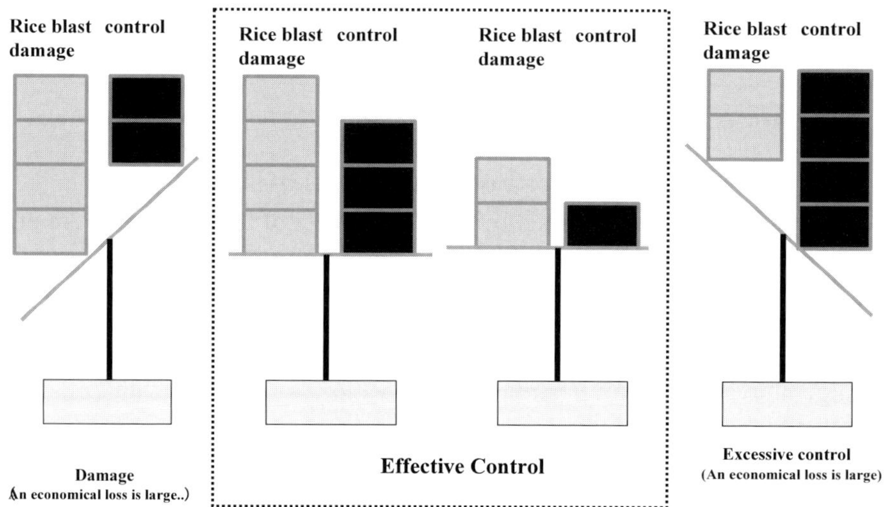

Fig.6. Effective Control of Rice Blast disease based on the Forecasting by a Computer Simulation Model "BLASTL"

REFERENCES

1. Hashimoto, A., Hirano, K. and Matsumoto, K. (1984) Studies on the Forecasting of Rice Leaf Blast Development by Application of the Computer Simulation. *Special Bulletin of Fukushima Prefecture Agricultural Experiment Station* 2, 1-104 [in Japanese, English summary].
2. Koshimizu, Y. (1988) A Forecasting Method for Occurrence of Rice Leaf Blast with AMeDAS Data. *Bulletin of the Tohoku National Agricultural Experiment Station.* 78:67-121 [in Japanese, English summary].
3. Nemoto, F., Hashimoto, A., Hanzawa, S. (1992) Control of Leaf Blast Forecasted by Computer Simulation Model (BLASTL). *Annual report of the Society of Plant Protection of North Japan.* 43:27-29 [in Japanese].
4. Nemoto, F. (1999) Computer-aided Forecasting of Rice Blast. *Farming Japan* Vol.33-4: 26-30.

SIMULATION MODELS OF RICE BLAST EPIDEMICS:

From a theoretical approach to a major component of IPM

Kiyoshi Ishiguro

National Agricultural Research Center for Tohoku region. Simo-Kuriyagawa, Morioka, Japan 020-0198

E-mail: ishiguro@affrc.go.jp

Abstract: In Japan, simulation models of rice blast epidemics have been developed as a tool for disease forecasting services in the public domain since the early 1970s. Later, the government changed the policy on disease management and started to emphasize lesser dependency on fungicides. Consequently, the disease forecasting had to introduce the concept of integrated pest management (IPM). An idea was proposed to use a simulation model for blast forecasting as a decision support system (DSS) based on IPM strategy. However, it is difficult to determine economic injury level (EIL) and action threshold (AT) of blast disease based on objective evidences. Also, accuracy and precision of simulation models of natural systems is innately insufficient. These two issues challenge the usefulness of simulation models for contemporary rice blast management. However, only if the EIL are subjectively determined by the decision-makers, forecasting with a simulation model can play a major role for IPM program for rice blast disease. The next step should be large scale field trials to test the usefulness of this system. The results certainly become strong evidences to encourage introducing the system into rice blast management programs.

Key words: IPM, Decision support system, Uncertainty, Bayesian estimate

1. MODELING RICE BLAST EPIDEMIC IN JAPAN

Attempts to develop simulation models of rice blast epidemics started in the early 1970s in Japan. It was almost immediately after the appearance in the USA of EPIDEM, the first-ever model to simulate plant disease

S. Kawasaki (ed.), Rice Blast: Interaction with Rice and Control, 297–302.
© 2004 *Kluwer Academic Publishers. Printed in the Netherlands.*

epidemics [1]. The development of rice blast models was considered possible due to the considerable accumulation of knowledge about the rice blast pathosystem until that time.

BLASTL, a simulation model of leaf blast epidemics [2], is one of the most well known and exploited models in Japan. The model was developed with a fund supported by Ministry of Agriculture, Forestry and Fisheries (MAFF). Interestingly, the original aim of BLASTL was different from that of EPIDEM. While EPIDEM was developed to test particular epidemiological hypotheses, BLASTL was constructed as a tool for disease forecasting services in the public domain. At that time, and still now to some extent, disease forecasting was considered as a tool to encourage growers to spray fungicides; in this respect, BLASTL was praised. BLASTL uses current and expected weather data as the input data to predict temporal disease progress on a rice plant of specific morphology and phenology. Historical weather data sets are used as reference for expected weather data.

2. RECENT POLICY CHANGE FOR DISEASE MANAGEMENT

As a tool for disease forecasting service, BLASTL has been used for many years. Until recently, whenever BLASTL predicted the risk of a severe leaf blast epidemic during the rice growing period, extension service officers would automatically recommend that growers spray fungicides repeatedly. This automatic response probably reflected the government's policy of avoiding any risk of shortage in domestic rice production, because rice is traditionally a major staple of the Japanese.

However, in the late 1990s, the government changed its disease management policy and started to emphasize a lesser dependency of rice production on fungicides and other agro-chemicals use. This drastic change probably resulted from two reasons. First, Japan rice production ability has improved to make even surplus rice since the 1970s due to introduction of high input and high yield cultivation methods. Second, more recently, the public concern about environment sound agriculture has increased.

The substitute policy introduced the concept of integrated pest management (IPM), in which fungicide spraying is considered necessary only when the cost of reduced yield is expected to exceed the cost of spraying. The concept of IPM includes two key words, economic injury level (EIL) and action threshold (AT): EIL is a level that the yield loss due to a disease is equal to the cost to control the disease; AT is a disease severity level that yield loss would exceed EIL unless adequate control measures are introduced at that time.

3. DISEASE FORECASTING FOR IPM

As a result of the policy shift of disease management, the main objective of disease forecasting service had to be changed. The disease forecasting has to play an important role for IPM. Is BLASTL, used to be only a tool for disease forecasting, also applicable to a component of IPM program? Ishiguro and Nanseki [3] have proposed a possibility to use BLASTL for a component of an IPM program of rice leaf blast. The procedures are as follows: (a) At any time during rice growing period, BLASTL predicts a range of plausible disease progress curves of leaf blast using current and several historical weather data sets (Figure 1). This is the same procedure as the conventional disease forecasting. (b) Once a plausible disease progress curve exceeds the EIL, then the corresponding AT is calculated using BLASTL model; (c) If the AT is expected to come within a few days, fungicide spray should be recommended. The duration between the forecasting and AT relies on how fast the decision-maker could prepare the spraying until he/she decides to spray.

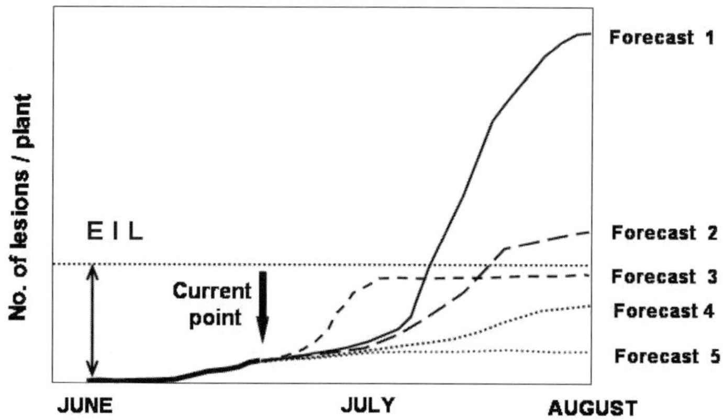

Figure 1. *The concept of forecasting with BLASTL model for leaf blast epidemics. Typical historical weather data sets are used as expected weather data.*

In an example shown in Figure 1, there are considerably different forecasts, and some of them exceed EIL. It is reasonable to consider that the actual disease progress curve would fall into the range between the most optimistic and the most pessimistic forecasts. Ishiguro and Nanseki [3] applied the maxmin criterion for a decision-making of such an uncertain situation. That is, choosing the best countermeasure against the worst expected outcome. In this example, a countermeasure in case of 'Forecast 1' should be considered.

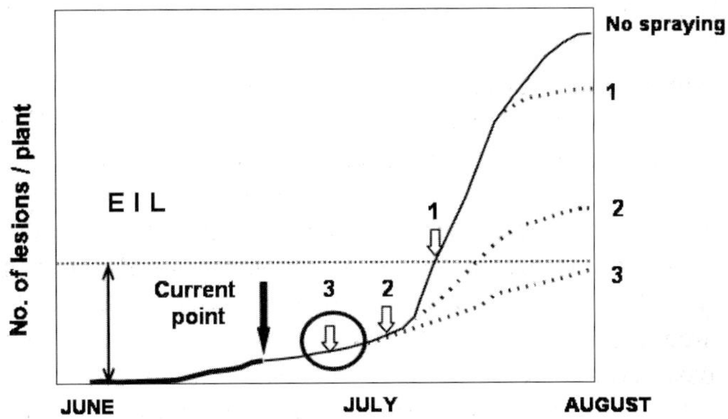

Figure 2. *The concept of calculation process of the action threshold (AT) of fungicide spraying with BLASTL model. Each spraying (an arrow with a number) results in the corresponding disease progress curve (a dot line).*

In the next step, spray timings of fungicide are explored to control the disease progress curve under EIL. This procedure can be also carried out using BLASTL. The manner of fungicide deposition on rice leaves, the behaviors of fungicide residues on the leaves and their modes of action against rice blast fungus were investigated in detail, and they were introduced into BLASTL model as parameters [4]. Figure 2 shows that spraying at the point 1 or 2 are too late but that spraying at the point 3 could control the disease severity under EIL. The latest timing of fungicide spraying that could control the disease severity under EIL is a critical point. The disease severity at the critical point is corresponding to AT. As predicted disease progress curves change every day, both the AT and the day when the disease progress curve reaches AT are not constant.

4. PROBLEMS IN INTRODUCING IPM

At first, the concept of IPM seemed to be easily introduced to the rice blast management program in which BLASTL consisted of a major component. However, two problems challenged the introduction of IPM concept to the rice blast management program. The first problem relates to the difficulties in determining EIL and AT of leaf blast severity based on objective evidences, where this is not a specific problem of simulation models but that of rice blast disease. The second problem is that accuracy and precision of simulation models to predict actual disease progress curves is generally inadequate. The results of quantitative validation of the BLASTL model demonstrated this problem [4, 5]. However, this limitation is unavoidable and common in simulation approaches that deal with natural systems [6].

To overcome these problems, Ishiguro and Nanseki [3] made several assumptions, while some of them seemed to be unrealistic. That is, the conventional approach to estimate EIL usually starts from quantitative experiments, and then estimates the EIL parameter from the results with the most likelihood method. However it is hard to estimate EIL of leaf blast epidemics. It is partially because leaf blast epidemics rarely affect the yield losses directly. It is also because leaf blast seems to play a role as an inoculum source for the subsequent panicle blast epidemics that mainly cause yield loss. Additionally there is still only poor correlation between severity of leaf blast and panicle blast epidemics within the same fields. This is probably due to the presence of some confounding factors. Thus, another effort to break through the bottleneck was needed and another approach was explored.

The EIL is estimated with the Bayesian estimation method instead of the most likelihood method. First, disease progress curves for several previous seasons are simulated with the BLASTL model using historical weather data sets at the local area. The decision-makers can use them to understand the actual risk of blast outbreaks before the beginning of the growing season. Second, each season's epidemic is categorized. Some seasons should be severe epidemic seasons (s), some should be mild (m), and others should be practically nothing (n). Based on the results, the probability of each category (P_s, P_m, P_n) is determined. Each probability is mutually exclusive and collectively exhaustive. Then, the decision-makers subjectively determine which categories fungicide spraying should be necessary. This decision is made based on their long time experiences. For example, they may feel that fungicide spraying should have been unnecessary in mild epidemic seasons but definitely necessary in severe ones. If so, the EIL must be between both categories. Once the EIL is agreed, the decision-support system (DSS) implemented by BLASTL can set an appropriate timing for fungicide spraying.

5. FUTURE PERSPECTIVE

We are currently examining the feasibility of the DSS for growers employing the concept of IPM. We have found many technical problems to be solved. Which kind of meteorological factors we really require, which kind of dew sensors are appropriate, and how the weather data are introduced efficiently to the computer in which the DSS is installed, should be considered. These issues should be cleared sooner or later.

However, the most important issue is to examine the effectiveness of the DSS on the rice blast management programs. Actually, the DSS includes several uncertainties. The estimated EILs have their own peculiar range of error. BLASTL model cannot always predict the real disease progress curve precisely and accurately. Essentially, future weather of the season cannot be always accurately predicted. The criterion for fungicide spraying is subjectively determined by the decision-makers, and it does not necessarily guarantee reasonable disease control. Thus, it should be crucial to collect sound evidences to demonstrate that DSS makes rice blast management program more effective and efficient.

These evidences can be obtained through extensive field tests comparing the IPM plots using DSS and conventional management plots with a statistically sound design. Meta analyses and systematic reviews of these results must provide stronger evidence of effectiveness of DSS. Based on the evidences, extension specialists can do justice to DSS and recommend it to the rice growers. This procedure would correspond to the idea of the evidence-based medicine (EBM) proposed in the clinical medicine [7].

REFERENCES

1. Waggoner, P. E. and Horsfall, J.G. 1969. EPIDEM, a simulator of plant disease written for computer. Conne. Agric. Exp. Stn. Bull. 698, 1-80.
2. Hashimoto, A., Hirano, K. and Matsumoto, K. 1984. Studies on the forecasting of rice leaf blast development by application of the computer simulation. Special Bulletin of Fukushima Prefecture Agricultural Experiment Station 2:1-104.
3. Ishiguro, K. and Nanseki, T. 2000. Exploring optimum application programs of fungicide using a simulator for leaf blast epidemics. in Advances in Rice Blast Research, (eds. D. Tharreau et al.) Klwer, Dordrecht, The Netherlands 203-208.
4. Ishiguro, K. 2001. Improving management strategies for rice blast disease using a simulation model of rice leaf blast epidemics. Bul. Tohoku Natl. Agric. Exp. Stn. 99, 1-110.
5. Ishiguro, K. 1994 Using simulation models to explore better strategies for the management of blast disease in temperate ricepathosystems. CAB International 435-449
6. Oreskes, N., Shrader-Frechette, K. and Belitz, K. 1994. Verification, validation, and confirmation of numerical models in the earth sciences. Science 263, 641-646.
7. Sackett, D. L., Haynes, R. B., Tugwell, P. and Guyatt, G. H. 1991. Clinical epidemiology: a basic science for clinical medicine. Little Brown, Boston, p441.